Urban Drainage and Storage Practices

Urban Drainage and Storage Practices focuses on the latest developments in urban stormwater design methods using drainage and storage approaches for both water quality and quantity control. It covers both the conventional approaches to flood mitigation and low-impact methods for stormwater quality enhancement. Theory, practice, and modeling methods are presented to illustrate how to build a holistic stormwater drainage and storage system using urban open space and parks through multiple land use.

Each chapter provides background theory, numerical experiments, illustrations, and Excel spreadsheets that outline design and calculation procedures. All urban watersheds are modeled as a series of cascading planes to drain stormwater from upstream roofs and parking lots onto downstream grass areas and vegetal beds. The drainage system is designed as a three-layer cascading system with various low-impact units for micro events, conveyance elements for minor events, and storage facilities for macro events. This book:

- presents the theory and practice of designing and building a stormwater system
- explains green approaches to designing and managing urban stormwater systems.

This text is ideal for senior and graduate students studying urban hydrology, hydraulic engineering, and water resource management. It will also be useful for engineers requiring a technical book with hands-on practical examples.

Urban Drainage and Storage Practices

James C.Y. Guo, Wenliang Wang, and Junqi Li

CRC Press
Taylor & Francis Group
Boca Raton London New York

CRC Press is an imprint of the
Taylor & Francis Group, an **informa** business

Cover image: James Guo

First edition published 2023
by CRC Press
6000 Broken Sound Parkway NW, Suite 300, Boca Raton, FL 33487-2742

and by CRC Press
4 Park Square, Milton Park, Abingdon, Oxon, OX14 4RN

CRC Press is an imprint of Taylor & Francis Group, LLC

© 2023 James C.Y. Guo, Wenliang Wang, and Junqi Li

Library of Congress Cataloging-in-Publication Data
Names: Guo, James C.Y. (James Chwen-Yuan), author. | Wang, Wenliang (Civil engineer), author. |
Li, Junqi (Civil engineer), author. Title: Urban drainage and storage practices /
James C.Y. Guo, Wenliang Wang, Junqi Li.
Description: First edition. | Boca Raton, FL : CRC Press, 2023. |
Includes bibliographical references and index.
Identifiers: LCCN 2022010106 | ISBN 9781032256122 (hbk) |
ISBN 9781032256146 (pbk) | ISBN 9781003284239 (ebk)
Subjects: LCSH: Storm sewers. | Drainage. Classification: LCC TD665 .G865 2023 |
DDC 628/.212–dc23/eng/20220722
LC record available at https://lccn.loc.gov/2022010106

ISBN: 9781032256122 (hbk)
ISBN: 9781032256146 (pbk)
ISBN: 9781003284239 (ebk)

DOI: 10.1201/9781003284239

Typeset in Sabon
by Newgen Publishing UK

Contents

Preface

Urban development leads to more pavements and impervious surfaces. The sharply increased storm runoff volumes and flows generated from urban areas change the spatial and temporal distributions of surface and sub-surface runoff in the hydrologic cycle. The major negative impacts of urban development on storm runoff are two-fold: (1) an increase in peak flows (*peak-flow Q-problem*) from extreme events, and (2) an increase in runoff volumes (*storage V-problem*) from both frequent events. Increased peak flows are concerning for public safety, whereas urbanization-induced runoff volumes deteriorate the water quality in the urban environment. This book is written to focus on the mitigation of peak flows associated with extreme events and the reduction of runoff volumes induced from frequent events. In the last two decades, the concept of low-impact-development (LID) has been developed to uphold the ultimate goal of preserving the watershed regime. This book presents a strategy using a green stormwater approach to preserve the pre-development flow-frequency relationship along a waterway after development.

A natural waterway has been shaped over geologic time by a full spectrum of runoff flows, from trickle flows to flood flows. In this book, a complete hydrologic database is recommended to determine the distribution of the population of runoff flows. In general, the lowest 80% of runoff flows are less than 6-month events and are classified as micro events. The top 4% of runoff flows are categorized as major events. The flows in-between are classified as minor events.

In this book, a man-made urban waterway is a corridor to pass stormwater and can be divided into three segments: (1) *on-site runoff disposal facilities* at the source of rainwater, such as roofs and parking lots, (2) *thru-site conveyance facilities* to pass stormwater, such as street gutters, sewers, and roadside ditches, and (3) *outfall detention facilities* to reduce peak flows and enhance stormwater quality at the system exit.

In this book, an urban waterway is shaped and sized to have three layers to pass micro, minor, and major (3M) storm events. The concept of the 3M *cascading flow system* is derived to design the three segments along a waterway: (1) the bottom *micro drainage system*, such as infiltrating beds and porous pavers that intercept flows up to the 6-month event, (2) the mid *minor drainage system*, such as storm drains and street inlets, that collect 2- to 5-yr peak runoff flows into underground sewers, and (3) the top *major drainage system*, such as street gutters, flood channels, and detention basins, that convey and store the 50- to 100-yr peak runoff flows. The concept of the 3M *Cascading Flow System* is applicable to both regional master drainage planning for new developments or/and retrofitting existing drainage systems for urban renewal.

This book is composed of 20 chapters. Chapter 1 presents a review of conveyance and storage hydrologic principles. The hydrologic frequency analyses in Chapter 2 cover traditional approaches to define minor and major events, while Chapter 18 analyzes the complete hydrologic data to quantify the range of micro events. In practice, urban catchments are small. For conveyance, engineers may choose peak-flow–based approaches or/and hydrograph approaches to size the hydraulic structures. Therefore, Chapters 3 to 7 are expanded from conventional peak-flow calculations to complete runoff hydrograph predictions. Chapters 8, 9, 10, and 13 cover the urban stormwater drainage designs, and Chapters 11 to 19 cover stormwater storage designs, including roadway storage basins using crossing culverts as outlets, stormwater detention basins, LID water-quality control basins, and energy dissipation basins. In this book, all LID water-quality basins are designed as an overtopping storage facility to intercept runoff flows up to the water quality control volume, while all detention basins are sized as a temporary storage facility to reduce the peak flows associated with extreme events. A complete review of the evolution of water quality control volume is presented in Chapter 18, and how to design a multiple-layer detention basin is summarized in Chapters 14 and 15. As a continual process in urban development, local drainage systems are always updated according to urban renewal. As a result, the concept of excess water detour is an important skill when connecting the updated local flow to the downstream existing trunk line. Chapter 20 presents the designs of interim storage facilities to accommodate the temporary drainage and storage conditions.

The ultimate goal of this book is to present the latest techniques in urban drainage engineering and to serve as a textbook for the senior to graduate level in the areas of water resources, hydrologic engineering, roadway drainage, highway and water environment, flood mitigation and control, hydraulic design, and related fields. Many numerical examples and real-world cases are incorporated into the illustrations of hydrologic principles and numerical schemes. In this book, examples and homework problems are formatted in spreadsheets to enhance the iterative calculations and routine numerical schemes. Numerous photos were taken from construction sites to better illustrate the real-world applications of the theories and principles in the book.

The authors believe that all these efforts will assist junior engineers with learning the basics of these subjects, will give senior engineers the latest information to improve their designs, and will help both to build a safe world for tomorrow.

James C.Y. Guo
University of Colorado, Denver

About the Authors

James C.Y. Guo received his PhD in water resources from the University of Illinois at Champaign-Urbana. Since 1982, he has been a member of the teaching and research faculty at the University of Colorado Denver, and he serves as Director of the Hydrology and Hydraulics Graduate Program. Guo is also a registered licensed professional engineer and actively participates in real-world design and planning projects. He has over 40 years of experience in hydrology, hydraulics, water resources, and groundwater modeling. Guo's approach is to apply the concept of the system to the development of hydrologic methods for engineering designs and analyses. He has also incorporated these new algorithms into practical design charts, procedures, and computer software models. Many of his publications have been adopted by the drainage manuals used in Colorado, Nevada, and California for stormwater designs.

Guo is the winner of national awards presented by the American Society of Civil Engineers, and is the author of numerous technical articles, training notes, and conference proceedings. He is in demand as a speaker at technical meetings and training seminars in the areas of flood mitigation and stormwater management.

Wenliang Wang is Associate Professor at Beijing University of Civil Engineering and Architecture and a Doctor of the Engineering Division. He is Vice President of the Sponge City and Stormwater Management Institute. He is a founding member of the China Association for Engineering Construction Standardization's Sponge City Committee. Wang also serves as Chief Editor of the Sponge City Construction Effectiveness Assessment Standard.

Junqi Li is Vice President, Professor, and Doctoral Supervisor at the Beijing University of Civil Engineering and Architecture and the Director of the Ministry of Education's Key Laboratory for urban stormwater system and water environment research. For the past 20 years, Li has focused on urban stormwater management and sponge city designs and methodology developments, as well as advancements in the areas of water environment ecological technology and environmental policy and management.

Chapter 1

Stormwater systems

1.1 URBAN STORMWATER DRAINAGE SYSTEMS

Stormwater is a natural resource for urban areas. Fresh water from precipitation renews water bodies in lakes, streams, greenbelts, and wetland areas. Facing the random processes in rainfall and runoff, the challenge in stormwater management is to cope with seasonal variations and uneven spatial distributions. A stormwater drainage system is one of the basic infrastructure systems in an urban area. From the aspect of stormwater quantity control, a stormwater system should be designed not only to quickly remove stormwater from urban areas but also to carefully store the runoff volume over an extended period for reuse. On the other hand, for the purpose of stormwater quality control, it is important to promote hydraulic efficiency when planning a stormwater drainage system, while the preservation of the water environment prefers slow flow movements to minimize surface erosion and disturbance of topsoil. As a result, the green stormwater approach has evolved from a conventional *conveyance approach* to a *conveyance-storage system*, and from a conveyance-storage system to an *infiltration-conveyance-storage system*. A green stormwater system balances stormwater quantity and quality management and preserves the pre-development watershed regime.

Urban streets are designed to satisfy the primary function defined by traffic needs. However, during a storm event, streets also serve as waterways to collect and deliver the excess storm runoff. As the runoff flow is accumulated on the street, the depth and speed of the water flow can impose potential hazards to traffic. For instance, a thin layer of water between tires and the road surface can cause hydroplaning. A wide spread of storm runoff compromises the level of service of roads and streets. Consequently, it is necessary to install a *street inlet* at locations where the water spread on the street exceeds the allowable width. As illustrated in Fig. 1.1, street inlets are tied into a storm sewer system or a roadside ditch. A storage facility is placed for flow release control at the exit of a storm sewer system. A stormwater storage facility can be a small *retention-wetland pond* for water quality control or a sizable *detention-dry basin* for regional flood control.

A major waterway, such as a stream or river, collects the outflows from sewers, ditches, culverts, detention basins, and wetland ponds. The stability of the waterway's morphology depends on the flow loadings. Fast urbanization process along a waterway can significantly increase runoff loadings and cause serious deterioration of the waterway regime. Therefore, regional drainage planning must be performed to

DOI: 10.1201/9781003284239-1

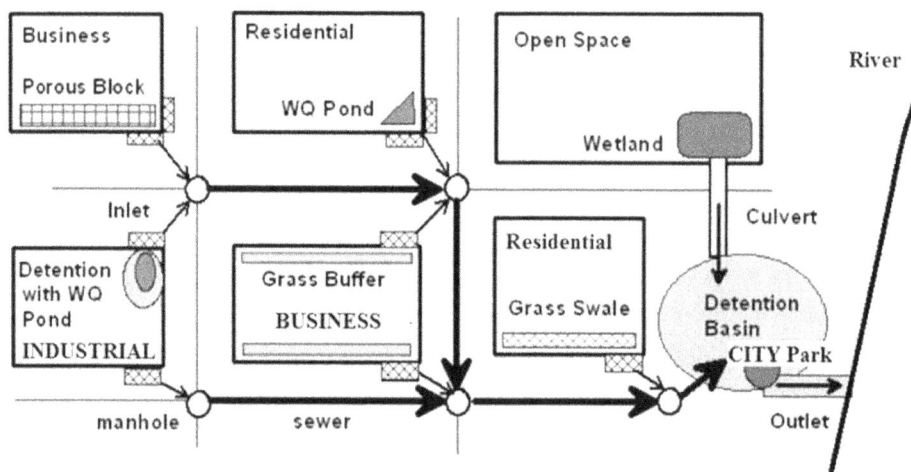

Figure 1.1 Stormwater drainage system

take effective mitigation measures into consideration. The main purpose of waterway drainage planning is not only to alleviate the flood threat from extreme events, but also to protect stream stability and wildlife habitats.

1.2 STORMWATER DRAINAGE FACILITIES

Storm runoff on the street is produced from the adjacent tributary areas. Street gutters guide stormwater to flow towards the inlets. An inlet is connected to a manhole. Between two adjacent manholes is a sewer line. Storm sewer lines are aligned with streets and eventually drain into downstream natural water bodies. Urban stormwater drainage systems are classified into four categories based on their functionality:

1. *water-quality facilities* for stormwater quality enhancement,
2. *collection facilities* to transfer surface stormwater into the system,
3. *conveyance facilities* to deliver stormwater to the system exit, and
4. *storage facilities* to control flow releases into the natural waterbody.

Although these drainage facilities are designed and built for different purposes, they jointly achieve similar goals of sustaining a healthy and functional urban water environment.

1.2.1 Water quality facilities

Surface erosion is inevitable when stormwater runs off a pervious surface. In an urban area, stormwater flows are often loaded with trash, debris, solids, chemicals, and petroleum pollutants. It is important to intercept 70–80% of solids before the stormwater enters the drainage system. As a result, a *water-quality control basin* (WQCB) is often

(a) Porous Basin for Filtering and Infiltration

(b) Permeable Pavement for Infiltration

(c) Vegetal Bed for Infiltration

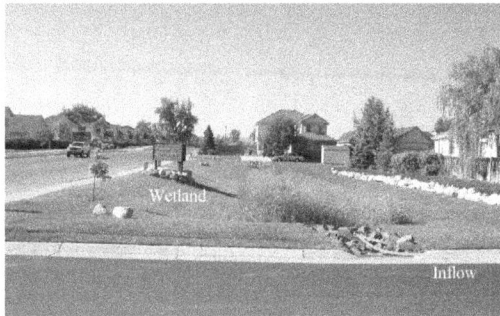

(d) Wetland Basin for Filtering

Figure 1.2 Water quality facilities

placed upstream of a street inlet for the purpose of pollutant source control. A WQCB is designed to remove the pollutants and solids from stormwater by filtering and sedimentation processes. Examples of WQCBs shown in Fig.1.2 are an *infiltration basin, porous pavement, vegetation bed, and wetland*. A WQCB is typically small and flat because it was designed to intercept early stormwater from frequent storms such as 3- to 6-month events. Infiltration basins and infiltrating beds are often embedded into the landscaping settings around buildings. Parking lots may be covered with pervious pavements; and wetlands are placed at locations where surface runoff and groundwater are sufficiently available.

From the hydrologic aspect, a WQCB is built to have a shallow basin with a sand–gravel sub-base for infiltration. The basin is built to capture the on-site runoff, while the sand–gravel media is used to filter the seepage flow for water quality enhancement. In general, the design volume for a WQCB is expressed in mm or inch/watershed. The water loading and depletion cycle in a WQCB can be as long as 12 to 24 hours. Of course, the longer the residence time, the more the sediment is removed.

Example 1.1: In Fig. 1.3, a pervious area of 2 hectares has been developed into a building complex. The proposed development reduces hydrologic losses, including soil infiltration and depression. Under the post-development conditions, a circular WQCB is proposed to compensate a hydrologic loss of 10 mm. Knowing that this WQCB is

Watershed Area	2.00	hectares
Volume in Depth	10.00	mm/watershed
Basin Volume	200.00	cubic m
Basin Geometry		
Basin Brimful Depth	0.30	m
Circular Basin Area	666.67	sq m
Basin Radius	14.57	m

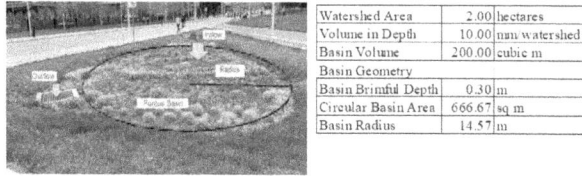

Figure 1.3 Example of a porous basin

(a) Curb-opening Inlet   (b) Grate Inlet

Figure 1.4 Example of street inlets

planned to have a water depth of 30 cm, determine the radius in meters for this circular basin.

Solution: In this case, the basin's volume is specified in the depth per watershed. The volume of 10 mm for 2 hectares is calculated as:

$$\text{Basin volume} = 10/1000 \times 2 \ (100 \times 100) = 200 \ \text{m}^3$$

1.2.2 Collection facilities

Excess stormwater overtops a WQCB into the streets. *Collection facilities* such as street inlets and culverts are laid out to intercept the runoff flows carried in street gutters. The flow interception capacity at an inlet depends on if the inlet operates like an orifice under deep water or a weir under shallow water. As shown in Fig. 1.4b, a steel grate is preferred at places where trash and urban debris are concentrated, otherwise a curb-opening inlet, as shown in Fig. 1.4a, is employed because it is not only hydraulically efficient but also safer for bikers.

1.2.3 Conveyance facilities

A *stormwater conveyance facility* is designed to deliver the runoff flows collected from upstream inlets. Examples of stormwater conveyance facilities include roadside ditches,

(a) Channel like River Park   (b) Channel with Flood Water

Figure 1.5 Conveyance facilities

flood channels, street gutters, storm sewers, and grass swale. Water flows delivered in a conveyance facility are open-channel flow in nature.

An urban flood channel system shall be designed like a river park, as shown in Fig. 1.5a. During an extreme storm event, the entire floodplain shown in Fig. 1.5b becomes wet and bank-full. As a conveyance facility, a channel is sized to pass the peak flow for the selected design event. Often, the cross-section of a flood channel is sized using Manning's formula to pass the 100-yr peak flow. Manning's formula is written as:

$$U = \frac{K}{N} R^{\frac{2}{3}} \sqrt{S_o} \tag{1.1}$$

$$R = \frac{A}{P} \tag{1.2}$$

$$Q = U A \tag{1.3}$$

where U = cross-sectional average velocity in [L/T], N = Manning's roughness coefficient, A = flow area, P = wetted perimeter, R = hydraulic radius in feet or meters, S_o = slope of conveyance element in ft/ft or meter/meter, Q = discharge in cubic feet/second (cfs) or cubic meter/second (cms), and K = 1 for the SI system or K = 1.486 for the English system. The Manning's roughness coefficient in Eq. 1.1 is a sensitive parameter. The recommended values of Manning's roughness coefficient are 0.012–0.014 for concrete linings, 0.030–0.035 for grass linings, and 0.04–0.045 for grouted riprap linings.

Example 1.2: After a storm event, the water mark on the bank walls in Fig. 1.6 shows that the flow area was a 5- by 10-ft rectangle. Knowing that the channel is laid on a slope of 0.01 ft/ft with a Manning's coefficient of 0.025, determine the discharge, Q.

Figure 1.6 Flow in a rectangular channel

Solution: For this case, the flow depth is: $y = 5$ ft in a 10-ft wide rectangular channel.

Flow area: A = 5 × 10 = 50 ft²

Wetted perimeter: P = 5 + 5 + 10 = 20 ft

$$Hydraulic\ radius: R = \frac{A}{P} = \frac{50}{20} = 2.50\ ft$$

$$Average\ flow\ velocity\ U = \frac{1.486}{0.025} \times 2.50^{2/3} \sqrt{0.01} = 11.0\ fps$$

Discharge in channel: Q = UA = 50.0 × 11.0 = 550.0 cfs

1.2.4 Storage facilities

From the hydrologic aspect, an urban catchment is mainly characterized by its land uses. Various land uses are often quantified by the area percentage of imperviousness, including roofs, pavements, roadways, and parking areas. The amount of storm runoff generated by a rainfall event is sensitive to the impervious area in the catchment. When the catchment is overly developed, excess stormwater must be mitigated on the site before being released to downstream properties.

A stormwater storage facility is often placed at a system outfall or slightly upstream of the entrance into a downstream natural waterbody. The main purpose of a storage facility is to temporarily store stormwater in order to reduce peak flows and delay the time to the peak. Examples of storage facilities are retention ponds and detention basins. A retention pond, as shown in Fig. 1.7a, has a permanent wet pool that is connected to the local groundwater, while a detention basin, as shown in Fig. 1.7b, is a dry pool until it rains.

(a) Retention Pond with a Wet Pool   (b) Detention Basin with a Dry Pool

Figure 1.7 Stormwater storage facilities

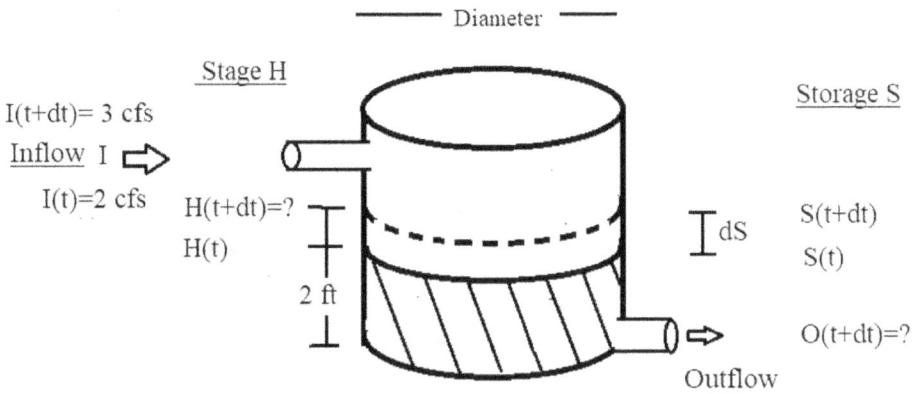

Figure 1.8 Volume balance through a storage facility

A stormwater detention basin is designed to have a large storage volume to mitigate extreme flood events. The large, depressed basin area may serve as a park during dry days or as a floodwater storage area during wet days. The primary design parameters for a stormwater storage facility are storage volume and drain time. Water flowing through a storage facility is unsteady and non-uniform. At the beginning of an event, the storage volume continues increasing as long as the inflow is greater than the outflow. During the recession, the stored volume is depleted.

An unsteady flow can be numerically modeled by the finite difference technique. The continuous flow process is discretized by a series of time intervals, and the flow within each time interval is assumed to be quasi-steady. The average flow within a time interval represents the flow condition. As illustrated in Fig. 1.8, the principle of continuity among inflow, outflow, and storage volume is described as:

$$I - O = \frac{\Delta S}{\Delta t} \tag{1.4}$$

Eq. 1.4 is balanced with inflow and outflow over a time interval, Δt or dt. Numerically, the finite difference equation for Eq. 1.4 is written as:

$$\frac{I(t)+I(t+\Delta t)}{2} - \frac{O(t)+O(t+\Delta t)}{2} = \frac{S(t+\Delta t)-S(t)}{\Delta t} \tag{1.5}$$

where I = inflow rate in [L³/T], O = outflow rate in [L³/T], S = storage volume in [L³], ΔS = change in storage volume in [L³], and Δt or dt = time interval in [T]. At each time step, the initial condition in terms of $I(t)$, $I(t + \Delta t)$, $O(t)$, and $S(t)$ are known. To solve for the two unknowns, $Q(t + \Delta t)$ and $S(t, t + \Delta t)$, in Eq. 1.5, we need a second equation that can be derived from the basin geometry to represent the storage-outflow relationship.

Example 1.3: A circular 10-feet in-diameter tank has already filled with a water depth of 2 feet. The inflow to the tank is changed from 2 cfs to 3 cfs over a time interval of 5 minutes. The outlet is formed with a circular 6-inch in-diameter pipe. Knowing that the outflow rate from the tank can be modeled by the orifice formula with an orifice coefficient, C_o, of 0.65, determine the water depth after five minutes of operation.

Solution: According to the orifice formula, the flow release from the tank is calculated as:

$$O = C_o \sqrt{2gH} = 0.65 \frac{\pi \times 0.5^2}{4} \sqrt{2.0 \times 32.2} = 1.03\sqrt{H} \ (\text{stage} - \text{outflow curve}) \tag{1.6}$$

where O = outflow, A = drain cross-sectional area in [L²], H = water depth in [L] above the center of pipe, and g = gravitational acceleration in [L/T²]. Eq. 1.6 is called the *stage-outflow curve* and is the relationship between the water depth, H, and the outflow, O. The water storage volume, S, in the tank is a function of water depth as:

$$S = HA = H\frac{\pi \times 10^2}{4} = 78.5H \ \text{cubic ft (stage} - \text{storage curve)} \tag{1.7}$$

Equation is called the *stage-storage curve* and is the relationship between water depth, H, and storage volume, S. Using Eqs 1.6 and 1.7, the *storage-outflow curve* for this circular tank is derived as:

$$S = 74.0 \ O^2 \ (\text{storage-outflow curve}) \tag{1.8}$$

The *initial condition* is defined with $H = 2$ ft. We have:

$$O(t) = 1.03\sqrt{2} = 1.45 \, cfs$$

$$S(t) = 78.5 \times 2 = 157.0 \, ft^3$$

Using Eqs 1.5 and 1.8, we have two equations as:

$$\frac{2+3}{2} - \frac{1.45 + Q(t+\Delta t)}{2} = \frac{S(t+\Delta t) - 157.0}{5 \times 60} \tag{1.9}$$

$$S(t+\Delta t) = 74.0O(t+\Delta t)^2 \tag{1.10}$$

By trial and error, the solutions are:

$O(t+\Delta t) = 2.20$ cfs, $S(t+\Delta t) = 359.0$ ft³, and $H(t+\Delta t) = 4.57$ ft

The *stage-storage curve* represents the basin's geometry, while the *stage-outflow curve* is derived from the outlet work. Eqs 1.9 and 1.10 are repeatedly applied to each time step to process the inflow hydrograph through the detention process. In practice, the volume units used in stormwater management are often expressed in *acre-ft*, which are equal to 43,560 cubic feet, hectare-meter (10,000 cubic meters), or *cfs per day (cfd)*, which is equal to the flow rate in ft³/sec times 86,400 seconds per day.

1.3 STORMWATER PLANNING

Urban drainage design is an attempt to control stormwater movement while maintaining the integrity of natural flow paths and existing legal relationships arising from land ownership. Stormwater planning for a region can affect all governmental jurisdictions and all parcels of property in a watershed. The characteristics of stormwater require coordination among different entities and cooperation between both the public and private sectors. *Drainage planning* is an approach that integrates both local and regional efforts to identify drainage conveyance and storage facilities based on *hydrologic optimization* and *cost minimization*, individually and collectively.

1.4 STORMWATER REGIONAL AND LOCAL PLANNING

A *regional master drainage plan* (RMDP) sets forth the current structural and regulatory means for improving existing flooding conditions within an area taking into account the possible effects of future development. An RMDP must provide basic drainage information, including:

1. locations of regional drainage facilities,
2. inflow and outflow information at design points,
3. flooding problems and future improvements, and
4. estimated costs for various alternatives.

An RMDP can be modified and/or revised from time to time to reflect the changes desired by the local entities, as long as the intent and integrity of the regional master plan is not compromised.

A *Local Drainage Plan* sets forth the site requirements for a new development project and also identifies the required public improvements. All local flood mitigation facilities must be designed in a manner that will collectively achieve the regional goal stated in the RMSP.

1.5 CONCLUSIONS

Urban stormwater drainage planning and design take a risk-based approach. The selection of the design event is based on many factors, including public perception, federal regulations, economic considerations, and public safety. It is important that a region maintains consistency in selecting design events. With a consistent underlying risk level, the relationship between the magnitude of storm runoff and tributary area can be established throughout the entire region. Since urban drainage systems are designed for the quick collection and fast delivery of stormwater, stormwater releases from man-made drainage systems have continuously produced shock loads of pollutants to the receiving water bodies downstream. Over the years, stormwater can lead to the significant deterioration of aquatic and wildlife habitats. This is because man-made drainage systems eliminate the natural infiltrating and filtering processes. In 1987, the *Federal Clean Water Act* triggered a shift in stormwater management from flood control to stormwater quality enhancement. Many new technologies and design methods have been developed using porous basins to promote on-site runoff volume disposal, and detention basins to reduce flow releases.

It was also recommended that stormwater be stored and reused for irrigation and entertainment purposes. It is imperative that both flood control and stormwater quality enhancement be designed using the concepts of multiple land use for multiple events and multiple functions for multiple purposes.

HOMEWORK

Q1-1   The channel in Example 1.2 has a trickle flow through the bottom. As shown in Fig. Q1-1, the trickle flow section has a depth of 1 ft and a width of 4 ft. Determine the discharge when the channel lining has a Manning's coefficient $N = 0.025$ and the channel bottom is laid on a slope of 1.0%.

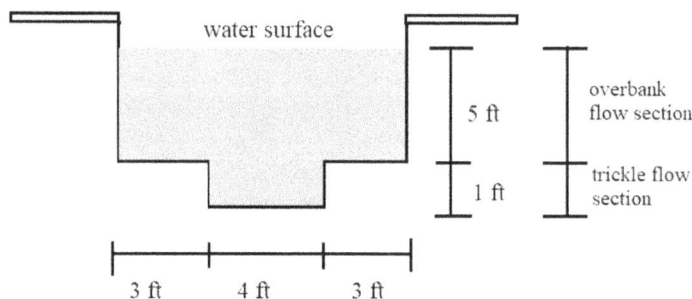

Figure Q1-1 Channel with overbank and trickle flow sections

Q1-2   Referring to the tank in Example 1.3, the outflow is increased as the water depth in the tank is increased. Estimate the time of equilibrium (i.e. when the outflow is equal to the inflow) under the specified operation.

Q1-3 The *stage-storage curve* (S,V) represents the relationship between the water depth (S) and water volume (V) in a storage facility. Consider the tank in Example 1.3. Construct the relationship between water depth and water volume in the tank. (Hint: The stage refers to the elevation above mean sea level while the depth refers to the vertical distance from the bottom of the tank.)

Q1-4 The *stage-outflow curve* (S,O) represents the relationship between the water depth and outflow from the storage facility. Assume that the orifice formula is applicable to the tank outlet in Example 1.2. With an orifice coefficient of 0.65, plot the stage-outflow curve for the tank.

Q1-5 The hydraulic characteristics of a storage facility are described by its stage-storage and stage-outflow curves. These two curves can be further merged into a single curve called the *storage-outflow curve* (V,O). Derive the storage-outflow curve for the tank in Example 1.2.

REFERENCES

City and County of Denver, Colorado (2001). *Urban Storm Water Drainage Criteria Manual*, Volumes 1, 2, and 3, Urban Drainage and Flood Control District, Denver, CO.

City and County of Sacramento, California (2015). *Hydrologic Standards*, Brown and Caldwell, Walnut Creek, CA.

City of Tucson, Arizona. (2010). *Standards Manual for Drainage Design and Floodplain Management*, Simons, Li and Associates, Fort Collins, CO.

Clark County, Nevada (2020). *Hydrologic Criteria and Drainage Design Manual*, Montgomery Watson, Las Vegas, NV.

Guo, James C.Y. (2017). *Urban Stormwater Management and Flood Mitigation*, CSC Company, Oxford.

Chapter 2

Design rainfall distribution

2.1 HYDROLOGIC CYCLE

The *hydrologic cycle* is the circulation of water through the atmosphere, land, lakes, and oceans. Fig. 2.1 illustrates that the various types of precipitation are formed in the atmosphere. Raindrops and snowflakes grow in size through collisions in the wind. Near the ground, all precipitation may be intercepted by buildings and trees before reaching the ground.

Soils on the ground present two major hydrologic losses: depression storage and infiltration depth. After subtracting interception, depression, and infiltration losses, the rainfall excess is converted into overland flows that drain into channels and lakes to refresh the natural water bodies and recharge the groundwater. At the sea, the large water surface area is a major source of evaporation.

During a dry season, the groundwater table is the source of the base flows in rivers and lakes. *Rainfall excess* is also called *runoff depth*, which is the amount of precipitation that survives hydrologic losses and produces overland flows to drain into streams, creeks, rivers, lakes, and oceans.

2.2 RAINFALL MEASUREMENT

Rain gauges are common instruments for measuring the incremental and total rainfall depths during a storm event. Rainfall depths in an event change temporally and spatially because the rainfall intensity decays from the center of the storm. Fig. 2.2a presents a mini-weather station, which is composed of wind gauge, rain gauge, temperature gauge, and solar panel.

The tipping bucket gauge in Figs 2.2b and 2.3 has a top orifice to collect raindrops, a bucket to measure the rainfall amount, and a reservoir to store the accumulated rainfall volume. The rainfall caught by the orifice opening is first funneled into the bucket that will tip and pour the accumulated water into a reservoir when the depth reaches the capacity of the bucket. This process will trigger an electron pen to continuously register the accumulation of rainfall depth on a roll of paper loaded on the revolving drum.

DOI: 10.1201/9781003284239-2

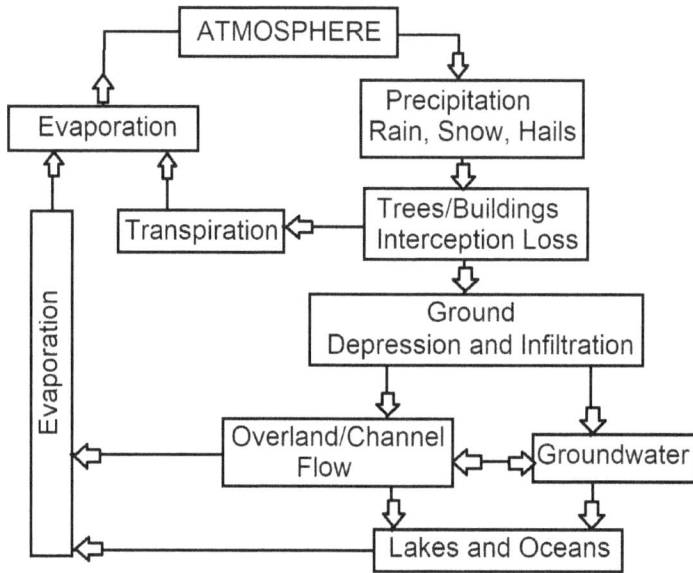

Figure 2.1 Illustration of the hydrologic cycle

(a) Weather Station   (b) Tipping Bucket

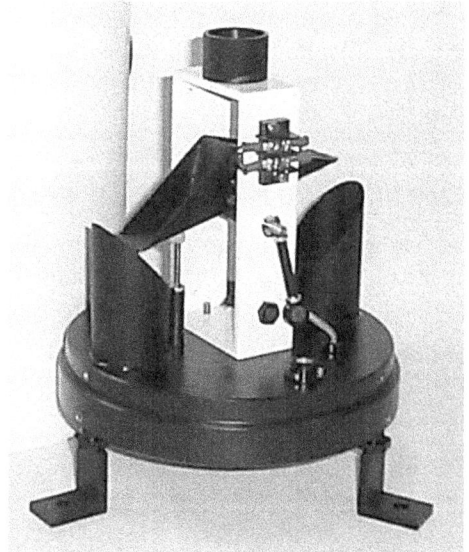

Figure 2.2 Rain gauge in the field

2.3 RAINFALL ANALYSIS

2.3.1 Continuous record

A rainfall record is continuous in time, as shown in Fig. 2.3. How to define the beginning and the end of a rainfall event depends on the *minimum interevent time* used to separate events. A minimal interevent time is defined as the minimum period of time with no rain. Referring to Fig. 2.4, using a minimal interevent time of 6 hours, Groups A and B shall be considered to be a single event, as are Groups D and E.

With a continuous record separated into individual events, the *event duration* is defined as the period of time from the beginning to the end of the event. An event rainfall volume is expressed in *depth per unit area*. The distribution of the incremental rainfall depths (rainfall blocks) with respect to time is called *hyetograph* or rainfall distribution. A rainfall event is often recorded by the incremental amounts in time and then analyzed to obtain: (1) a *mass curve*, and (2) an *intensity–duration curve*.

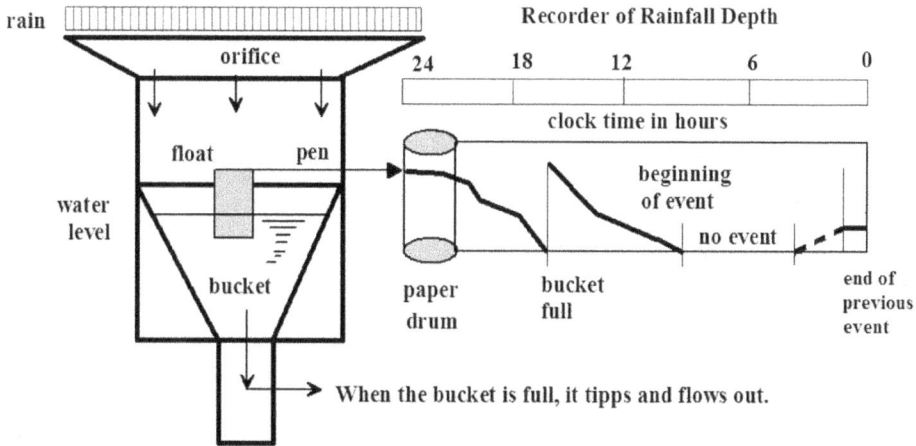

Figure 2.3 Tipping bucket rain gauge

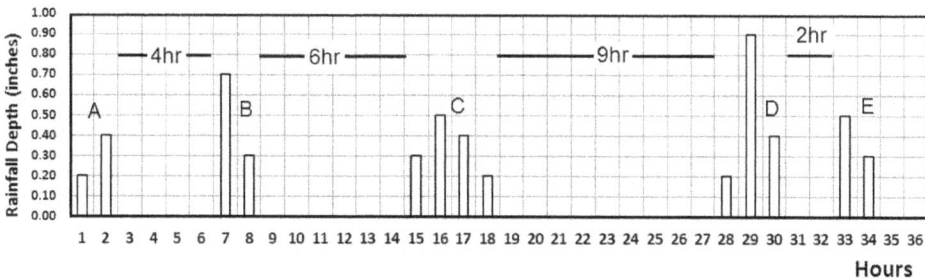

Figure 2.4 Continuous rainfall records

2.3.2 Rainfall depth–duration and intensity–duration curves

A *rainfall time-distribution* represents the sequential incremental rainfall amounts recorded according to clock time. The *cumulative rainfall distribution* is the mass curve for the event. To further analyze a rainfall distribution, we may divide a continuous record into segments using the selected duration. The duration is a window width in time to pick the most intense amount during a rainfall event. For instance, there are six 5-min increments in a 30-min rainfall distribution. Considering a duration of 10 minutes, then every two adjacent 5-min rainfall blocks is a 10-min segment. As a result, a total of five 10-minute segments can be derived from this 30-min rainfall distribution. Among these five 10-min segments, the most intense one is selected as the representative 10-min precipitation depth observed in this storm event. Similarly, using duration of 20 minutes as a basis, a total of three 20-min segments can then be formed from this 30-min storm, and the most intense one is selected to represent this storm event. Under the assumption that all of the segments are independent, an observed storm event can be dissected into many small segments to augment the database for conservative designs. A storm event can be converted from its time distribution into rainfall depth–duration (P–D) pairs. The plot of rainfall depth (P) versus duration (D) is called a P–D *curve*. A P–D curve is an increasing function of time, starting from the highest 5-min depth to the total depth for the entire event. It is important to understand that a rainfall distribution is plotted according to clock time, and a P–D curve is plotted using durations. The ratio of rainfall depth to duration gives the rainfall intensity. The average rainfall intensity is defined as:

$$I = \frac{P}{T_d} \tag{2.1}$$

where I = average intensity in inch/hr or mm/hr, P = precipitation depth in inches or mm, and T_d = duration in hours. The plot of rainfall intensity (I) versus duration (D) is called an I–D *curve*. An I–D curve is a decay curve with respect to duration, starting with the highest 5-minute intensity to the event average intensity.

Example 2.1: Table 2.1 presents a rainfall event recorded from 5:00 PM to 6:00 PM. The total precipitation depth for this event is 0.79 inches for a period of 60 minutes. The mass curve for this event is derived in Column 3. The rainfall mass curve can be further normalized with the event duration, D, of 1.0 hour and event precipitation of 0.79 inch. In this case, the highest 5-minute rainfall depth is 0.22 inch observed at 5:25 PM. The highest 10-minute rainfall depth is the sum of 0.22 and 0.18. Similarly, the sum of the three blocks, 0.22 + 0.18 + 0.13 = 0.53 inches, represents the 15-min rainfall depth for this case. Repeating the same procedure generates the P–D curve listed in Column 8. The P–D curve can be converted into its I–D curve using Eq. 2.3. For instance, the 5- and 10-min rainfall intensities are calculated as:

$$I_5 = \frac{0.22}{5} \times 60 = 2.64 \, \text{inch/hour}$$

Table 2.1 Rainfall-duration analysis

Clock time	Time t min	Rainfall Increment dP(t) inches	Cumualtive Rain Depth P(t) inches	Normalized Time t/Td	Normalized Rainfall Mass Curve P(t)/P	Duration D min	Depth P(D) inch	Intensity I(D) inch/hr
5:00 PM	0.00	0.00	0.00	0.00	0.00			
5:05 PM	5	0.01	0.01	0.08	0.01	5.00	0.22	2.64
5:10 PM	10	0.02	0.03	0.17	0.04	10.00	0.40	2.40
5:15 PM	15	0.04	0.07	0.25	0.09	15.00	0.53	2.12
5:20 PM	20	0.13	0.20	0.33	0.25	20.00	0.63	1.89
5:25 PM	25	0.22	0.42	0.42	0.53	25.00	0.67	1.61
5:30 PM	30	0.18	0.60	0.50	0.76	30.00	0.71	1.42
5:35 PM	35	0.10	0.70	0.58	0.89	35.00	0.73	1.25
5:40 PM	40	0.04	0.74	0.67	0.94	40.00	0.75	1.13
5:45 PM	45	0.02	0.76	0.75	0.96	45.00	0.76	1.01
5:50 PM	50	0.01	0.77	0.83	0.97	50.00	0.77	0.92
5:55 PM	55	0.01	0.78	0.92	0.99	55.00	0.78	0.85
6:00 PM	60	0.01	0.79	1.00	1.00	60.00	0.79	0.79

Rainfall Event Duration   Td = 60.00 min

Rainfall Event Depth   P = 0.79 inch

$$I_{10} = \frac{0.40}{10} \times 60 = 2.40 \text{ inch/hour}$$

Repeating the above produces the I–D curve in Column 9.

It is noticed that the mass curve in Fig. 2.5 sharply increases before the peak and then becomes flatter as time increases. Fig. 2.5c and 2.5d present the P–D and I–D curves. Both the mass and the P–D curves appear as cumulative values. It is important to understand that the mass curve follows clock time, while the P–D curve follows a window width in time that often starts from the peak and then symmetrically expands in both directions.

2.4 RAINFALL FREQUENCY–DEPTH ANALYSIS

Hydrologic frequency analysis is a statistical technique applied to long-term continuous rainfall or runoff records collected at a gauge station. The watershed hydrologic conditions must be carefully examined to ensure that no major changes have occurred during the period of the records. Only records that represent relatively steady watershed hydrologic conditions should be used for flood flow-frequency analysis. Predictions from such a database are only applicable to similar hydrologic conditions. In this section, the hydrologic frequency technique is introduced and applied to the rainfall records.

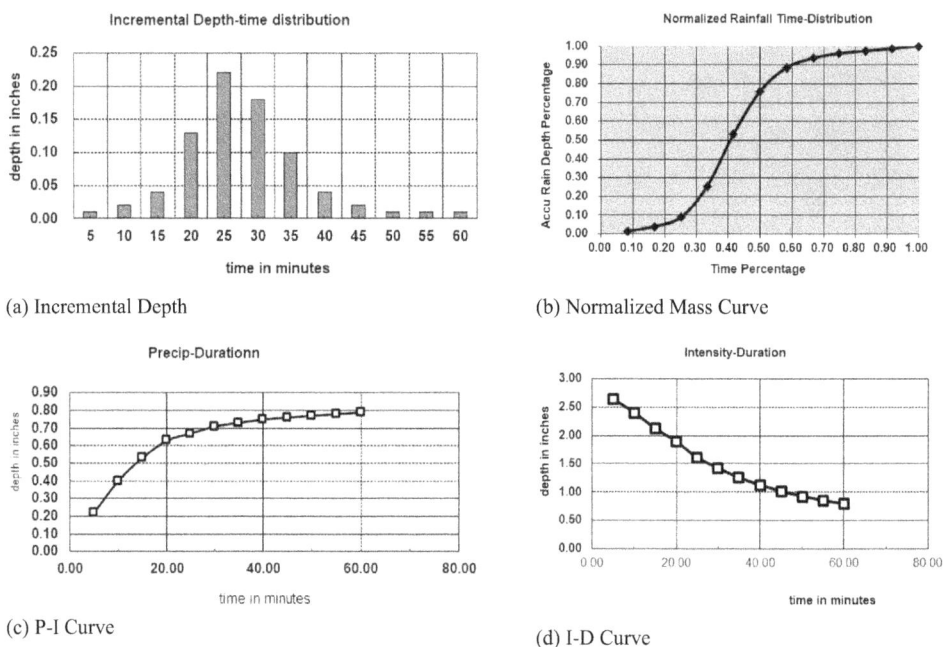

(a) Incremental Depth

(b) Normalized Mass Curve

(c) P-I Curve

(d) I-D Curve

Figure 2.5 Rainfall-time and rainfall-duration distributions

2.4.1 Rainfall annual database

The reliability of a frequency analysis depends on the length of the record. A record of 10 years is barely acceptable for hydrologic predictions because most hydraulic designs are governed by 50- or 100-year events. It is important to know how to interpret the sample with confidence to extrapolate the derived relationship into extreme events. For this purpose, there are two types of databases developed from the *complete data series* (CDS) for hydrologic frequency analysis: namely, the *annual maximum series* (AMS) and the *annual exceedance series* (AES).

An AMS uses a period of 365 days as a hydrologic year. The representative flood for each period is the largest one observed within 365 days. As a result, an N-year record will only include N events, one from each year, that is then used in the analysis. It often happens that the second highest event in a wet year exceeds the largest floods observed in some dry years. However, by definition, an AMS only considers the highest magnitude and ignores the rest.

An AES accepts hydrologic data as a continuous record. All observed events are ranked in descending order of magnitude. An AES consists of the top N events from an N-year record. This allows the inclusion of more than one event from a wet year and none from a dry year. An AES is a more conservative approach than an AMS.

Example 2.2: Table 2.2 presents a 10-yr record of the 10-min rainfall depths, P_{10}, collected at a rain gauge station between the years 2000 and 2009. The AMS and AES for this database are selected and constructed as shown in Table 2.2.

Table 2.2 Databases for AMS and AES

Year	inch	P-10 inch	Depth inch	inch	Highest Value	Data Base AMS	AES
2000	1.50	0.98	1.15	0.75	1.50	1.64	1.64
2001	1.35	0.85	1.64		1.64	1.50	1.50
2002	0.75	1.25			1.25	1.25	1.35
2003	0.68	0.89	0.52	0.85	0.89	0.96	1.25
2004	0.11	0.58			0.58	0.89	1.15
2005		0.95	0.85	0.96	0.96	0.75	0.98
2006	0.45	0.75	0.50	0.67	0.75	0.75	0.96
2007	0.65	0.55	0.75		0.75	0.65	0.95
2008	0.55	0.62			0.62	0.62	0.89
2009	0.55	0.52	0.65		0.65	0.58	0.85

Solution: The AMS consists of the highest value from each year. The second largest, 0.98 inch, in the year 2000, was excluded from the AMS but was included in AES. The AES is more conservative than the AMS.

Regardless of how the data series is selected, an AMS or AES serves as a sample to estimate the statistical parameters of the P_{10} population. Predictions of certain events can then be modeled by the selected probability distribution. The validity and applicability of a database depends directly on the characteristics of the sample data used to estimate the model parameters. An AES may be considered if the flood damage is caused by repeated floods, such as traffic interruptions by flooded highways. In other cases where the design is controlled by the most critical condition such as spillway design, an AMS may be used.

2.4.2 Sample statistics

The variable $p(i)$ can be represented by its mean and the departure of the variable from the mean. Such a departure can be expressed by a fraction of the standard deviation. The mean, standard deviation, and skewness of the distribution of the variable $p(i)$ are calculated as:

$$\bar{P} = \frac{1}{n}\sum_{i=1}^{i=n} p(i) \tag{2.2}$$

$$S = \sqrt{\frac{1}{n}\sum_{i=1}^{i=n}\left(p(i)-\bar{P}\right)^2} \tag{2.3}$$

$$g = \frac{1}{S^3}\frac{n}{(n-1)(n-2)}\sum_{i=1}^{i=n}\left(p(i)-\bar{P}\right)^3 \tag{2.4}$$

where $\bar{P}$ = mean, $p(i)$ = i-th value, n = number of years of the rainfall record, S = unbiased standard deviation, and g = unbiased skewness coefficient. Often, logarithmic values can narrow the differences between the predicted and observed values

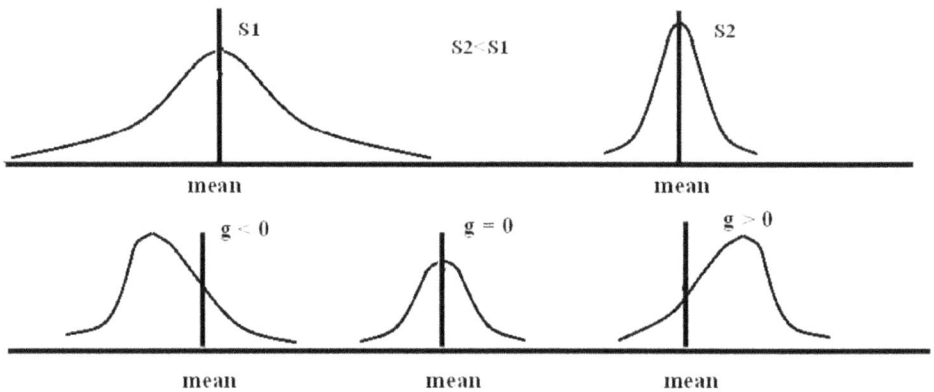

Figure 2.6 Statistic parameters for various distributions

in the analysis. Using logarithmic values, log $p(i)$, will convert Eqs 2.2 to 2.4 to log-value distributions.

As illustrated in Fig. 2.6, a standard deviation represents the spread of $q(i)$. A higher standard deviation implies a wider spread and a lower standard deviation implies a lower spread. A skewness coefficient indicates the shape of the distribution. When $g = 0$, it is a symmetric distribution; $g > 0$ means there are more events greater than the mean; while $g < 0$ means there are more events smaller than the mean.

2.4.3 Plotting position

The primary purpose of hydrologic frequency analysis is to determine the recurrence intervals and magnitudes of extreme events. Referring to the AES in Example 2.2, the magnitude $P_{10} = 1.50$ inch was exceeded once and also equaled once, or on average, this magnitude was exceeded or equaled once every 5 years. Therefore, the 5-year 10-min precipitation depth is estimated to be 1.5 inches. Statistically, the exceedance probability is $\text{Prob}(P \geq 1.5) = 1/5$ and the non-exceedance probability is $\text{Prob}(P < 1.5) = 4/5$.

Graph papers are made with the vertical axis to be the magnitude of the variable and the horizontal axis to be the non-exceedance probability. During data analysis, the N-year data series should be ranked in order of decreasing magnitude. The plotting position for the m-th event is determined by the total number of observations, n, and its rank, m. The empirical formula for determining the non-exceedance probability is:

$$Prob(P < P_{10}) = 1 - \frac{m+a}{n+b} \tag{2.5}$$

where $\text{Prob}(P < P_{10})$ = non-exceedance probability, P_{10} = 10-min precipitation depth, P = precipitation depth variable, n = length of record in years, m = rank of an event in order of decreasing magnitude, and a and b = empirical constants. The value of $a = 0.00$ for the uniform distribution, $a = 0.375$ for a Normal distribution, and $a = 0.44$ for a Gumbel distribution. Table 2.3 shows the recommended values for the variables a and b.

The selection of variables of a and b is a matter of local hydrology characteristics. In general, the California formula is recommended for an AES analysis and the Weibull formula is recommended for an AMS analysis.

The fundamental definition in statistical hydrology is the relationship between *return period* and *exceedance probability*. They are defined as:

$$Prob(P \geq P_{10}) = 1 - Prob(P < P_{10}) \tag{2.6}$$

$$T_r = \frac{1}{Prob(P \geq P_{10})} \tag{2.7}$$

where $Prob(P \geq P_{10})$ = exceedance probability, and T_r = return period in years. The return period is defined as the average recurrence interval. However, it does not mean that an exceedance always occurs once every T_r years, but it means that the average time between two adjacent exceedances is T_r years.

Example 2.3: For comparison, both *California* and *Hazen* formulas are tested for the AES of 10-min precipitation depth, P_{10}, in Example 2.2. Using Eqs 2.2, 2.3, and 2.4, the statistics for the AES are: $\overline{P_{10}}$ = 1.15 inches, S = 0.27 inches, g = 0.6836. As shown in Table 2.4, the Hazen formula assigns a return period of 20 years to the highest value, but the California formula assigns a return period of 10 years. The difference

Table 2.3 Empirical constants for plotting formulas

Empirical Formula	a	b	Prob (p < P)
California (1923)	0.0	0.0	$1 - m/n$
Hazen (1914)	−0.5	0.0	$1 - (m - 0.5)/n$
Weibull (1939)	0.0	1.0	$1 - m/(n + 1)$
Beard (1943)	−0.3	0.4	$1 - (m - 0.3)/(n + 0.4)$

Table 2.4 Plotting positions by California and Hazen formulas

AES inch	Rank Order m	California Prob(P<P10)	Formula Tr in Years	Hazen Prob(P<P10)	Formula Tr in Years
1.64	1.00	0.90	10.00	0.95	20.00
1.50	2.00	0.80	5.00	0.85	6.67
1.25	3.00	0.70	3.33	0.75	4.00
0.96	4.00	0.60	2.50	0.65	2.86
0.89	5.00	0.50	2.00	0.55	2.22
0.75	6.00	0.40	1.67	0.45	1.82
0.75	7.00	0.30	1.43	0.35	1.54
0.65	8.00	0.20	1.25	0.25	1.33
0.62	9.00	0.10	1.11	0.15	1.18
0.58	10.00	0.00	1.00	0.05	1.05

between these two formulas diminishes as the rank, m, increases. This implies that the Hazen formula is more suitable when dealing with outliers.

2.4.4 Probability distributions

The generalized form for probability models is that the departure of a variable, $P(T_r)$, from its mean can be related to the standard deviation and frequency factor as (Chow et.al 1988):

$$P = \bar{P} + Z(T_r)S \tag{2.8}$$

in which P = variable of precipitation depth, $\bar{P}$ = mean, $Z(T_r)$ = frequency factor for the return period T_r, and S = standard deviation. As revealed in Eq. 2.8, the event for the mean is defined as:

$$P = \bar{P} \quad \text{When } Z(T_r) = 0 \tag{2.9}$$

The value of the frequency factor in Eqs 2.8 and 2.9 depends on the underlying probability distribution and the return periods. Theoretical formulas have been developed for the following distributions.

Gumbel distribution

The frequency factor for the Gumbel distribution is defined as:

$$Z_g(T_r) = -\frac{\sqrt{6}}{\pi}\left\{0.5772 + \ln\left[\ln\frac{T_r}{(T_r - 1)}\right]\right\} \tag{2.10}$$

where $Z_g(T_r)$ = Gumbel frequency factor for return period, $T_r \geq 1$, and π = 3.1416. The Gumbel distribution is applicable only to an annual database such as T_r = 1 to 200 years. According to Eq. 2.10, the return period for the average magnitude is the one that has Z_g = 0.0. Setting Eq. 2.10 equal to zero, the return period for the average magnitude is T_r = 2.33 years for the Gumbel distribution.

Exponential distribution

The frequency factor for the Exponential distribution is defined as:

$$Z_e(T_r) = \frac{\sqrt{6}}{\pi}(\ln T_r - 0.5772) \tag{2.11}$$

where Z_e = exponential frequency factor for the return period, $T_r \geq 0$. The Exponential distribution can be used with include monthly data such as T_r = 0.5 year. Setting $Z_e(T_r) = 0$ in Eq. 2.11, the return period for the average magnitude is 1.78 years for the Exponential distribution.

Normal distribution

The Normal distribution is symmetric with skewness = 0.0. The frequency factors for the Normal distribution are closely approximated by:

$$B = \sqrt{\ln\left(\frac{1}{p^2}\right)} \tag{2.12}$$

where the variable p is the exceedance probability. Having found the variable B in Eq. 2.12, the value of z for the Normal distribution is (Abramowitz and Stegun, 1965):

$$z = B - \frac{2.515517 + 0.802853\ B + 0.010328\ B^2}{1 + 1.432788\ B + 0.189269\ B^2 + 0.001308\ B^3} \tag{2.13}$$

Eq. 2.12 is also valid for the Log-Normal distribution when logarithmic values are used. Eq. 2.12 indicates that the return period for the average magnitude is 2 years under a Normal distribution.

Pearson Type III distribution

Pearson Type III distributions are a set of family curves. Each curve is identified by the mean, standard deviation, and skewness coefficient. The frequency factors for a Pearson or Log-Pearson Type III distribution depend on the return period and skewness coefficient as well. When the skewness coefficient is zero, the Pearson Type III distribution is reduced to the Normal distribution. When the skewness coefficient is between 9.0 and –9.0 and the exceedance probability is between 0.0001 and 0.9999, the frequency factors of the Pearson Type III distribution can be calculated using the value of Z_p in Eq. 2.14 as (Harter, 1971)(Kite, 1977):

$$Z_p = \frac{2}{g}\left\{ \left[(z-k)k+1\right]^3 - 1 \right\} \tag{2.14}$$

where Z_p = frequency factor for Pearson Type III distribution, and k is defined as:

$$k = \frac{g}{6} \tag{2.15}$$

Example 2.4: Continued from Examples 2.2 and 2.3. Predict the 50-yr 10-min precipitation depth, $P10_{50}$, using the Gumbel and Normal distributions.

Solution: Let us convert T_r into its non-exceedance probability as:

$$P(P < P10_{50}) = 1 - \frac{1}{T_r} = 0.98 \,(\text{non - exceedance probability})$$

$$P(P1 \geq P10_{50}) = \frac{1}{T_r} = 0.02 \,(\text{exceedance probability})$$

Table 2.5 Prediction of P_{10} using Gumbel and normal distributions

Return Period Tr year	Non-Exceed Probability P(P<P10)	Gumbel Freq Z Z	10-min Depth P10 inch	Normal Variable B	Normal Freq Z Z	10-min Depth P10 inch
2.00	0.500	− 0.164	1.11	1.177	0.000	1.15
5.00	0.800	0.719	1.35	1.794	0.841	1.38
10.00	0.900	1.305	1.51	2.146	1.282	1.50
25.00	0.960	2.044	1.71	2.537	1.751	1.63
50.00	0.980	2.592	1.86	2.797	2.054	1.71
100.00	0.990	3.137	2.01	3.035	2.327	1.79

Using the Gumbel distribution, we have:

$$Z_g(50) = -\frac{\sqrt{6}}{\pi}\left\{0.5772 + \ln\left[\ln\frac{50}{(50-1)}\right]\right\} = 2.592$$

$$P10_{50} = \overline{P_{10}} + Z_g(50)S = 1.15 + 2.592 \times 0.27 = 1.86\,\text{inch}$$

Using the Normal distribution, the exceedance probability for a 50-year event is 0.02. With $p = 0.02$, Eq. 2.12 yields:

$$B = 2.7971$$

Substituting $B = 2.7971$ into Eq. 2.13 yields $z = 2.054$. Then, the 50-year 10-min precipitation depth predicted by the Normal distribution is:

$$P10_{50} = 1.15 + 2.054 \times 0.27 = 1.71\,\text{inch}.$$

Repeat the same process to predict a set of 10-min precipitation depths as shown in Table 2.5.

In comparison, the Normal distribution underestimates the 25-, 50-, and 100-yr magnitudes because it is not a skewed curve.

2.4.5 Confidence limits

The length of record affects the accuracy of predictions. For instance, any 30-year continuous record out of a 50-year record may constitute a sample set. Therefore, at least 20 sets of a 30-year sample can be derived from the 50-year continuous record. Each sample produces its mean and standard deviation. As a result, we will have 30 pairs of (*mean, standard deviation*). As illustrated in Fig. 2.7, these 30 data points form a Normal distribution. *Confidence limits* are expressed as the reliability band above and below the frequency curve. For instance, the 5% (upper limit) and 95% (lower limit) confidence interval gives a confidence level of 90% for the predicted value to be within these two limits.

Magnitude Q

Normal Distribution for Q

95% Limit

95%

Q_u

Best-Fitted Line

Q

5% Limit

Q_d

5%

Tr

Pearson III probabilistic Scale

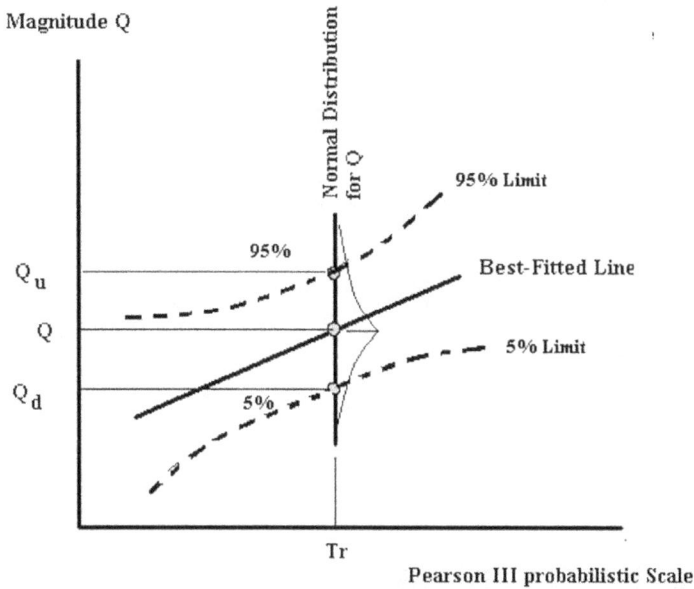

Figure 2.7 Illustration of confidence limits

The Water Resources Council (Bulletin 17 in 1982) suggests that these two confidence limits be determined as:

$$P_u = \bar{P} + SZ_u \tag{2.16}$$

$$P_d = \bar{P} - SZ_d \tag{2.17}$$

$$Z_u = \frac{1}{c}\left(Z + \sqrt{Z^2 - cd}\right) \tag{2.18}$$

$$Z_d = \frac{1}{c}\left(Z - \sqrt{Z^2 - cd}\right) \tag{2.19}$$

$$c = 1 - \frac{z_*^2}{2(n-1)} \tag{2.20}$$

$$d = Z^2 - \frac{z_*^2}{n} \tag{2.21}$$

where P_u = upper limit, P_d = lower limit, n = years in the record, Z = frequency factor of the underlying probability distribution for the specified event such as the 100-year magnitude by the Gumbel distribution, z_* = frequency factor of the Normal distribution for the limits selected, and c and d = constant, determined for the specified event.

Table 2.6 Predicted 10-min precipitation depths with confidence limits

Return Period Tr year	Non-Exceed Probability P(P<P10)	Gumbel Freq Z Z	10 min Depth P10 inch	Value of d	Upper Freq F Zu	Lower Freq F Zd	Upper Limit P10 inch	Lower Limit P10 inch
2.00	0.500	−0.164	1.11	−0.244	0.376	−0.763	1.25	0.94
5.00	0.800	0.719	1.35	0.247	1.500	0.194	1.56	1.21
10.00	0.900	1.305	1.51	1.431	2.356	0.715	1.80	1.35
25.00	0.960	2.044	1.71	3.907	3.496	1.315	2.11	1.51
50.00	0.980	2.592	1.86	6.449	4.362	1.740	2.35	1.63
100.00	0.990	3.137	2.01	9.568	5.230	2.153	2.58	1.74

Example 2.5: Continued from Examples 2.2 to 2.4. Considering the 5% and 95% confidence limits applied to the AES in Example 2.2, determine the values of Z_u and Z_d for the 100-year event described by the Gumbel distribution.

To determine the confidence limits between 5% and 95% requires $z_* = 1.645$, which is the frequency factor in Eq. 2.13 for the Normal distribution with an exceedance probability of 5% or a non-exceedance probability of 95%. For a 100-year event, the frequency factor of the Gumbel distribution is $Z_g (100) = 3.137$ using Eq. 2.10. In this case, using Eqs 2.20 and 2.21, we have $c = 0.85$ and $d = 0.27$. Eqs 2.18 and 2.19 provide $Z_u = 5.230$ and $Z_d = 2.153$. The corresponding confidence limits are 2.58 to 1.74 inches, determined by Eqs 2.16 and 2.17. This means that the most likely value for the 100-yr P_{10} is 2.01 inches, and the chance of it being between 1.71 and 2.58 inches is 90%. The same procedure can be applied to 2-, 5-, 10-, and 50-year events. Table 2.6 presents the confidence limits for a range of values of P_{10}.

As discussed above, the sample AES or AMS database collected from the 10-year record can be expanded using hydrologic frequency analyses to estimate extreme events. The reliability of the predicted magnitude is evaluated with confidence limits. This technique expands the precipitation-duration curves derived from a single event to yield the precipitation-duration-frequency (P–D–F) curve for engineering applications.

2.5 DESIGN RAINFALL INFORMATION

Rainfall data are available from the US Weather Bureau in two forms: *raw data* and *published data*. Raw data are in the form of continuous records of rainfall events with various incremental time intervals. Raw data can be analyzed using the depth–duration–frequency approach and published in forms of isohyetal maps. Each map shows contours of equal precipitation depth for specified duration and recurrence intervals.

2.5.1 Technical Paper 40 (TP40)

Weather Bureau Technical Paper No. 40 (TP 40), the results of previous Weather Bureau investigations of the precipitation-frequency regime of the US continent were combined into a single publication. TP 40 has been accepted as the standard source for rainfall P–D–F information in the United States since 1961. A sample of isohyetal

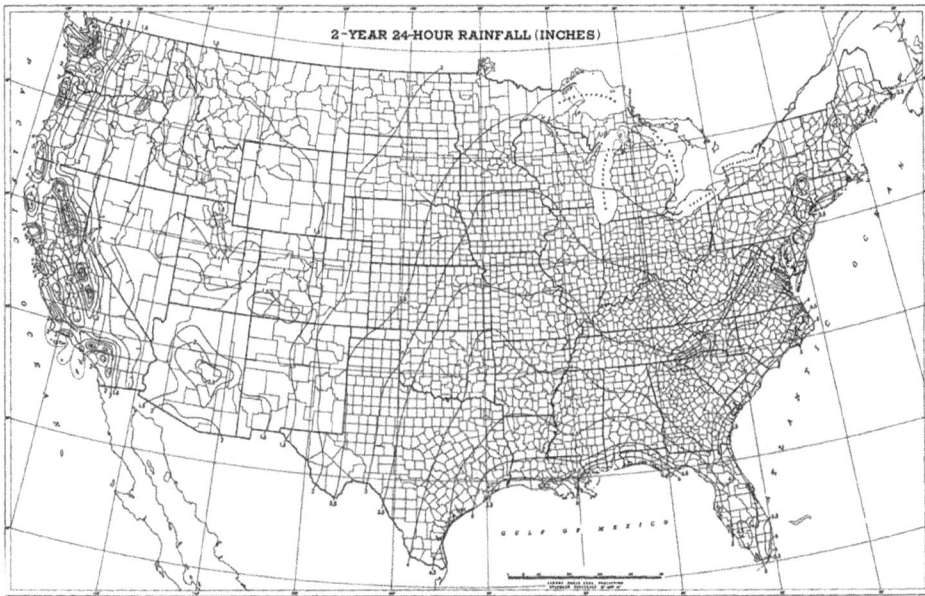

Figure 2.8 Rainfall isohyetal map (2-year 24-hour) from TP 40

maps is presented in Fig. 2.8. TP 40 includes isohyetal maps with return periods of 1, 2, 5, 10, 25, 50, and 100 years, and duration periods of 30 minutes, 1, 3, 6, 12, and 24 hours.

2.5.2 NOAA Atlas 14

NOAA Atlas 14 contains precipitation-frequency estimates for the United States and US affiliated territories with associated 90% confidence intervals and supplementary information on the temporal distribution of heavy precipitation, analysis of seasonality and trends in the AMS database. It includes pertinent information on development methodologies and intermediate results. The results are published through the Precipitation Frequency Data Server (PFDS) (http://hdsc.nws.noaa.gov/hdsc/pfds). The PFDS is a point-and-click interface in Fig. 2.9 developed to deliver NOAA Atlas 14 precipitation-frequency estimates and associated information. Upon clicking a state on the map, a user can identify the project site for which the precipitation-frequency estimates are needed. Estimates and their confidence intervals are displayed directly as P–D–F tables.

Example 2.6: Download Atlas 14 intensity–duration–frequency (IDF) table for Stapleton Airport, Denver, Colorado.

Figure 2.10 presents a plot of the IDF curves in Table 2.7.

As illustrated in Fig. 2.10, the rainfall intensity in an *I–D* curve decreases as the duration increases. The best-fitted formula for IDF curves is a hyperbolic equation as:

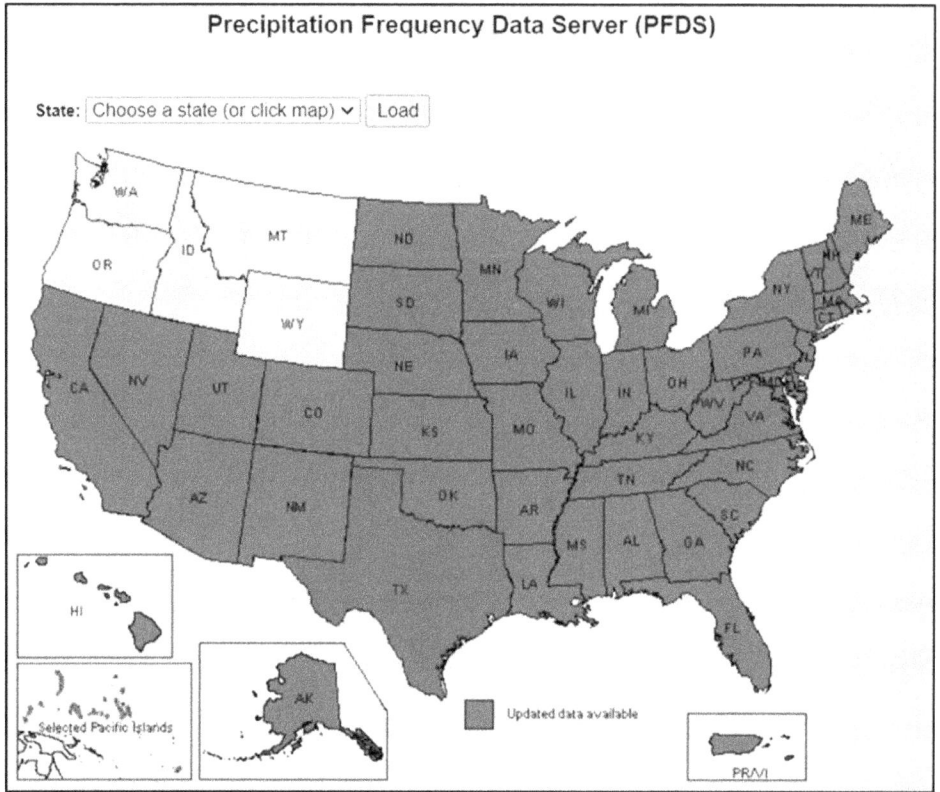

Figure 2.9 Precipitation frequency data server for Rainfall Atlas 14

$$I = \frac{C_1 P_{Index}}{\left(C_2 + T_d\right)^{C_3}} \tag{2.22}$$

where I = rainfall intensity in [inch/hr or mm/hr], T_d = duration in minutes, P_{Index} = index precipitation such as the 1-hr or 6-hr precipitation depth for the selected return period and C_1, C_2, and C_3 are local empirical constants. For instance, the IDF curves in Fig. 2.10 can be approximated as:

$$I = \frac{28.5 P_1}{\left(10 + T_d\right)^{0.789}} \quad \text{for Stapleton Airport, Denver, Colorado only} \tag{2.23}$$

in which P_1 = 1-hr precipitation depth in inches. For instance, use the 2-yr 1-hr precipitation, P_1 = 0.77 inch, to estimate the 2-yr rainfall intensities at Stapleton Airport as:

$$I = \frac{28.5 \times 0.77}{\left(10 + T_d\right)^{0.789}} \quad \text{for the 2-yr } I\text{--}D \text{ curve in Fig. 2.10.} \tag{2.24}$$

Table 2.7 IDF table at Stapleton Airport, Denver, Colorado

AMS-based precipitation frequency estimates with 90% confidence intervals (in inches/hour)

Duration					Annual exceedance probability (1/years)				
	1/2	1/5	1/10	1/25	1/50	1/100	1/200	1/500	1/1000
5-min	**2.95** (2.41-3.62)	**4.22** (3.44-5.18)	**5.29** (4.28-6.53)	**6.85** (5.41-8.86)	**8.15** (6.26-10.6)	**9.55** (7.07-12.7)	**11.0** (7.82-15 0)	**13.2** (8.95-18.3)	**14.9** (9.80-20.8)
10-min	**2.16** (1.77-2.65)	**3.09** (2.52-3.80)	**3.87** (3.14-4.78)	**5.02** (3.96-6.49)	**5.97** (4.58-7.78)	**6.99** (5.17-9.29)	**8.08** (5.73-11.0)	**9.64** (6.55-13.4)	**10.9** (7.18-15.2)
15-min	**1.76** (1.44-2.16)	**2.51** (2.05-3.09)	**3.15** (2.55-3.89)	**4.08** (3.22-5.27)	**4.85** (3.73-6.32)	**5.68** (4.20-7.55)	**6.57** (4.66-8.94)	**7.84** (5.33-10.9)	**8.86** (5.84-12.4)
30-min	**1.23** (1.01-1.51)	**1.75** (1.43-2.16)	**2.19** (1.77-2.70)	**2.83** (2.23-3.65)	**3.35** (2 57-4.36)	**3.92** (2.90-5.20)	**4.52** (3.20-6.14)	**5.37** (3.65-7.47)	**6.07** (3.99-8.48)
60-min	**0.770** (0.630-0.945)	**1.08** (0.884-1.33)	**1.35** (1.09-1.67)	**1.74** (1.37-2.25)	**2.06** (1 59-2.69)	**2.41** (1.79-3.21)	**2.79** (1 97-3.79)	**3.32** (2.26-4.61)	**3.75** (2.47-5.24)
2-hr	**0.461** (0.380-0.562)	**0.646** (0.530-0.788)	**0.802** (0.654-0.984)	**1.03** (0.821-1.33)	**1.23** (0.948-1.59)	**1.43** (1.07-1.89)	**1.66** (1 18-2.24)	**1.97** (1.35-2.73)	**2.23** (1.48-3.10)
3-hr	**0.335** (0.276-0.407)	**0.466** (0.383-0.567)	**0.577** (0.472-0.705)	**0.743** (0.592-0.950)	**0.880** (0.683-1.14)	**1.03** (0.769-1.35)	**1.19** (0.852-1.60)	**1.42** (0.974-1.95)	**1.60** (1.07-2.22)
6-hr	**0.199** (0.165-0.240)	**0.275** (0.227-0.332)	**0.339** (0.278-0.411)	**0.432** (0.346-0.548)	**0.510** (0.397-0.651)	**0.593** (0.446-0.773)	**0.682** (0.492-0.911)	**0.809** (0.560-1.11)	**0.912** (0.612-1.25)
12-hr	**0.121** (0.101-0.145)	**0.166** (0.138-0.199)	**0.203** (0.168-0.245)	**0.256** (0.206-0.321)	**0.299** (0.234-0.378)	**0.345** (0.261-0.445)	**0.393** (0.285-0.520)	**0.461** (0.321-0.624)	**0.516** (0.348-0.702)
24-hr	**0.074** (0.062-0.088)	**0.100** (0.084-0.119)	**0.121** (0.101-0.145)	**0.151** (0.122-0.187)	**0.175** (0.137-0.219)	**0.199** (0.151-0.255)	**0.225** (0.164-0.294)	**0.260** (0.182-0.349)	**0.288** (0.196-0.390)

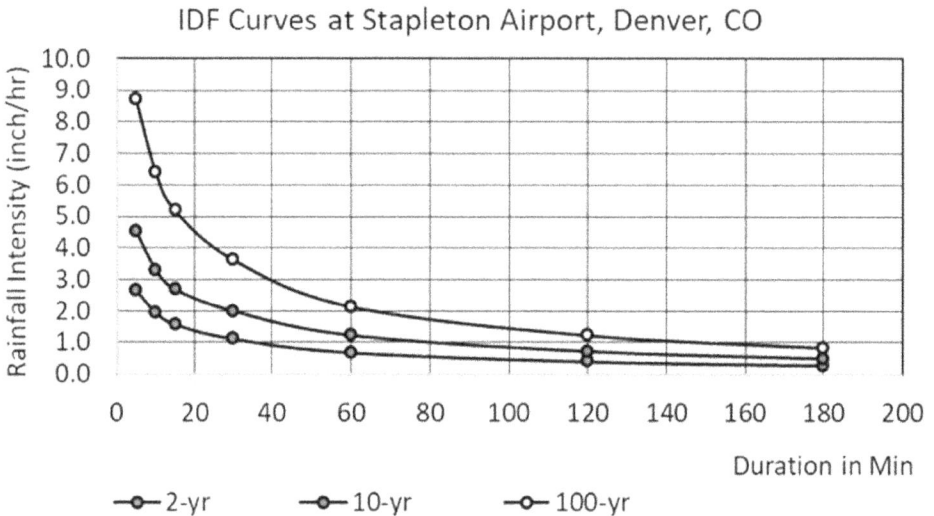

Figure 2.10 Rainfall IDF curves at Stapleton Airport, Denver, Colorado

Similarly, using the 100-yr one-hr, $P_1 = 2.41$ inches as the index precipitation depth, the 100-yr I–D curve is estimated as:

$$I = \frac{28.5 \times 2.41}{\left(10 + T_d\right)^{0.789}} \quad \text{for the 100-yr } I–D \text{ curve in Fig. 2.10} \tag{2.25}$$

2.5.3 Continuous precipitation data

Under the concept of low-impact-development (LID), hydrologic designs for extreme events need to be evaluated by all events observed on a long-term basis in order to detect the impact on the watershed regime. Usually, 1-hr continuous precipitation data series are recommended for continuous numerical simulations. The resultant flow-frequency and duration curves serve as a basis for evaluating the performance of a hydraulic facility. Such long-term data records are provided by the National Climate Center (NCC). The service centers of NCC are presented in Fig. 2.11.

2.6 DESIGN RAINFALL DISTRIBUTION

2.6.1 24-hour rainfall distribution curves

Hershield's studies in 1962 led to a set of 24-hour rainfall time-distributions. Each rainfall distribution covers a period from the clock time of 0:00 to 24:00, with the peak intensity at 12:00 or the center of the period of 24 hours. The *Natural Resources Conservation Service* (NRCS) developed several Soil Conservation Services (SCS) 24-hr rainfall curves to distribute the 24-hr precipitation depths recommended by the

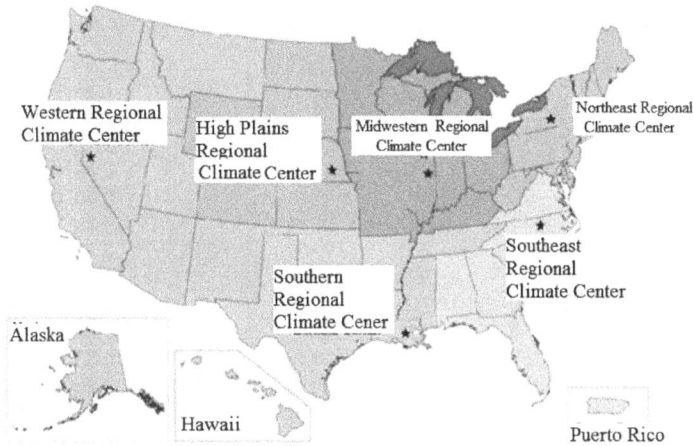

Figure 2.11 Centers of the National Climate Center

National Weather Service's precipitation-frequency map. As shown in Table 2.8, these SCS synthetic 24-hr rainfall distributions include Type I, Type IA, Type II, and Type IIA. The SCS 24-hour distributions are expressed as $p(t)/P_{24}$ ratios where $p(t)$ is the cumulative amount at time t, and P_{24} is the 24-hour precipitation depth for any selected frequency. These distributions are characterized by a sharp rising curve with a major portion of design rainfall depth precipitated in the central two hours. The SCS has also developed a non-dimensional rainfall distribution using 6-hour rainfall depth as listed in Table 2.8. The peak hours of the 6-hour distribution do not rise as sharply as the 24-hour distributions. Regional applicability of these curves is shown in Fig. 2.12.

As discussed before, the duration is the period of the burst that carries the highest intensity during a storm event. A synthetic 24-hr rainfall distribution consists of bursts of 2-, 6-, and 12-hr durations. A 6-hr bust is the central 6 hours from 9:00 to 15:00 around the peak time at 12:00. The 2-hr and 12-hr bursts are similarly centered around the peak time at 12:00. Therefore, a 24-hr rainfall curve also provides the most intense 2-hr, 6-hr, and 12-hr time-distributions. Fig. 2.13 presents the concept of 2-, 6-, 12-, and 24-hour rainfall distributions derived from the SCS Type I curve. The rainfall duration selected for hydrologic designs has to be compatible with the time of concentration of the watershed. It is critically important that a watershed is completely covered under the design storm. If not, then only a portion of the watershed is the tributary area. As a result, using a rainfall distribution shorter than the time of concentration of the watershed will underestimate the peak flow.

2.6.2 2-hour design rainfall distributions

In the front range of the Rocky Mountains, the average size of watersheds is approximately 5 to 10 square miles because of the hilly conditions. As a result, a 2-hr design storm is long enough to cover the entire watershed as a tributary area to the peak flow

Table 2.8 SCS rainfall distributions

Time (hours)	p(t)/P$_{24}$		Ratios		6-hr Curve
	Type I	Type II	Type IA	Type IIA	
0	0	0	0	0	0
0.50	0.0080	0.0050	0.0025		0.0350
1.00	0.0170	0.0110	0.0050		0.0800
1.50	0.0260	0.0160	0.0075		0.1350
2.00	0.0350	0.0220	0.0100	0.0500	0.2300
2.50	0.0450	0.0280	0.0150		0.6000
3.00	0.0550	0.0350	0.0200		0.7050
3.50	0.0650	0.0410	0.0250		0.7800
4.00	0.0760	0.0460	0.0300	0.0750	0.8350
4.50	0.0870	0.0560	0.0500	0.1400	0.8800
5.00	0.0990	0.0630	0.0600	0.1600	0.9250
5.50	0.1220	0.0710	0.1000	0.1900	0.9650
6.00	0.1250	0.0800	0.7000	0.2200	1.0000
6.50	0.1400	0.0890	0.7500	0.2500	
7.00	0.1560	0.0980	0.7800	0.2750	
7.50	0.1740	0.1090	0.800	0.3650	
8.00	0.1940	0.1230	0.8200	0.4500	
8.50	0.2190	0.1330	0.8300	0.485	
9.00	0.2540	0.1470	0.8400	0.5250	
9.50	0.3030	0.1630	0.8500	0.5500	
10.00	0.5150	0.1810	0.855	0.575	
10.50	0.5830	0.2040	0.8600	0.6050	
11.00	0.6240	0.2350	0.8650	0.6250	
11.50	0.6540	0.2830	0.8850	0.6500	
12.00	0.6820	0.6630	0.8900	0.6750	
12.50	0.7050	0.7350	0.9000	0.6900	
13.00	0.7270	0.7720	0.9050	0.7100	
13.50	0.7460	0.8000	0.9100	0.7250	
14.00	0.7670	0.8200	0.9150	0.7400	
14.50	0.7840	0.8400			
15.00	0.8000	0.8540			
15.50					
16.00	0.8300	0.8800	0.9400	0.8000	
16.50	0.8440	0.8910			
17.00	0.8570	0.9020			
17.50					
18.00	0.8820	0.9200			
18.50	0.8930	0.9290			
19.00	0.9050	0.9370			
19.50	0.9160	0.9450			
20.00					
20.50	0.9360	0.9590			
21.00	0.9460	0.9650			
21.50	0.9550	0.9720			
22.00					
22.50	0.9740	0.9840			
23.00					
23.50	0.9920	0.9950			
24.00	1.0000	1.0000	1.0000	1.0000	

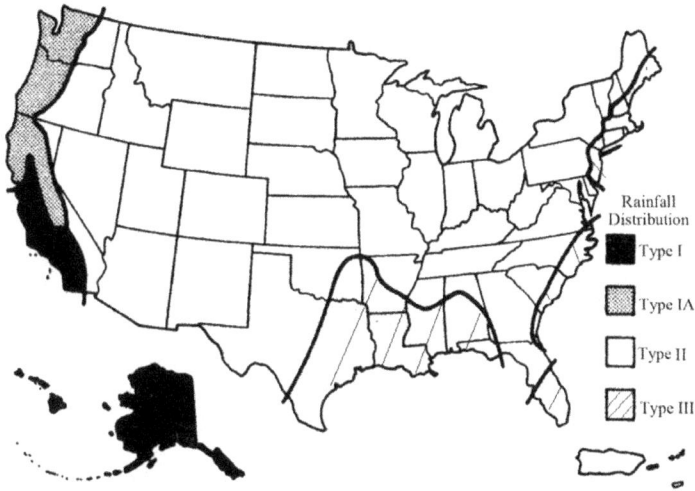

Figure 2.12 Applicability of SCS 24-hour rainfall distributions

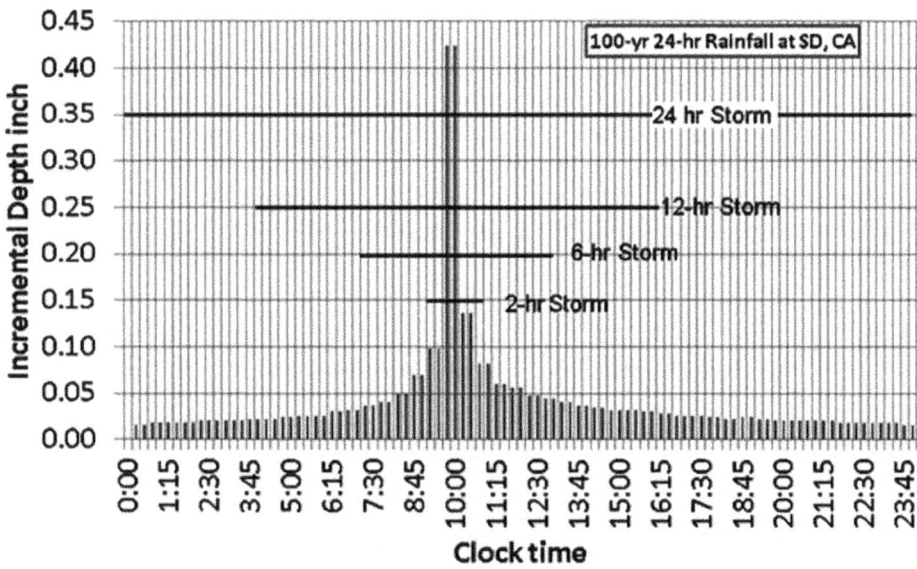

Figure 2.13 Duration of a design storm

at the outlet. Using the 1-hour rainfall depth, P_1, as the index precipitation, the IDF curve derived for the Denver area is:

$$I = \frac{28.5P_1}{(10+T_d)^{0.789}}$$
(2.26)

where I = rainfall intensity in inch/hr, P_1 = 1-hr precipitation depth in inches (see Table 2.7), and T_d = rainfall duration in minutes. The value of P_1 represents the design return period. Set the range of duration from 10 to 120 minutes with an increment of 10 minutes. The 12 incremental depths can be calculated as:

$$P(T_d) = I\, T_d, \text{ where } T_d = 10 \text{ to } 120 \text{ minutes with } \Delta t = 10 \text{ minutes} \qquad (2.27)$$

$$\Delta P = P(T_d + \Delta t) - P(T_d) \qquad (2.28)$$

Among these 12 incremental depths, the highest one is $P(10)$, which can be placed at the center of the duration of 120 minutes, and the rest of 11 incremental depths can be alternately decreased from the center to both directions.

Example 2.7: Considering P_1 = 2.41 inch, derive the 2-hr rainfall time-distribution for the City of Denver, Colorado.

Solution: As summarized in Table 2.9, apply Eqs 2.26, 2.27, and 2.28 to produce $I(T_d)$, $P(T_d)$, and $\Delta P(T_d)$, where T_d varies from 10 to 120 minutes with Δt = 10 minutes. Re-arrange the 12 values of $\Delta P(T_d)$ using a symmetric curve with its center at t = 60 minutes. For this case, the event precipitation P = 2.962 inches and the event duration T_r = 120 minutes. The last two columns present the normalized clock-time rainfall time-distribution.

Similarly, the two-hour rainfall distributions expressed as $p(t)/P_1$ in percentage versus clock time were derived for the City of Denver, Colorado, and are summarized in Table 2.10. The peak time for these synthetic rainfall distributions is set at t = 25 or 30 minutes. These 2-hr rainfall time-distributions are recommended for hydrologic designs in the front range of the Rocky Mountains. In fact, these 2-hour rainfall distributions are comparable to the sharp rise from 11:00 to 13:00 on the SCS Type IIA distribution.

Table 2.9 Example of rainfall time-distribution derived from IDF curve

Duration Td min	I(Td) inch/hr	P(Td) inch	ΔP(Td) inch	Clock t min	ΔP(t) inch	P(t) inch	Normalized Curve P(t)/P	t/Td
10	6.46	1.08	1.08	12:10	0.08	0.083	0.028	0.083
20	4.69	1.56	0.49	12:20	0.10	0.188	0.063	0.167
30	3.74	1.87	0.31	12:30	0.14	0.330	0.111	0.250
40	3.14	2.09	0.22	12:40	0.22	0.550	0.186	0.333
50	2.72	2.26	0.17	12:50	0.49	1.038	0.350	0.417
60	2.40	2.40	0.14	13:00	1.08	2.115	0.714	0.500
70	2.16	2.53	0.12	13:10	0.31	2.420	0.817	0.583
80	1.97	2.63	0.10	13:20	0.17	2.593	0.875	0.667
90	1.81	2.72	0.09	13:30	0.12	2.713	0.916	0.750
100	1.68	2.81	0.08	13:40	0.09	2.806	0.947	0.833
110	1.57	2.88	0.08	13:50	0.08	2.882	0.973	0.917
120	1.48	2.95	0.07	14:00	0.08	2.962	1.000	1.000

Table 2.10 Design storm distribution used in Colorado

Time min	2-yr	5-yr	10-yr	50-yr	100-yr
			p(t)/P₁		
0.00	0.00	0.00	0.00	0.00	0.00
5.00	2.00	2.00	2.00	1.30	1.00
10.00	10.00	4.00	3.70	3.70	3.50
15.00	15.00	8.40	8.00	5.60	4.00
20.00	16.00	8.00	8.70	8.00	8.00
25.00	25.00	25.00	25.00	25.00	14.00
30.00	14.00	13.00	12.00	12.00	25.00
35.00	8.30	6.30	5.80	5.60	14.00
40.00	5.00	5.00	4.40	4.30	8.00
45.00	3.00	3.00	3.60	3.80	6.20
50.00	3.00	3.00	3.60	3.20	5.00
55.00	3.00	3.00	3.20	3.20	4.00
60.00	3.00	3.00	3.20	3.20	4.00
65.00	3.00	3.00	3.20	3.20	4.00
70.00	3.00	3.00	3.20	2.40	2.00
75.00	2.50	2.50	3.20	2.40	2.00
80.00	2.20	2.20	2.50	1.80	1.20
85.00	2.20	2.20	1.90	1.80	1.20
90.00	2.20	2.20	1.90	1.40	1.20
95.00	2.20	2.20	1.90	1.40	1.20
100.00	1.50	1.50	1.90	1.40	1.20
105.00	1.50	1.50	1.90	1.40	1.20
110.00	1.50	1.50	1.90	1.40	1.20
115.00	1.50	1.50	1.70	1.40	1.20
120.00	1.50	1.50	1.30	1.40	1.20

Note: The total percentage is 115% that is approximately the ratio of P_2/P_1.

2.6.3 Derivation of localized design rainfall distribution

The time distribution of an observed rainfall event can be normalized using the event precipitation depth and duration. Fig. 2.14 presents five randomly selected storm events recorded at Stapleton Airport, Denver, Colorado. Comparisons with the SCS 24-hr Type I and II curves suggest that the design rainfall curves are constructed using a low enveloping curve for the leading portion and a high enveloping curve for the tail portion, with a sharp rise in-between. The steeper the sharp rise, the higher the peak discharge. If the local rainfall data is inadequate, the conservative approach is to combine the low and high enveloping curves with a sharp rise through the center of the rainfall distribution.

In comparison, the SCS Type I curve has a milder central rise. As expected, a Type I rainfall curve produces lesser peak flows than the Type II curve. Therefore, Type I is recommended for a harbor-based, hilly city like San Diego, CA, while Type II is more suitable for inland cities where the watersheds have a large tributary area. A Type I curve is also useful to depict the distribution of winter storms, which are long and mild, whereas Type II is better suited for thunderstorms, which are short and intense.

$t=Td$

$$\sum_{t=0} p(t)/P$$

Rainfall --Time Distribution

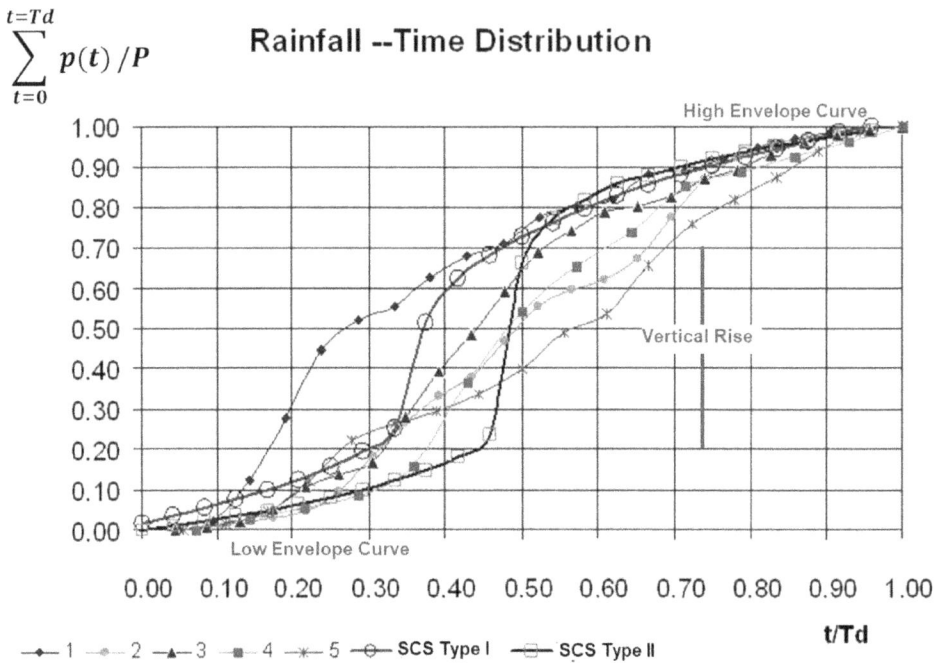

Figure 2.14 Normalized rainfall cumulative depth–time distributions

Table 2.11 Hourly incremental rainfall depths in inches for two storms in Taiwan

Date		Time (hour)													
	1	*2*	*3*	*4*	*5*	*6*	*7*	*8*	*9*	*10*	*11*	*12*	*13*	*14*	*15*
6/20	0.00	0.10	0.25	0.75	1.00	1.30	0.50	0.25	0.20	0.15	0.10	0.00			
11/23	0.00	0.10	0.15	0.20	0.30	0.40	0.70	1.20	1.50	2.00	1.50	1.20	0.80	0.30	0.00

Example 2.8: As always, the basic challenge in hydrologic design is not enough data. Table 2.11 presents the hourly incremental rainfall depths in two storm events observed on 6/20 and 11/23 in 2008 in Taiwan. Recommend a conservative design rainfall distribution.

Solution: Convert these two rainfall distributions into their mass curves. Normalize each mass curve by the total rainfall depth for the depth axis and the duration for the time axis. Plot these two normalized mass curves. Identify the low and high envelopes and the steepest connection. As shown in Fig. 2.15, the solid line represents the most severe rainfall distribution that will produce the highest flood flow, with 70% of the total rainfall amount blasted within 10% of the rainfall duration.

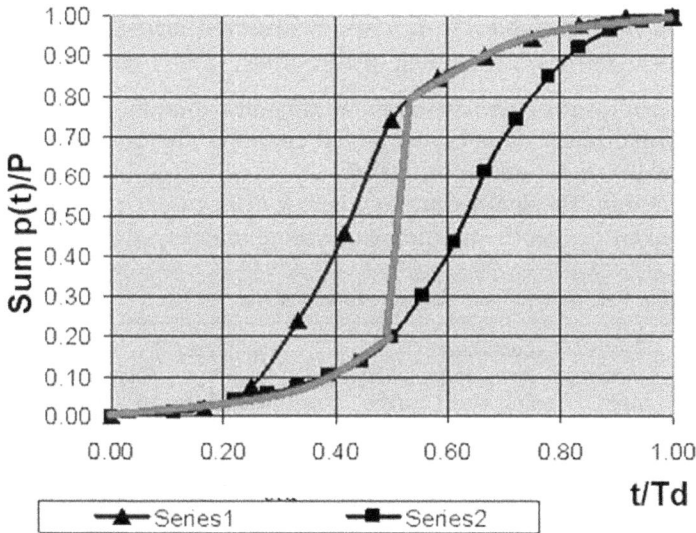

Figure 2.15 Conservative design rainfall distributions

2.7 CONCLUSIONS

It is important to understand the difference between design rainfall statistics and observed rainfall distributions. Rainfall statistics are preserved in the form of IDF tables or formulas. Although we can quantify the precipitation depths using IDF formulas such as 5-yr 10-min and 5-yr 15-min depths, these two depths are not the same as the observed depths. The former are the results of long-term frequency analyses, whereas the latter follows clock time. Therefore, a design rainfall distribution is a sequence of statistical values that represent the most likely rainfall time-distribution to come.

The IDF formula is most useful in dealing with small watershed hydrology (<150 acres), whereas SCS 24-hr distributions or any normalized rainfall curves are required for numerical rainfall-runoff simulations. Although there are many design rainfall distributions recommended for stormwater simulation studies, the proper design rainfall distribution for a specific project site should be selected according to the design rainfall duration. As a rule of thumb, the rainfall duration should be longer than the flow time through the waterway in the study area. A design rainfall distribution is a composite curve that is composed of leading and trailing envelope curves that are connected with a sharp peak. A large number of rainfall events are not required to derive a local design rainfall distribution. A thunderstorm should be modeled with a steep rise that carries 60 to 70% of the total rainfall amount, whereas a winter storm should have a mild rise carrying 40 to 50% of the total rainfall amount.

HOMEWORK

Q2-1 As presented in Table Q2-1, a set of observed incremental rainfall depths, $\Delta p(t)$, was recorded, according to clock time, t. Your tasks are:

1. Determine the cumulative rainfall time-distribution, pairs of $[P(t),t]$.
2. Normalize the cumulative rainfall curve by the event duration, t/T_d, and total precipitation depth, $P(t)/P$.
3. Determine the depth–duration pairs for this case.
4. Convert the depth–duration curve to an intensity–duration curve.

Table Q2-1 Rainfall depth–intensity–duration analysis

Clock Time t (min)	Rainfall Increment $\Delta p(t)$ (inch)	Cumulative Rain Depth P(t) (inch)	Normalized Time t/T_d	Normalized Rainfall Mass Curve P(t)/P	Duration D (min)	Depth P(D) (inch)	Intensity I(D) (inch/hr)
5.00	0.01		0.08		5.00		
10.00	0.02		0.17		10.00		
15.00	0.04		0.25		15.00		
20.00	0.13		0.33		20.00		
25.00	0.22		0.42		25.00		
30.00	0.18		0.50		30.00		
35.00	0.10		0.58		35.00		
40.00	0.04		0.67		40.00		
45.00	0.02		0.75		45.00		
50.00	0.01		0.83		50.00		
55.00	0.01		0.92		55.0		
60.00	0.01		1.00		60.00		

Q2-2 For the given rainfall intensity–duration, we can construct the rainfall duration–intensity plot for a range of durations from 5 to 50 minutes.

$$I(\text{inch/hr}) = \frac{74.1}{(10+D)^{0.789}}, \text{ where } D = \text{duration in minutes}$$

In order to reproduce the rainfall time-distribution (the temporal distribution), a symmetric distribution is chosen to have the highest $\Delta p(D)$ placed at the center of the rainfall event. Complete Table Q2-2 to reproduce the temporal distribution for the rainfall event.

Table Q2-2 Rainfall I–D curve and time distribution

Duration, D (min)	I(D) (inch/hr)	P(D) (inch)	$\Delta p(D)$ (inch)	Clock time, t (min)	$\Delta P(t)$ (inch)	P(t) (inch)	P(t)/P	t/T_d
5	8.75	0.73	0.73	5	0.11	0.11	0.11	0.11
10	6.97	1.16	0.43	10	0.15	0.15	0.15	0.15
15	5.85	1.46	0.30	15	0.23	0.23	0.23	0.23
20				20	0.43			
25				25	0.73			
30				30	0.30			

Table Q2-2 (Continued)

Duration, D (min)	I(D) (inch/hr)	P(D) (inch)	Δp(D) (inch)	Clock time, t (min)	ΔP(t) (inch)	P(t) (inch)	P(t)/P	t/T_d
35				35	0.18			
40				40	0.13			
45				45	0.10			
50				50	0.09			

Q2-3 You are given a record of 60-min precipitation depths for a continuous period of 11 years. Fill in your answers to complete the frequency analyses.

REFERENCES

Beard, L.R. (1962). *Statistical Methods in Hydrology*, US Army Corps of Engineers, Sacramento, CA.

Bulletin 17 (1983), "Guidelines for Determining Flood Flow Frequency", Interagency Advisory Committee on Water Data, Bulletin #17B of the Hydrology Subcommittee, OWDC, US Geological Survey, Reston, VA.

Chow, Ven T., Maidment, D.R., and Mays, L.M. (1988). "*Applied Hydrology*", McGraw Hill, New York.

Gumbel, E.J. (1954). "The Statistical Theory of Theory of Droughts", *Proceedings of the American Society of Civil Engineers*, vol. 80, pp. 419–439.

Guo, James C.Y. (1986). *Software for Hydrologic Frequency Analysis*, Fourth National Conference on the Application of Microcomputer to Civil Engineering, Orlando, FL, November, pp. 308–310.

Han, Charles T. (1977). *Statistical Method in Hydrology*, Iowa State University Press, Ames, IO.

Harter, H. Leon (1969), "A New Table of Percentage Points of the Pearson Type III Distribution", *Technometrics*, vol. 11, no. 1, pp. 177–187.

Harter, H. Leon (1971), "More Percentage Points of the Pearson Distribution", *Technometrics*, vol. 13, no. 1, pp. 203–204.

HEC SSP (2005). *Hydrologic Statistical Package*, Hydrologic Engineering Center, US Army Corps of Engineers, Davis, CA.

HEC-1 Flood hydrograph Package (1985). Hydrologic Engineering Center, US Army Corps of Engineers, Davis, CA.

Hershfield, D.M. (1961). "Rainfall Frequency Atlas of the United States for Durations from 30 Minutes to 24 Hours and Return Periods from I to 100 Years", Tech. pap. 40, US Department of Commerce, Weather Bureau, Washington, DC, May.

Hershfield, D.M. (1962). "Extreme rainfall relationships", *Journal of the Hydraulics Division – American Society of Civil Engineers*, vol. 88 (HY6), pp. 73–92.

Kite, G.W. (1977). *Frequency and Risk Analysis in Hydrology*, Water Resources Publications, Littleton, CO.

Peak FQ Program (2005). User's Manual for Computer Program PEAK FQ, Office of Surface Water, USGS. http://water.usgs.gov/osw/bulletin17b/bulletin_17B.html.

Wallis J.R., Matalas, N.C., and Slack, J.R. (1974). "Just a Moment", *Water Resources Research*, vol. 10, no. 2.

Chapter 3

Runoff hydrology

3.1 WATERSHED LAND USES

After subtracting hydrologic losses, excess rain is a major source for storm runoff. Flood flows impose many challenges to traffic safety and flood inundation in urban areas. Mitigating stormwater has become a primary directive in reducing negative impacts on aging and undersized stormwater drainage systems. Many innovative green concepts have been developed to take advantage of grass filtering, soil infiltration, wetland settlement, and detention storage to enhance stormwater quality control and to reduce post-development peak runoff flows for quantity control. In this chapter, the major effort is to understand numerical algorithms to quantify hydrologic losses and to identify the characteristics of the runoff hydrograph.

Ground surface textures are composed of *impervious* and *pervious* areas. Soil infiltration losses only occur in pervious areas. Impervious areas are subject to almost no infiltration losses. The runoff volume generated from an urban catchment is highly sensitive to the percentage of the area that is impervious. The greater the impervious area, the faster the runoff flow. The more concentrated the surface runoff, the greater the runoff flow rates and volumes. the increase of impervious areas is an inevitable trend in urbanized areas. Many flood prediction methods directly correlate the impact of urbanization on storm runoff to the percentage of impervious area in the watershed.

3.2 HYDROLOGIC TYPES OF SOILS

An infiltration rate reflects the ability of the soil medium to absorb water. This parameter is usually given in inches per hour or millimeters per hour. The US Natural Resources Conservation Service (NRCS 1964) has developed a set of SCS classifications of soils. In general, all soils are categorized into four hydrologic types, A, B, C, and D, based on their infiltrating nature:

Type A soil: soils with high infiltration rates of between 0.30 and 0.45 inch/hour if thoroughly wetted, consisting mainly of moderately deep, well to excessively drained sand and gravel.
Type B soil: soils with moderate infiltration rates of between 0.15 and 0.30 inch/hour if thoroughly wetted, consisting mainly of moderately deep to deep, moderately well to well-drained soils with moderately fine to moderately coarse textures, such as loamy soils.

DOI: 10.1201/9781003284239-3

Type C soil: soils with slow infiltration rates of between 0.05 and 0.15 inch/hour if thoroughly wetted, consisting mainly of soils with a layer that impedes the downward movement of water or soils with moderately fine to fine textures.

Type D soil: soils with very slow infiltration rates of between 0.01 and 0.05 inch/hour if thoroughly wetted, consisting mainly of clay soils with swelling potential, highly saturated soils, soils with a clay pan or clay layer at or near the surface, and shallow soils over nearly impervious materials.

Infiltration rates are described by a decay function changing from a high rate at the beginning of the event when the soil is dry to a low rate when the soil becomes saturated. When the watershed has several different types of soils, the representative infiltration rate can be determined as the area-weighted value.

3.3 HYDROLOGIC LOSSES

3.3.1 Interception losses

Interception is the portion of the precipitation that is retained by leaves and stems of vegetation or other obstructions that prevent raindrops from falling on the ground. *Interception loss* is also called *initial loss* and is expressed in millimeters or inches per watershed. Interception loss is proportional to the total precipitation in an event as (Horton 1933):

$$L_a = 0.04 + 0.18P \tag{3.1}$$

where L_a = interception loss in inches or mm, and P = precipitation in inches or mm in Eq. 3.1. In an urban area, the interception loss is a relatively small fraction of the precipitation. In practice, this amount is subtracted from the earliest precipitation amount at the beginning of the event.

3.3.2 Infiltration losses

Infiltration is the process of transferring surface water into the top layer of soil. A soil layer becomes saturated when its pores are filled up with water. Seepage flows through soil layers may move laterally into streams and lakes or vertically into groundwater aquifers. The infiltration rate varies with respect to the topsoil texture and the water content of the soil. The soil porosity, θ_s, of the soil column is defined as:

$$\theta_s = void\ volume/column\ volume \tag{3.2}$$

The initial water content, θ_o, in the soil column is defined as:

$$\theta_o = water\ volume/void\ volume\ \text{at the beginning of a storm event} \tag{3.3}$$

The initial water content in a soil column cannot exceed θ_s. Soil porosity ranges from 0.25 for loam to 0.35 for sand and 0.45 for gravel. For convenience, the infiltration volume is expressed as a depth per unit area, such as mm or inches. The *infiltration rate*, $f(t)$, is expressed in inch/hr or mm/hr. As illustrated in Fig. 3.1, the initial

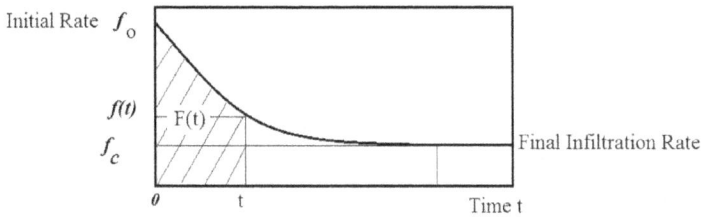

Figure 3.1 Variation of the soil infiltration rate

Table 3.1 Recommended soil infiltration rates

Soil Type	Initial Rate (inch/hr)	Final Rate (inch/hr)	Decay Coeff (1/hr)
A	4.50	0.60	6.48
B	4.00	0.55	6.48
C	3.00	0.50	6.48
D	3.00	0.50	6.48

infiltration rate, f_0, at the beginning of a storm event tends to be higher because the soil column is drier. As time goes on, the infiltration rate decays to its final rate, f_c, after the soils become saturated. The *infiltration volume or amount*, $F(t)$, over a period of time, t, is the shaded area under the curve of infiltration rate.

Horton's Formula

Under a constant headwater, Horton's formula was developed as an exponential decay model using three parameters to depict the water volume transfer rate from the surface into the top layer of a soil column. The infiltration rate varies, with the highest rate representing dry conditions in the soil column and the lowest rate after the soil column becomes saturated. Horton's formula states:

$$f(t) = f_c + (f_0 - f_c)e^{-kt} \tag{3.4}$$

where $f(t)$ = infiltration rate in [L/T] at time t, f_c = final rate in [L/T], f_0 = initial rate in [L/T], and k = decay constant in [1/T] such as 1/hour or 1/second. As a continuous decay curve as Eq. 3.4, it implies that the rate of arriving rainfall is always higher than the infiltration capacity. Otherwise, the actual infiltration rate depends on the availability of the arriving rainfall depth. The initial and final infiltration capacity in Horton's formula depend primarily on the soil's type and initial moisture content, and on surface vegetation conditions. The recommended design values for the three parameters in Eq. 3.4 are summarized in Table 3.1 (UDFCD 2010):

Using the finite difference approach, the incremental infiltration amount, $\Delta F(t)$, over a time interval Δt is calculated as:

$$\Delta F(t + \Delta t) = \frac{1}{2}\left[f(t) + f(t + \Delta t)\right]\Delta t \tag{3.5}$$

Integrating Eq. 3.4 yields the cumulative water volume, $F(t)$, infiltrated into the soils over a period of time, t, as:

$$F(t) = \int_{t=0}^{t} f(t)\,dt = f_c t + \frac{(f_0 - f_c)}{k}\left(1 - e^{-kt}\right) \approx \sum_{t=0}^{t=t} \Delta F(t) \qquad (3.6)$$

For engineering practice, the integral may be replaced with the numerical summation in Eq. 3.6. When the operational time is long enough, Eq. 3.6 is reduced to:

$$F(t) = f_c t + \frac{f_0 - f_c}{k} \quad (t => \text{sufficiently long}) \qquad (3.7)$$

Eqs 3.4 to 3.7 are only applicable to infiltration processes on the land surface or top layer of soils. As the water volume is percolated into the soils, the movement of the seepage flow is dictated by groundwater hydraulics. Therefore, the prediction of seepage flow requires the soil hydraulic conductivity and hydraulic gradient.

Green and Ampt

The *Green and Ampt model* was developed to estimate the seepage flow through soil pores as a diffusion process. The continuity principle for moisture diffusion is written as:

$$\frac{\partial \theta(t)}{\partial t} + \frac{\partial V(t)}{\partial z} = 0 \qquad (3.8)$$

where V = seepage flow in [L/T], z = downward vertical distance in [L], and t = elapsed time in [T]. Re-arranging Eq. 3.8, the downward movement of the wetting front, Δz, is calculated as:

$$\Delta F(t) = \Delta V(t)\Delta t = \Delta \theta(t)\Delta z(t) \qquad (3.9)$$

As illustrated in Fig. 3.2, the moisture deficit, $\Delta \theta$, in the soil column is the difference between the current moisture level and the soil porosity as:

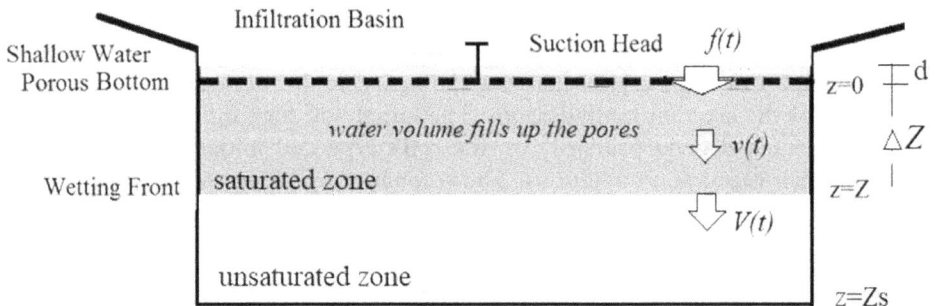

Figure 3.2 Illustration of infiltration flow and seepage flow

Table 3.2 Saturated soil hydraulic conductivity

Soil Type	Capillary Suction Head ψ mm of water	Moisture Deficit $\Delta\theta = \theta_s - \theta_o$	Hy Conductivity cm/sec
Gravel 3/4 inch	2.5	0.36	10^{-2} to 10^0
Sand	4.00	0.34	10^{-3} to 1.0^{-1}
Sandy Loam	8.00	0.33	
Silt Loam	12.00	0.32	10^{-5} to 10^{-3}
Loam	8.00	0.31	
Clay Loam	10.00	0.26	
Clay	7.00	0.21	10^{-5} to 10^{-3}

$$\Delta\theta = \theta_s - \theta_o \tag{3.10}$$

The seepage flow is dictated by the soil hydraulic conductivity and the energy gradient. The hydraulic head associated with the seepage flow is composed of the vertical travel distance and the suction head due to soil capillary effects. A suction head is often expressed as an additional headwater depth in [L] above the water surface. According to Darcy's law, the vertical velocity, $V(t)$, of the wetting front at $z(t)$ is calculated as:

$$V(t + \Delta t) = K_s i = K_s \left(\frac{z(t) + d + \psi}{z(t)} \right) \approx K_s \left(1 + \frac{\psi}{z(t)} \right) \tag{3.11}$$

where I = hydraulic gradient in [L/L], K_s = saturated hydraulic conductivity in [L/T], $z(t)$ = saturated downward distance in [L] from $z(t = 0) = 0$ to Z_s which is the height in [L] of the soil column, d = water depth in [L] as a sheet flow on the land surface, ψ = suction head in [L]. Table 3.2 lists the recommended soil parameters for various types of soils. In practice, the shallow water depth, d, is numerically negligible.

3.3.3 Movement of the wetting front

Like a culvert, the flow capacity is dictated by the inlet control at the entrance and the outlet control through the length of the conduit. The movement of the wetting front in a soil column is dictated by the water infiltrating rate, $f(t)$, on the land surface and the seepage flow rate, $V(t)$, through the soil medium. The smaller value dictates the wetting front movement as:

$$v(t) = \min(f(t), V(t)) \tag{3.12}$$

where v = downward velocity in [L/T] for the wetting front. At the elapsed time, t, the infiltration water fills up the soil pores to advance the wetting front to the location, $z(t)$, as:

$$\Delta F(t) = 0.5(v(t) + v(t - \Delta t))\Delta t \tag{3.13}$$

$$\Delta z(t) = \frac{\Delta F(t)}{(\theta_s - \theta_o)} = \frac{\Delta F(t)}{\Delta \theta} \tag{3.14}$$

$$F(t) = F(t - \Delta t) + \Delta F(t) \tag{3.15}$$

$$Z(t) = Z(t - \Delta t) + \Delta z(t) \tag{3.16}$$

The movement of the wetting front is the loading process to saturate the soil column. After $Z(t) \approx Zs$, the soil column becomes saturated. The seepage flow will become steady to percolate through the soil column to recharge the local groundwater table.

Example 3.1: As shown in Table 3.3, the water infiltration on a sandy soil surface is described by Horton's formula with f_o = 3.0 inch/hr, f_c = 0.50 inch/hr, and the decay coefficient = 0.0018 per second. The soil has a porosity of 0.35, an initial water content of 0.050, a suction head of 0.32 inches, and a hydraulic conductivity of 0.42 inch/hr. Determine the soil saturation process for Zs = 12 inches as the wetting front moves downward.

Solution: As shown in Table 3.3, the seepage flow fills up soil pores until t = 360 min. At t = 360 min, the 12-inch soil column becomes saturated. After t = 360 min, the constant seepage flow becomes steady to percolate through the soil column.

3.3.4 Depression losses

Depression losses account for the early runoff volume trapped in lowland areas, puddles, and potholes without running off. These volumes are eventually dispersed through soil infiltration and evaporation. It is important to quantify the *depression storage capacity* before estimating runoff flows. The depression loss is expressed as the sum of the entrapped water volume in inch or mm per watershed. Depression storage may be treated as a calibration parameter, particularly to adjust runoff volumes generated from pervious and/or impervious areas. Table 3.4 is a summary of the recommended depression capacity for use in computer modeling.

Values in Table 3.4 represent the *maximum depression storage volume*, D_m, available in a lowland area. During an event, the depressed area will not be filled until the rainfall depth exceeds the soil infiltration capacity. After the low land is filled up to D_m, the overland flows will overtop the depressed area to produce concentrated runoff flows.

3.3.5 Surface detention volume

As illustrated in Fig. 3.3, the *surface storage capacity* includes two portions: (1) *depression losses* and (2) the *surface detention* volume under the water surface profile of the overland flow. Depression losses are counted as a part of hydrologic losses, whereas the surface detention volume is temporarily stored as the water flow depth, which will be gradually released to form the recession hydrograph.

Table 3.3 Soil saturation process through 12-inch soil column

Suction Head	ψ =	0.32 inches		Soil Porosity	θs =	0.38
Infiltration	initial rate fo =	3.00 inch/hr		Initial Water Content	θo =	0.05
	final rate fc =	0.50 inch/hr		Moisture Deficit	$\Delta\theta$ =	0.33
Decay factor	k =	0.0018 1/sec		Beginning Time	to =	0.00 min
Hy Conductivity	Ks =	0.42 inch/hr		Incremental Time	Δt =	60.00 min
Soil Depth	Zs =	12.00 inches				

Elapsed Time t min	Surface Flow f(t) inch/hr	Seepage Flow V(t) inch/hr	Actual Flow v(t) = min(f,V) inch/hr	Incremental Infiltrating Volume ΔF(t) = Avg v(t) x dt inches	Cumulative Water Volume F(t) inches	Movement of W Front ΔZ(t) = ΔF/Δθ inches	Downward Distance Z inches
1	2	3	Min(2 and 3)	5	Sum (Col 5)	7	Sum (Col 7)
0	3.000	3.000	3.000	0	0.000	0.000	0.000
60	0.504	3.000	0.504	1.752	1.752	5.309	5.309
120	0.500	0.445	0.445	0.475	2.254	1.438	6.747
180	0.500	0.440	0.440	0.443	2.754	1.341	8.088
240	0.500	0.437	0.437	0.438	3.254	1.328	9.416
300	0.500	0.434	0.434	0.435	3.754	1.320	10.736
360	0.500	0.433	0.433	0.433	4.254	1.313	12.049
420	0.500	0.431	0.431	0.432	4.754	0.431	12.480
480	0.500	0.431	0.431	0.431	5.254	0.431	12.911
540	0.500	0.430	0.430	0.431	5.754	0.430	13.341
600	0.500	0.430	0.430	0.430	6.254	0.430	13.772

>Zs

Table 3.4 Depression losses for various land uses

Land cover	Range(inches)	Design Value(inches)
Large Paved Area	0.05-0.15	0.10
Flat Roofs	0.10-0.30	0.10
Sloped Roofs	0.05-0.10	0.05
Lawn Grass	0.20- 0.50	0.03
Wooded Area	0.20- 0.60	0.40
Open Fields	0.20-0.60	0.40
Sandy Area		0.02
Loams		0.15
Clay		0.10

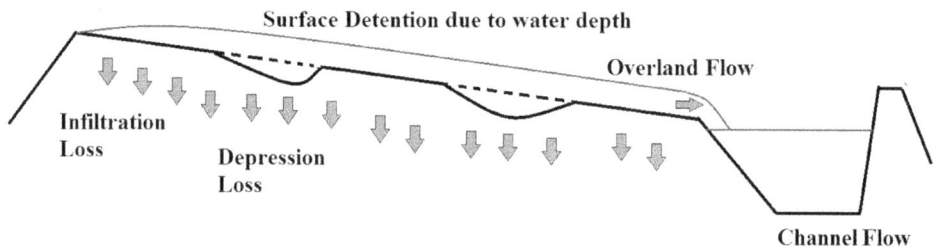

Figure 3.3 Surface storage

The flow depth associated with an overland flow represents the surface detention volume. An overland flow is a one-dimensional sheet flow that can be calculated as:

$$q = \frac{k}{N} y^{\frac{5}{3}} \sqrt{S_o}$$

(3.17)

where q = unit-width overland flow in [L²/T], k = 1 when using SI units or 1.486 when using English units, N = Manning's roughness coefficient, y = flow depth in [L], and S_o = ground slope in [L/L]. Overland flows are sensitive to the ground surface roughness. Values of the Manning's roughness coefficient are not as well known for overland flow as for channel flow because of the considerable variability in ground cover, transitions between laminar and turbulent flow, very small depths, etc. Table 3.5 presents estimates of the Manning's roughness coefficient for overland flows on various surface textures.

3.4 EXCESS RAINFALL

As mentioned above, excess rainfall is the amount of rainfall that survives hydrologic losses and then turns into overland flows. Excess rainfall is also called net rainfall or the runoff depth. Soil infiltration and depression losses are the major reductions from the observed rainfall distribution. Obviously, the natural rainfall reduction process

Table 3.5 Recommended Manning's coefficient

Surface Texture	Recommended Value for N	Range of N
Concrete or asphalt	0.011	0.01 - 0.013
Bare sand	0.01	0.01 - 0.016
Gravelled surface	0.02	0.012 - 0.033
Bare clay-loam (eroded)	0.02	0.012 - 0.033
Range (natural)	0.13	0.01 - 0.32
Bluegrass sod	0.45	0.39 - 0.63
Short grass prairie	0.15	0.10 - 0.20
Bermuda grass	0.41	0.30 - 0.48

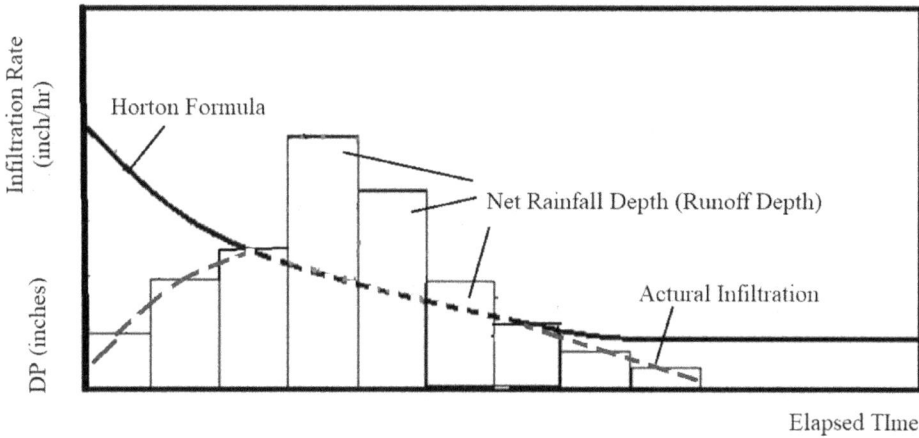

Figure 3.4 Seepage flow model to estimate net rainfall depth

is complicated and it takes time to transfer surface runoff into the sub-surface soil layers. Numerical algorithms are developed to rationalize the hydrologic loss process in a timely order. Although none of these numerical methods fully represent reality, they offer a way of quantifying the runoff depth for storm hydrograph prediction. Two methods are considered in this chapter. They are: (1) the seepage flow model and (2) the soil storage model.

3.4.1 Seepage flow model for rainfall reduction

After it rains, the decay of the soil infiltration rate follows Horton's formula. As illustrated in Fig. 3.4, it does not mean the Horton's infiltration rate can be sustained. In reality, the actual infiltration rate is also dictated by the availability of the amount of rainfall. Obviously, the infiltration amount cannot be higher than the available rainfall amount.

$$\Delta F_a(t) = \min\left[\Delta F(t), \Delta P(t)\right] \tag{3.18}$$

where $\Delta F_a(t)$ = actual infiltration amount in [L] at time t, $\Delta F(t)$ = predicted infiltration amount in [L] by an empirical formula such as Horton's formula, and $\Delta P(t)$ = incremental rainfall depth in [L]. After the incremental rainfall depth survives the infiltration loss, the initial flow will have to fill up the depressed area before overland flows can be produced. The net rainfall depth is also called the incremental runoff depth, $\Delta R(t)$, which is calculated as:

$$\Delta R(t) = \Delta P(t) - \Delta F_a(t) - \Delta D(t) \quad \text{or zero if } \Delta R(t) \le 0 \tag{3.19}$$

where $\Delta D(t)$ = incremental depression depth in [L] and $\Delta R(t)$ = runoff depth in [L] at time t. Each $\Delta R(t)$ applies to the unit hydrograph to convolute the storm hydrograph.

Example 3.2: A rainfall event is observed with a total precipitation depth of 1.363 inches. The rainfall distribution and soil infiltration rates are described in Table 3.6. If the depression loss is 0.42 inch, derive the amount of rainfall excess.

Solution: Table 3.6 presents the numerical rainfall reduction process based on the seepage flow model. Without considering the soil storage capacity, the numerical method leads to a net 10-min rainfall depth of 0.316 inch, which was likely to be produced from this rainfall event. All of these numerical methods need calibration

Table 3.6 Rainfall reduction without considering soil moisture deficit

Depression Losses		0.42 inch					
Infiltration	Initial Rate	3.00 inch/hr					
	Final Rate	0.50 inch/hr					
	Decay Coef	6.25 1/hr					

			Soil Infiltration				
	Incremental Rain Depth	Horton	Horton		Initial Flow Depth	Incremental Depression	Net Rain Depth ΔR
Time min	$\Delta P(t)$ inches	f inches	ΔF inches	Actual ΔFa inches	ΔP-ΔFa inches	ΔD inches	Depth ΔR inches
a	b	c	d	e = min (b,d)	f = b-e	g	h = f-g
0	0	3.000	0.000	0	0	0.000	0.000
10.00	0.151	1.38	0.365	0.151	0.000	0.000	0.000
20.00	0.585	0.81	0.183	0.183	0.402	0.402	0.000
30.00	0.452	0.61	0.118	0.118	0.334	0.018	0.316
40.00	0.095	0.54	0.096	0.095	0.000	0.000	0.000
50.00	0.055	0.51	0.088	0.055	0.000	0.000	0.000
60.00	0.025	0.50	0.085	0.025	0.000	0.000	0.000
Total	1.363			0.627	0.736	0.420	0.316

Column c = predicted by Horton formula
Column d = 0.5[f(t)+f(t–Δt)]Δt
Column e = min [$\Delta P(t)$,$\Delta F(t)$]
Column f = Column b – Column e
Column g = assigned incremental depression losses to meet the total of 0.42 inches.
Column h = Column f – Column g

using field data to choose the correct parameters for the calculations and then need verification using observations from monitoring systems.

3.4.2 Soil storage model for rainfall reduction

The soil column acts as a sponge, and it exhibits a moisture deficit at the beginning of a rainfall event. The area under the Horton infiltration curve presents the required cumulative infiltration, $F(t)$ at time t, to saturate the pore storage capacity in the soil column. At each time step, Δt, the soil moisture deficit, as marked in the shaded area in Fig. 3.5, is defined as the shortage between the cumulative rainfall depth, $P(t)$, and the required infiltration amount, $F(t)$. Initial flows will not be produced until $P(t) > F(t)$. Next, the initial flows have to fill up the depressed area before overland flows can be produced. Therefore, it is recommended that the numerical computation be conducted using the cumulative rainfall depth, $P(t)$, and cumulative infiltration amount, $F(t)$, to determine the *ponding time*, which is the required period of time to satisfy the soil moisture deficit before initial flow can occur.

Example 3.3: Continued from Example 3.2. A rainfall event is observed with a total precipitation depth of 1.494 inches. The rainfall distribution is described in Table 3.7. If the depression loss is 0.42 inch, derive the amount of rainfall excess.

Solution: As illustrated in Table 3.7, at $t = 10$ min, the soil infiltration is equal to the rainfall depth available. At $t = 20$ min, the cumulative infiltration amount defined under Horton's infiltration curve is:

$F(t = 20) = 0.365 + 0.183 = 0.548$ inch

In this case, the pond time is approximately $t = 20$ min. At $t = 20$ min, the actual infiltration amount is accumulated as:

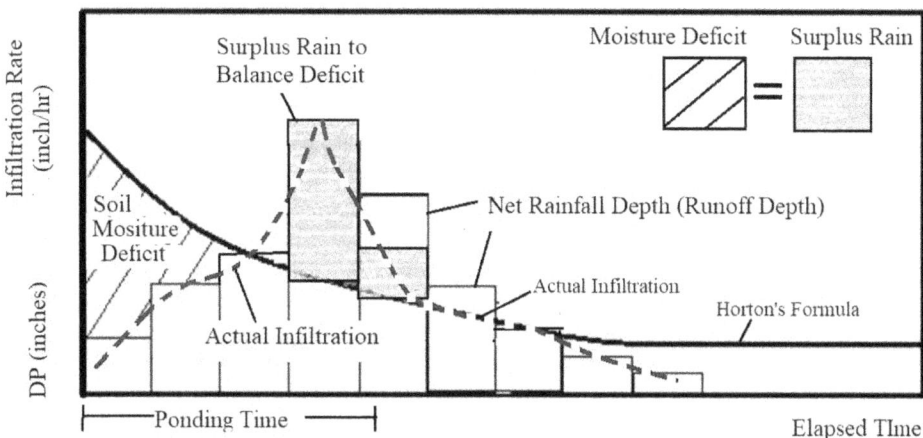

Figure 3.5 Soil storage model to estimate net rainfall depth

Table 3.7 Rainfall excess derived from hydrologic losses

	Depression Losses		0.42	inch
	Infiltration	Initial Rate	3.00	inch/hr
		Final Rate	0.50	inch/hr
		Decay Coef	6.25	l/sec

Time min a	Incremental ΔP(t) inches b	Soil Horton ΔF inches c	Infiltration Actual ΔFa inches d = min(b,d)	Initial Flow Depth ΔP−ΔFa inches e = b − d	Incremental Depression ΔD inches f	Net Rainfall Depth ΔR inches g = e − f
0.00	0.000	0.000	0.000	0.000		0.000
10.00	0.151	0.365	0.151	0.000	0.000	0.000
20.00	0.585	0.183	0.397	0.188	0.188	0.000
30.00	0.452	0.118	0.118	0.334	0.232	0.102
40.00	0.095	0.096	0.095	0.000	0.000	0.000
50.00	0.055	0.088	0.055	0.000	0.000	0.000
60.00	0.025	0.085	0.025	0.000	0.000	0.000
Total	1.363		0.841	0.522	0.420	0.102

$$Fa(t = 20) = \Delta fa(t = 10) + \Delta fa(t = 20) = 0.151 + \Delta fa(t = 20) = 0.548 \text{ inch}$$

So, $\Delta fa(t = 20) = 0.397$ inch, which is the amount transferred from the peak rainfall depth at $t = 20$ min and used to balance the soil moisture deficit.

Table 3.7 presents the numerical rainfall reduction process based on the soil storage model. Considering the soil moisture deficit, the numerical method indicates that a net 10-min rainfall depth of 0.102 inch was likely to be produced from this rainfall event. As expected, the latter produced less net rainfall than the former because of more infiltration losses in the soil storage model. Both numerical methods produce consistent predictions. Their accuracy can be refined until sufficient model calibrations are obtained.

3.5 RUNOFF HYDROGRAPH

Stream flows are recorded at a gauge station continuously in time. Prior to a storm event, the stream gauge registers the *base flow* in the river. During the storm event, the response of the watershed is registered as a *runoff hydrograph*, which shows the variation of flow rates with respect to time. A *single-event hydrograph* is often used to design hydraulic structures, and *long-term hydrographs* provide design information for water resources planning. Unlike small catchments, stormwater characteristics in a large watershed are complicated because of the differences in elevation, slope, soil, and vegetation cover. In the case that the delay of surface runoff becomes so significant that it interferences with subsurface and groundwater flows shall also be considered. In practice, the stream gauge network in a large watershed is not adequate to provide complete rainfall-runoff information for the entire drainage area. Therefore, a procedure such as a *synthetic hydrograph* needs to be developed so that the flood discharge at ungauged sites can be related to observations at nearby gauged sites. In general, such a method requires a large amount of data and extensive calibration.

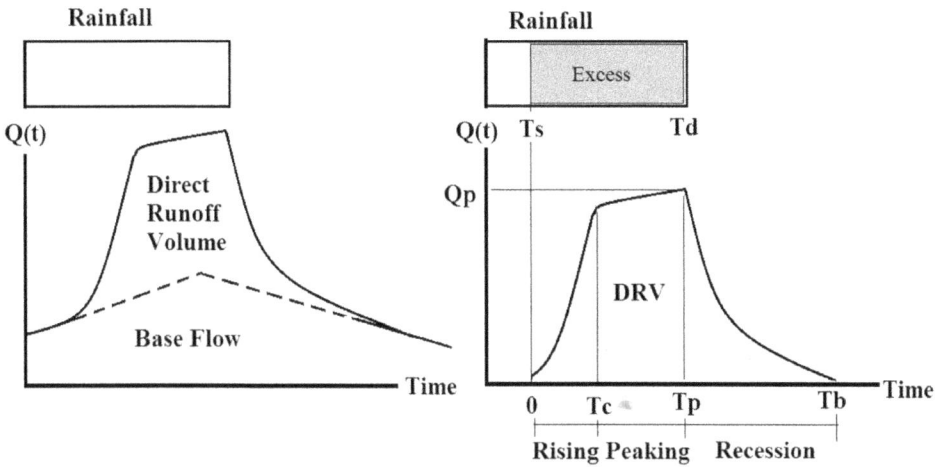

Figure 3.6 Runoff hydrograph

Before it rained, the waterway carried a base flow that came from the local ground-water table. An observed runoff hydrograph consists of both the base flow and the direct flow, which is generated from a storm event. Use the two tangent lines as illustrated in Fig. 3.6 to separate the base flow from an observed runoff hydrograph. The *direct runoff volume (DRV)* is then calculated as:

$$V = \int_{t=0}^{t=T_b} Q(t)\,dt = \Delta t \Big[\sum_{t=0}^{t=T_b} Q(t)\Big] \quad \text{where } 0 \le T \le T_b \tag{3.20}$$

where V = DRV in [L³], $Q(t)$ = direct runoff rate in [L³/T] at time t, T_s = ponding time to fill up potholes and soak the surface soils, T_b = base time, and Δt = time interval. Eq. 3.20 indicates the total DRV is equal to the sum of runoff ordinates multiplied by the time interval.

Example 3.4: Continued from Example 3.2. The observed storm hydrograph is represented in Fig. 3.7. Based on the conclusion in Example 3.2, derive the unit hydrograph.

Solution:

1. Derive the direct hydrograph by subtracting the groundwater base flow.
2. All flow ordinates on the direct hydrograph are divided by the net rainfall depth.
3. The hydrograph derived in (2) is a 10-minute unit hydrograph for this watershed.
4. Knowing that the hydrograph runoff volume = 1 inch × watershed area, find the watershed area. (Hint: runoff volume = 1 inch × watershed area)

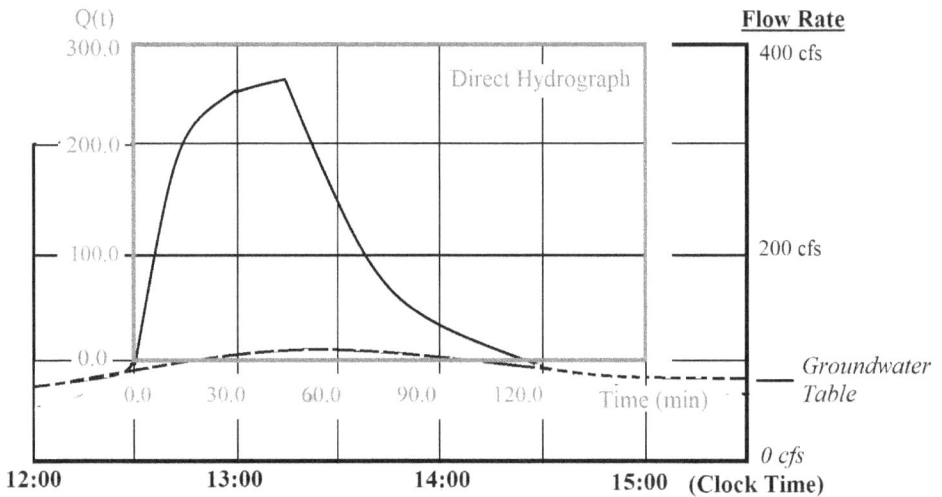

Figure 3.7 Direct runoff hydrograph derived from an observed storm hydrograph

3.5.1 Runoff hydrograph analysis

As illustrated in Fig. 3.8, a *direct runoff hydrograph* (DRH) can be represented by seven points, including:

1. the starting point at $(Q, t) = (0,0)$,
2. the peak time at $(Q, t) = (Q_p, T_p)$,
3. two points to define the time width, W_{50}, at 50% of peak runoff, $50\%Q_P$,
4. two points to define the time width, W_{75} at 75% of peak runoff, $75\%Q_P$, and
5. the end point at $(Q, t) = (0, T_b)$.

These seven points on the DRH are directly related to the following important time parameters on the DRH:

1. the *ponding time*, T_s, which is the period of time taken to compensate the soil moisture deficit,
2. the *duration of rainfall excess*, T_d, which is the period time to produce runoff flows,
3. the *time to peak*, T_p, which is the period of time from the beginning to the peak flow, and
4. the *lag time*, *T*-lag, which is the time span between the centers of rainfall excess and DRH.

Using these seven points to represent an observed hydrograph, a database can be built from a group of DRHs collected from a gauged watershed. Furthermore, regression analyses can be performed to generate best-fitted empirical formulas by which synthetic hydrographs can be produced for the purpose of hydrologic designs. For instance, the

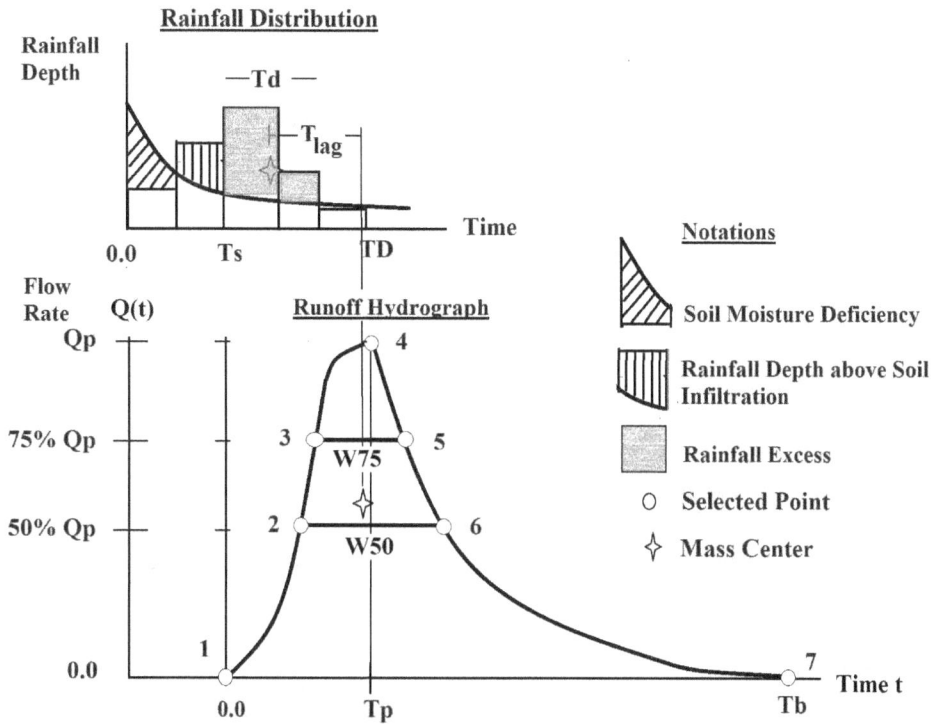

Figure 3.8 Seven points to represent a direct runoff hydrograph

Colorado urban hydrograph procedure is the typical *synthetic hydrograph prediction* method using regression equations for seven points on a hydrograph (CUHP 2005).

3.6 UNIT HYDROGRAPH AND THE S-CURVE

The *unit hydrograph* is defined as a *direct runoff hydrograph* (DRH) produced by *1 inch* or *1 cm* of rainfall excess uniformly distributed over the entire watershed with a *specified duration* (Sherman 1932). Since the duration of the 1-inch excess rainfall can vary, a watershed has many unit hydrographs that are identified by duration. For instance, a 10-minute unit hydrograph means that the watershed is subject to 1 inch of excess rainfall over 10 minutes. As a result, its rainfall intensity is equal to 6.0 inch/hr. Similarly, a 30-minute unit hydrograph is produced under a rainfall intensity of 2.0 inch/hr. As a result, the peak flow in the 10-min unit hydrograph is higher than that in the 30-min unit hydrograph. The major assumptions in the concept of the unit hydrograph are:

1. the linear relationship between the DRV and the runoff rates in the DRH,
2. the constant lag time between the centers of the rainfall excess and the DRH, and
3. the applicability of linear superimposition among DRHs.

A unit hydrograph for a watershed can be derived from: (1) an *observed hydrograph*, or (2) an *S-curve method*.

3.6.1 Unit hydrograph derived from an observed hydrograph

A dataset for deriving a unit hydrograph includes rainfall hyetographs, runoff hydrographs, base flows, and hydrologic losses. For each set of data, the duration, excess rainfall, and hydrologic losses have to be derived from balancing the rainfall and runoff volumes. For convenience, runoff and rainfall volumes in the unit hydrograph method are expressed in inches or cm per watershed area. The *volume ratio* between the unit hydrograph and the observed DRH is:

$$K_v = \frac{1}{\text{DRV}} \qquad (3.21)$$

where K_v = volume ratio, and DRV = volume under the observed DRH in [L] per watershed. Based on the assumption of the linearity between runoff rate and runoff volume, the unit hydrograph can be derived from an observed DRH by multiplying the DRH runoff ordinates by the volume ratio, K_v. Under the assumption that the lag time is independent of runoff volume, this conversion process does not change the timescale on the observed DRH.

For mathematical convenience, a unit hydrograph can be converted into its *mass curve*, which is the plot of the cumulative runoff ordinates under a unit hydrograph. A mass curve appears to be a shape of letter 'S'; therefore, it is called the *S-curve*.

Example 3.5: Based on the volume balance analysis, it is concluded that a storm event has a rainfall excess of 0.30 inch over duration of 5 minutes for the observed DRH listed in Table 3.7. Derive the unit hydrograph and S-curve from the observed DRH given in Table 3.7.

Solution: If the DRV = 0.3 inch for a duration of 5 minutes, then the volume ratio is calculated as:

$$K_v = \frac{1}{0.30} = 3.33$$

The 5-minute unit hydrograph is derived by multiplying the ratio of 3.33 with the runoff ordinates under the observed DRH, as shown in Table 3.8. The S-curve is then derived by accumulating the runoff ordinates for each time step. The total of runoff ordinates amounts to 1197.90 cfs. According to Eq. 3.20, the total runoff volume under this 5-min unit hydrograph is:

$$V = \Delta t [\sum_{t=T_S}^{t=T_b} Q(t)] = (5 \times 60) \times 1197.9 = 359370 \text{ ft}^3 \text{ under the 5-min unit hydrograph.}$$

Based on the definition of the unit hydrograph, the watershed area for this case has to satisfy:

$$V = 1.0 \text{ inch} \times \text{watershed area} = 359,370 \text{ ft}^3.$$

So the watershed area = 431,2440 sq ft = 99 acres.

Example 3.6: Use the unit hydrograph derived in Example 3.3 to determine the storm hydrograph for the three 5-minute rainfall excess blocks, as shown in Table 3.9.

Table 3.8 Unit hydrograph and S-curve

Time (min)	DRH (cfs)	5-minute Unit Hydrograph (cfs)	5-minute S-curve (cfs)
0.00	0.00	0.00	0.00
5.00	5.00	16.50	16.50
10.00	24.00	79.20	95.70
15.00	80.00	264.00	359.70
20.00	72.00	237.60	597.30
25.00	60.00	198.00	795.30
30.00	48.00	158.40	953.70
35.00	35.00	115.50	1069.20
40.00	20.00	66.00	1135.20
45.00	10.00	33.00	1168.20
50.00	5.00	16.50	1184.70
55.00	3.00	9.90	1194.60
60.00	1.00	3.30	1197.90

Table 3.9 Storm hydrograph predicted by unit hydrograph

Time minutes	5-minute Unitgraph cfs	DRH-1 cfs	DRH-2 cfs	DRH-3 cfs	Storm DRH cfs
			Rainfall Excess in inches		
		0.20	0.50	0.15	
0.00	0.00	0.00			0.00
5.00	16.50	3.30	0.00		19.80
10.00	79.20	15.84	8.25	0.00	103.29
15.00	264.00	52.80	39.60	2.48	358.88
20.00	237.60	47.52	132.00	11.88	429.00
25.00	198.00	39.60	118.80	39.60	396.00
30.00	158.40	31.68	99.00	35.64	324.72
35.00	115.50	23.10	79.20	29.70	247.50
40.00	66.00	13.20	57.75	23.76	160.71
45.00	33.00	6.60	33.00	17.33	89.93
50.00	16.50	3.30	16.50	9.90	46.20
55.00	9.90	1.98	8.25	4.95	25.08
60.00	3.30	0.66	4.95	2.48	11.39
65.00		0.00	1.65	1.49	3.14
70.00			0.00	0.50	0.50
75.00				0.00	0.00
80.00					0.00

Solution: Column 2 in Table 3.9 lists the unit graph. The first rainfall block of 0.2 inch produces the hydrograph listed in Column 3. The runoff rates in Column 3 are generated by the value in Column 2 times 0.20 inch and then shifted by 5 minutes. The fourth column is the hydrograph generated by Column 2 times 0.5 and then shifted by 10 minutes. The third hydrograph is generated by Column 2 times 0.15 and then shifted by 15 minutes. The predicted storm hydrograph in Fig. 3.9 is found by adding the three hydrographs together, according to the time step. In this case, the peak flow is found to be 429 cfs, which occurs at $t = 20$ minutes (see Fig. 3.9).

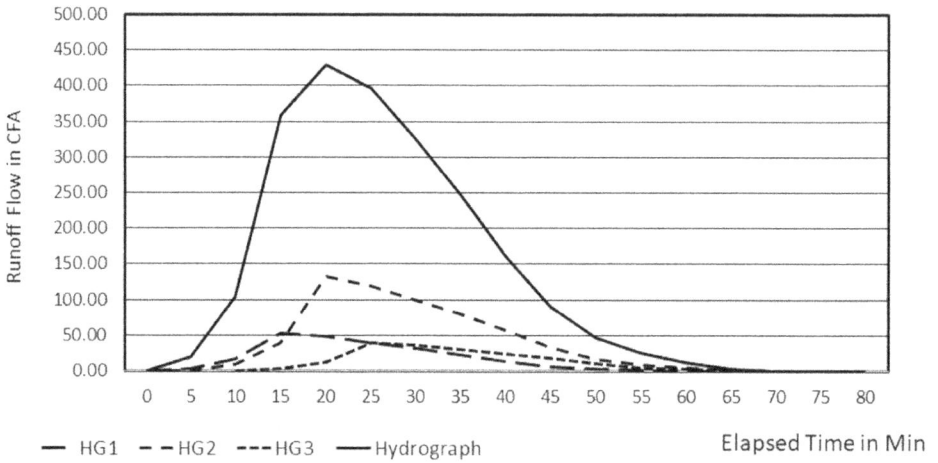

Figure 3.9 Convolution of hydrographs

3.6.2 Unit hydrograph derived from an S-curve

Unit hydrographs with different durations can be derived from the S-curve. For instance, how do we derive the 15-minute unit hydrograph from the 5-minute S-curve? First, we shall list two 5-minute S-curves with a time shift of 15 minutes. The ordinates on the 15-minute unit hydrograph are equal to the differences between the two 5-minute S-curves divided by the ratio of 15/5, which is determined as:

$$R = \frac{D_{New}}{D_{Base}} \tag{3.21}$$

where R = ratio of duration, D_{New} = duration of the new unit hydrograph to be derived, and D_{Base} = duration of the known S-curve.

Example 3.7: Derive a 15-minute unit hydrograph from the 5-minute S-curve in Example 3.5.

Solution: Table 3.8 lists two 5-minute S-curves with a time shift of 15 minutes. The difference between these two S-curves must be divided by the ratio as:

$$R = \frac{15}{5} = 3.0$$

Column 3 in Table 3.10 is the 15-minute unit hydrograph derived for this case. The sum of the runoff ordinates in Column 3 is verified to be 1197.9 cfs, or equivalent to one inch of water on the watershed of 99 acres. So, Column 3 is the 15-min unit hydrograph.

Table 3.10 15-minute unit hydrograph derived
            from 5-minute S-curve

5-minute S-curve (cfs)	Shifted S-curve by 15 minutes (cfs)	15-minute unit hydrograph (cfs)
(1)	(2)	[(1) − (2)]/3
0.00		0.00
16.50		5.50
95.70		31.90
359.70	0.00	119.90
597.30	16.50	193.60
795.30	95.70	233.20
953.70	359.70	198.00
1069.20	597.30	157.30
1135.20	795.30	113.30
1168.20	953.70	71.50
1184.70	1069.20	38.50
1194.60	1135.20	19.80
1197.90	1168.20	9.90
	1184.70	0.00
	1194.60	0.00
	1197.90	0.00

3.7 CONCLUSIONS

A runoff hydrograph is a combined response from the watershed to the arriving rain-fall and also the base flows from the local groundwater aquifer. The separation of the base flow is a practice using the engineer's best judgement and experience. To derive the unit hydrograph from an observed direct runoff hydrograph, the principle of volume balance is observed among the rainfall, runoff, infiltration loss, and depression loss. Recognizing that the soil infiltration rate predicted by Horton's formula was derived under the assumption of a constant headwater, the actual infiltration rate is also limited with the arriving rainfall as the source. The movement of water flow through soil pores depends on the soil hydraulic conductivity and energy grade. As indicated in the example, the movement of wetting front is a slow process to saturate the soil column. It implies that the compensation of the soil moisture deficit should take longer than the movement of surface runoff. Care must be taken when selecting a numerical procedure and empirical parameters to simulate a natural process. Above all, the reliability of a numerical modeling technique depends on in-depth calibrations with adequate field data. It is a continual effort from development, calibration, verification, and refinement to improve the numerical method.

The unit hydrograph method provides a practical and relatively easy-to-apply tool for quantifying the effect of a unit of rainfall on the runoff generation from a specified watershed. The unit hydrograph method assumes that a watershed's runoff response is linear and time-invariant, and that the effective rainfall occurs uniformly over the watershed. In the real world, none of these assumptions are strictly true. Nevertheless, applications of the unit hydrograph method typically yield a reasonable approximation of the runoff response from watersheds. A watershed can have many unit hydrographs, depending on the rainfall duration. In practice, the S-curve method is used to verify the consistency between two unit hydrographs derived from the same watershed.

HOMEWORK

Q3-1   Referring to Fig. Q3-1, a triangular hydrograph is generated from a 10-min uniform rainfall of 8.5 inch/hr on a square paved parking lot. The parking lot has a slope of $S_o = 1\%$ and a Manning's roughness $N = 0.03$. The rising hydrograph reaches a peak flow of $Q_p = 34$ cfs at $t = 10$ min. After the rain ceases, the recession hydrograph is continued to $t = 20$ min.

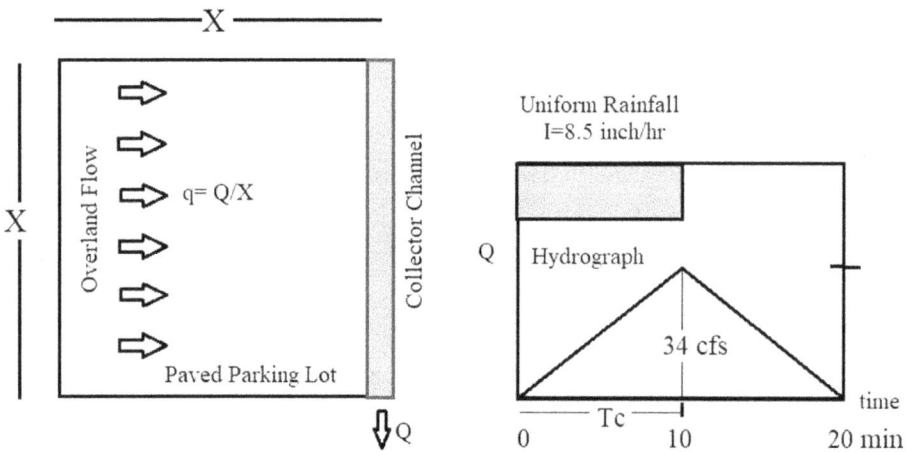

Figure Q3-1 Hydrograph from a square paved parking lot

1. Determine the tributary area. (Hint: runoff volume $= 0.5 \times 34 \times 20 \times 60$ cubic ft $= 8.5 \times 10/60 \times$ area $= X^2$)
2. Knowing that unit-width discharge $q = Q_p/X$, find the normal depth defined by the Manning formula.
3. Determine the runoff volume under the recession hydrograph.
4. The surface detention volume is defined as the water volume under the uniform flow profile. Prove that the surface detention volume is equal to the runoff volume under the recession hydrograph.

Q3-2   Knowing that the soil infiltration rate is constant at $f = 4.5$ inch/hr, derive the unit hydrograph from the observed hydrograph shown in Fig. Q3-2.

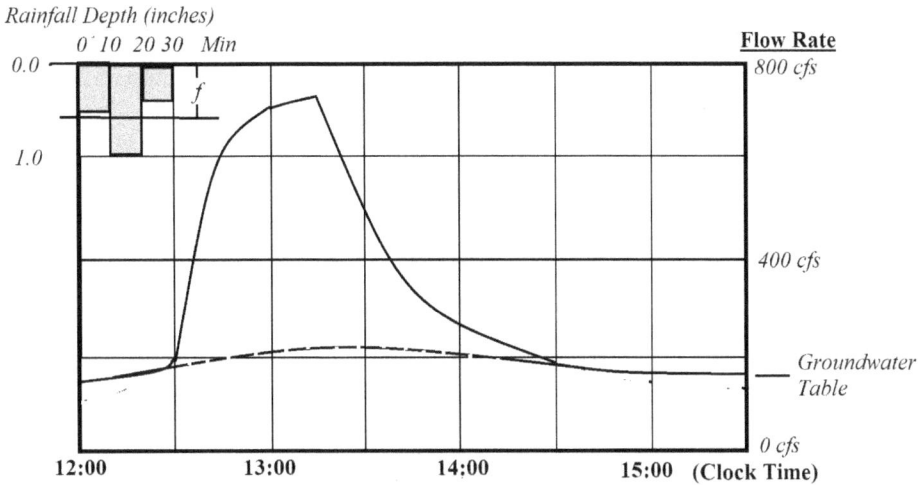

Figure Q3-2 Observed hydrograph

Q3-3   Continued from Q3.2. An observed rainfall distribution is given below. Complete the calculations of net rainfall depth. The depression loss for the watershed is 0.35 inches. The soil infiltration is described with $f_o = 3.0$ inch/hr, $f_c = 0.5$ inch/hr, and $k = 0.108$ per min.

Time t (min)	Precipit- ation $\Delta p(t)$ (inches)	Infiltration $f(t)$ (inch/hr)	Infiltration $\Delta F(t)$ (inches)	Runoff Depth (inches)	Incremental Depression (inches)	Cumulative Depression (inches)	Net Rain Depth (inches)
0	0.000	3.000	0.000	0.000	0.000	0.000	
10	0.110	1.349	0.362	0.000	0.000	0.000	
20	0.220	0.788	0.178	0.042	0.042	0.042	
30	0.330						
40	0.250						0.062
50	0.120						0.033
60	0.050						
70	0.030						
80	0.010						
Total	1.120		1.089	0.445	0.350		

Q3-4   Apply the net rainfall depths derived in Q3.3 to the unit hydrograph derived in Q3.2 to predict the storm hydrograph.

Q3-5   A porous bottom of a rain garden is shown in Fig. Q3-5. The basin is operated with a constant water depth of 2.0 inch on a 24-inch layer of sandy medium. The infiltration rate on the porous bottom at $Z = 0$ is described by Horton's formula with f_o = 4.5 inch/hr, f_c = 1.0 inch/hr, and the decay coefficient K = 6.25 per hour. The hydraulic conductivity for this sandy medium is 0.75 inch/hr and the sand moisture deficit is 0.33. Predict the movement of the wetting front and the time required to saturate the sandy medium of 24 inches.

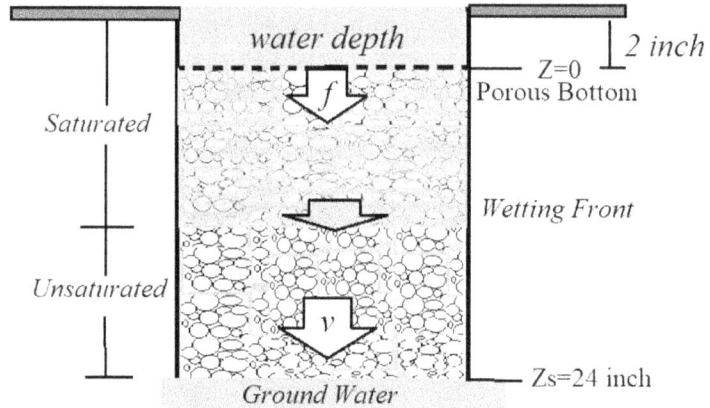

Figure Q3-5 Infiltration through the soil column

REFERENCES

American Society of Civil Engineers (1957). *Hydrology Handbook*, ASCE, New York.

Chow, Ven Te, Maidment, David R., and Mays, L.W. (1988). *Applied Hydrology*, McGraw Hill, New York.

CUHP (2005). "Colorado Urban Unit hydrograph Procedure", in *Runoff, Urban Storm Water Design Criteria Manual*, UDFCD, Denver, CO

Dooge, J.C.I. (1959). "A General Theory of the Unit Hydrograph", *Journal of Geophysical Research*, vol. 64, no. 2, pp. 241–256.

Eagleson, P.S. (1962). "Unit Hydrographs for Sewered Areas", *Journal of the Hydraulics Division*, vol. 88, no. HY2, pp. 1–25.

Green, W.H. and Ampt, G.A. (1911). "Studies of Soil Physics, I: The Flow of Air and Water Through Soils", *Journal of Agriculture Science*, vol. 4, no. 1, pp. 1–24.

Guo, J.C.Y. (1988). "Colorado Unit Hydrograph Procedures – Its Synthetic Unit Hydrograph Characteristics", Proceeding of ASCE International Conference on Hydraulic Engineering held at Colorado Springs, Colorado, in August.

Henderson, F.M. and F.A. Wooding (1964). "Overland Flow and Groundwater Flow from Steady Rainfall of Finite Duration", *Journal of Geophysical Research*, vol. 69, no. 8, pp. 1531–1539.

HMS (2015). "Hydrologic Analytic Model", *Flood Hydrograph Package Manual*, updated 1987, US Army Corps of Engineers, Davis, CA.

Horton, R.E. (1933). "The Role of Infiltration in the Hydrologic Cycle", *Transactions of the American Geophysical Union*, vol. 14, pp. 446–460

NRCS SCS (1964). *SCS National Engineering Handbook, Section 4: Hydrology*, updated 1972, US Department of Agriculture, Washington, DC.

NRCS SCS (1986). *Urban Hydrology for Small Watersheds*, 2nd ed., Tech. Release no. 55 (NTIS PB87-101580), US Department of Agriculture, Washington, DC.

Overton, D.E. and M.E. Meadows (1976). *Stormwater Modeling*, Academic Press, New York.

Sherman, L.K. (1932). "Stream Flow from Rainfall by Unit-graph Method", *Engineering News-Record*, vol. 108, April 7, pp. 501–505

Snyder, F.F. (1938). "Synthetic Unit hydrographs", *Transactions of the American Geophysical Union*, vol. 19, pp. 447–454.

Snyder, W.M. (1955). "Hydrograph Analysis by the Method of Least Squares", *Journal of the Hydraulics Division*, vol. 81, pp. 1–25.

Tauxe, G.W. (1978). "S-Hydrographs and Change of Unit Hydrograph Duration", *Journal of the Hydraulics Division*, vol. 104, no. HY3, pp. 439–444.

Taylor, A.B. and H.E. Schwartz (1952). "Unit Hydrograph Lag and Peak Flow Related to Basin Characteristics", *Transactions of the American Geophysical Union*, vol. 33.

UDFCD (2010). *Urban Stormwater Design Criteria Manual*, Urban Drainage and Flood Control District, Denver, CO.

Chapter 4

Rational method

4.1 RATIONAL METHOD

The Rational method was derived to predict the peak flows from a small urban catchment. The key variables in the Rational method are rainfall intensity, watershed tributary area, and runoff coefficient. The Rational method can be stated as (Kuichling 1889):

$$Q = K C I A \tag{4.1}$$

where Q = flow rate of runoff in cfs or cms, C = runoff coefficient for the design event, I = average intensity of rainfall over the watershed in inch/hr or mm/hr, A = tributary area in acre or hectare. The value of $K = 1$ if Q is in cfs, I is in inch/hr and A is in acre; or $K = 1/360$ if Q is in cms, I is in mm/hr, and A is in hectare. The *time of concentration* and *runoff coefficient* are the two major parameters that describe the drainage characteristics of the tributary watershed. The *runoff coefficient*, C, representing the percentage of rainfall excess, is determined by the soil infiltration and watershed's imperviousness. The time of concentration representing the contributing rainfall amount to the peak runoff is determined by the flow time through the waterway in the tributary watershed.

4.2 DESIGN RAINFALL INFORMATION

To predict a design event, local rainfall statistics should be used. A small watershed can be represented by the point rainfall statistics. The US Weather Bureau has published a set of rainfall statistics for the United States, such as Rainfall Technical Paper 40 and Rainfall Atlas 14. Table 4.1 is an example of the rainfall intensity–duration–frequency (IDF) relationship for the City of Denver, Colorado.

An IDF curve has a decay nature with respect to rainfall duration. It can be described by a hyperbolic function as:

$$I = \frac{C_1 P_*}{(C_2 + T_d)^{C_3}} \tag{2.22 in Chapter 2}$$

where I = intensity in inch/hr or mm/hr, P_* = index rainfall depth in inches or mm, T_d = rainfall duration in minutes, and C_1, C_2, and C_3 are constants. As shown in Fig. 4.1, the IDF information for Denver International Airport, Colorado, was obtained from

DOI: 10.1201/9781003284239-4

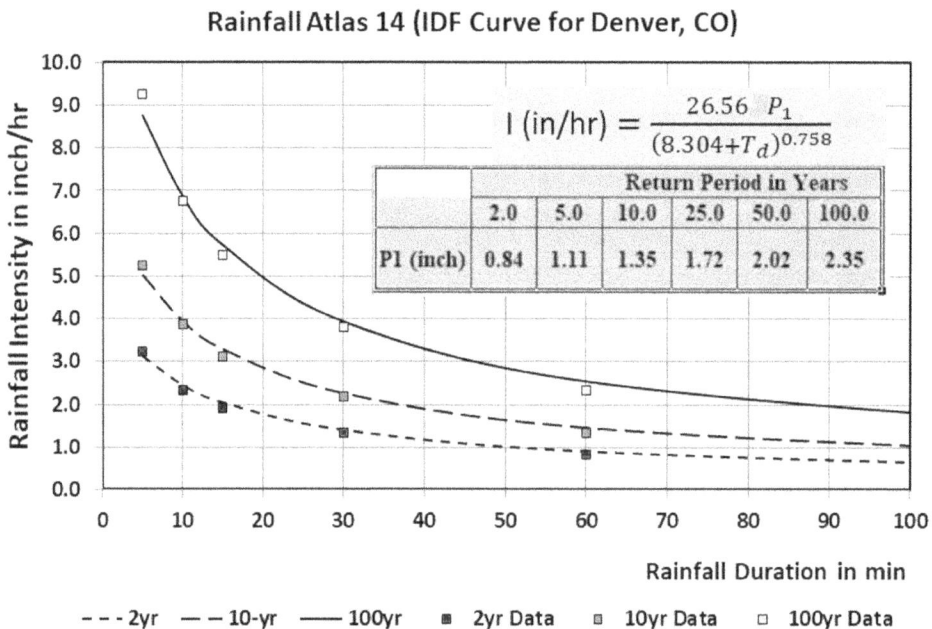

Figure 4.1 IDF curves and formula for Denver, Colorado

the NOAA website for Rainfall Atlas 14 (https://hdsc.nws.noaa.gov/hdsc/pfds/pfds_m ap_cont.html?bkmrk = pa). The best-fitted formula was derived for the Denver metro area to have $C_1 = 26.56$, $C_2 = 8.304$, and $C_3 = 0.758$ when using the 1-hr precipitation depth in inches and the rainfall duration in minutes.

$$I\left(\frac{inch}{hr}\right) = \frac{26.56 P_1}{(8.304 + T_d)^{0.758}} \text{ (Atlas 14 for Denver, Colorado)} \tag{4.2}$$

Eq. 4.2 represents the average rainfall intensity over the specified rainfall duration. In reality, the rainfall time-distribution is not uniform in nature. Therefore, the applicability of the Rational method is limited to small, homogenous urban catchments.

Example 4.1: The rainfall IDF curve developed for the City of Beijing, China, is given by:

$$I = \frac{1602(1 + 1.037 \log_{10} T_r)}{(11.594 + T_d)^{0.681}} \tag{4.3}$$

where I = rainfall intensity in liter per second per hectare, T_r = return period in years, and T_d = rainfall duration in minutes. Convert the IDF formula for Beijing to conform to Eq. 4.2

Table 4.1 Rainfall IDF relationship for the city of Beijing, China

Duration (min)	Year 2 mm/min	inch/hr	Rainfall 10 mm/min	Intensity Year inch/hr	50 mm/min	Year inch/hr	100 mm/min	Year inch/hr
5	1.859	4.390	2.885	6.815	3.912	9.240	4.354	10.285
10	1.553	3.669	2.411	5.696	3.269	7.723	3.639	8.596
15	1.348	3.184	2.093	4.943	2.837	6.702	3.158	7.459
30	0.994	2.348	1.543	3.645	2.092	4.942	2.329	5.501
60	0.687	1.622	1.066	2.518	1.445	3.414	1.609	3.800

Solution: Obviously, there are multiple solutions for this case. The value of the P-index should be chosen to represent the design frequency. For this case, it is reasonable to set the three variables as:

$$P_{Index} = (1 + 1.037 \log_{10} T_r)$$

$$C_2 = 11.594$$

$$C_3 = 0.681$$

$$C_1 = 1602 \left(\frac{liter}{sec. \ ha} \right) \frac{1,000,000 \ mm^3}{1 \ liter} \frac{60 \ sec}{1 \ min} \frac{1 \ ha}{10,000*10,000 \ mm^2}$$
$$= 9.593 \ mm/min$$

$$C_1 = 9.583 \left(\frac{mm}{min} \right) \frac{60 \ min}{1 \ hr} = 594.98 \frac{mm}{hr} = 22.63 \frac{inch}{hr}$$

The value C_1 depends on the unit for the rainfall intensity. Table 4.1 presents the IDF rainfall statistics for the City of Beijing, China.

4.3 VOLUME-BASED RUNOFF COEFFICIENT

The peak flow from an urban catchment in Fig. 4.2 depends on catchment's area, waterway length, and slope, imperviousness, and rainfall intensity and duration. Dimensional analysis leads to the following conclusion:

$$\frac{Q}{AI} = fct \left(\frac{L}{\sqrt{A}}, I_a, S_o, soil loss \right) \tag{4.4}$$

$$I_a = \frac{A_i}{A} \tag{4.5}$$

$$\beta = \frac{L}{B} \approx \frac{L}{\sqrt{A}} \tag{4.6}$$

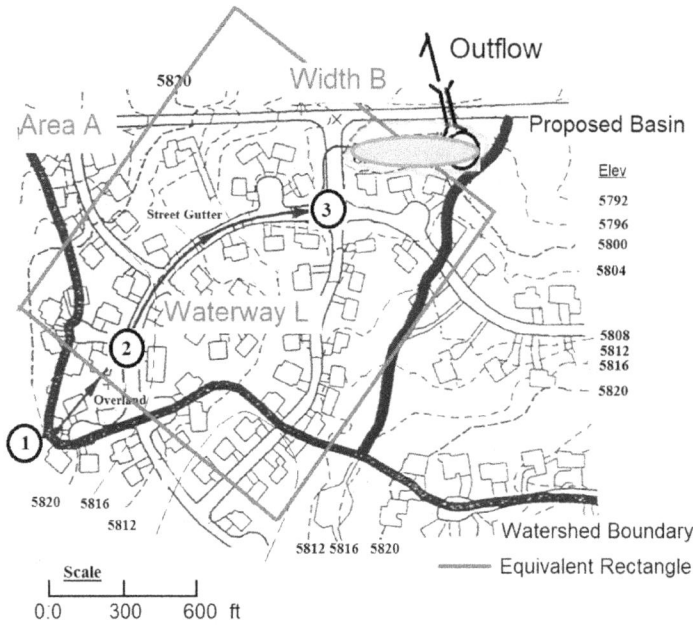

Figure 4.2 Urban watershed under dimensional analysis

where *fct* = functional relationship, L = length of waterway in [L], S_o = waterway slope in [L/L], I_a = catchment imperviousness between zero and unity, A_i = impervious area in [L²], A = total tributary area in [L²], B = width of catchment, and β = catchment shape factor defined as catchment's width-to-length ratio. Comparing with Eq. 4.1, Eq. 4.4 is re-written as:

$$C = \frac{Q}{IA} = fct(\beta, I_a, S_o, \text{soil loss})$$ (4.7)

Eq. 4.7 implies that the value of C varies with respect to the catchment shape, catchment development, I_a, flow movement in term of the catchment's slope, and soil loss. From the aspect of rainfall and runoff characteristics, Eq. 4.7 is further modified to include the flow time and rainfall duration as:

$$c = \frac{Q}{IA}\frac{T_c}{T_d} \approx \frac{V_F}{V_R} = \frac{\text{rain excess}}{\text{rain}}$$ (4.8)

Eq. 4.8 is a *volume-based runoff coefficient*, which is the ratio of the rainfall excess to rainfall depths or the runoff to rainfall volumes, i.e.:

$$C = \frac{V_F}{V_R} = \frac{\sum_{T=0}^{T=T_B} Q(T)\Delta T}{A\sum_{T=0}^{T=T_d} \Delta P(T)}$$ (4.9)

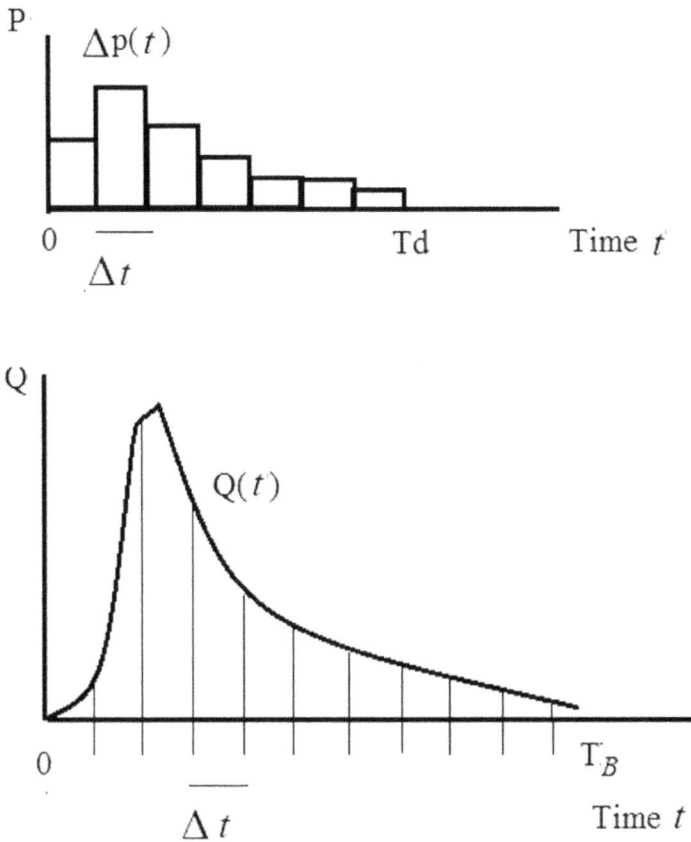

Figure 4.3 Rainfall and runoff volumes

Referring to Fig. 4.3, V_F = runoff volume in [L³] under the hydrograph, V_R = rainfall volume in [L³] under the hyetograph, ΔT = incremental time step on runoff hydrograph, such as 5 minutes, $Q(T)$ = runoff flow in [L³/T] at time T, $\Delta P(t)$ = incremental rainfall depth in [L] at time T, A = tributary area in [L²], T_d = rainfall duration, and T_B = base time of runoff hydrograph.

In theory, both Eqs 4.7 and 4.8 should yield identical runoff coefficients. In practice, the difference between the observed rainfall hyetograph and the rainfall IDF curve result in a numerically negligible gap. Generally, the volume-based runoff coefficient represents the average value for the entire event (Guo and Urbonas 2014).

Example 4.2: As illustrated in Fig. 4.4, an urban catchment of A = 10 acres (4 hectares) has been developed with I_a = 70%. Knowing that the soil infiltration loss is f = 0.5 inch/hr (1.27 cm/hr), and depression losses are D_{vp} = 0.4 inch (1.0 cm) for a pervious surface and D_{vi} = 0.1 inch (0.25 cm) for an impervious surface, determine the runoff coefficient, C, for a rainfall event with a precipitation depth of 3 inches (7.62 cm) over a duration of 1 hr.

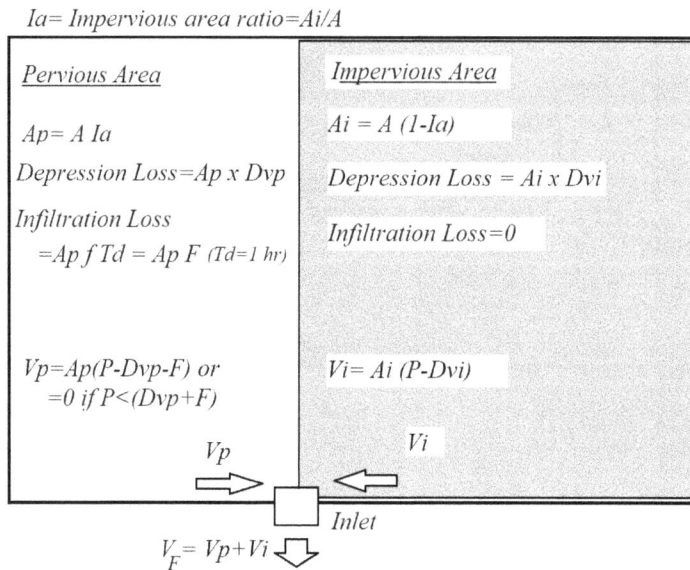

Figure 4.4 Example to calculate runoff coefficients

Solution:

In this case, A = 10 acres *and* I_a = 0.70

The total rainfall volume V_R = 3 inch × 10 acres = 30 inch-acre

Infiltration loss $F = f\, T_d$ = 0.5 0 inch/hr × 1.0 hr = 0.5 inch

For the pervious area: $A_p = A\,(1 - I_a)$ = 10 × (1 − 0.7) = 3.0 acres

Depression loss = $D_{vp}\, A_p$ = 0.4 × 3.0 = 1.2 inch-acre

Soil loss = $FA_p = f \times T_d \times A_p$ = 0.5 × 1 × 3 = 1.5 inch-acre

Runoff volume from pervious area: $V_p = A_p(P - D_{vp} - F)$ = 3.0 × (3.0 − 0.4 − 0.5) = 6.3 inch-acre > 0

For the impervious area: $A_i = A\, I_a$ = 10 × 0.7 = 7.0 inch-acre

Depression loss = $D_{vi}\, A_i$ = 0.1 × 7.0 = 0.7 inch-acre

Runoff volume from impervious area: $V_I = A_i(P - D_{vp})$ = 7.0 × (3.0 − 0.1) = 20.3 inch-acre > 0

Total runoff volume $V_F = V_P + V_I$ = 6.3 + 20.3 = 26.6 inch-acre

$$C = \frac{V_F}{V_R} = \frac{26.6}{30} = 0.88$$

When the precipitation depth is less than the total hydrologic losses or $P \leq (F_t + D_{vp})$, the runoff volume from the pervious area is nullified. In fact, the runoff volume in an urban area is mostly dominated by the impervious area.

Example 4.3: Repeat the Example 4.2 with the precipitation depth of 0.5 inch.

Solution:

The total rainfall volume V_R = 0.50 inch × 10 acres = 5.0 inch-acre

Runoff volume from pervious area: $V_p = A_p (P - D_{vp} - F) = 3.0 \times (0.5 - 0.4 - 0.5)$ < 0. So, set V_p = 0.

Runoff volume from impervious area: $V_I = A_i (P - D_{vp}) = 7.0 \times (0.5 - 0.1) = 2.8$ inch-acre > 0

Total runoff volume $V_F = V_P + V_I = 0.0 + 2.8 = 2.8$ inch-acre

$$C = \frac{V_F}{V_R} = \frac{2.8}{5.0} = 0.56$$

4.3.1 Runoff coefficients

From Examples 4.2 and 4.3, the runoff coefficient varies with respect to the precipitation, P, representing the design return period, the imperviousness, I_a, representing land use, and infiltration loss representing type of soil in the catchment. For convenience, the normalized equation for the runoff coefficient is derived as:

$$C = 1 - \frac{D_{vi}}{P}(1 - I_a) - m\left(\frac{D_{vp} + F}{P}\right)I_a \tag{4.10}$$

where F = 1-hr infiltration depth in inch or mm, and $m = 0$ when $P \leq (D_{vp} + F_t)$ or $m = 1$ when $P > (D_{vp} + F_t)$

Example 4.4: Consider D_{vi} = 0.10 inch, D_{vp} = 0.40 inch, and F = 0.88 inch over the first hour for Type C/D soils. A set of runoff coefficients are produced as summarized in Table 4.2 using the 1-hr rainfall depths shown in Table 4.1.

Using Table 4.2 as a template, runoff coefficients for Type B soils are produced with F = 1.0 inch in Table 4.3. Similarly, Table 4.4 is prepared for Type A soils with F = 1.8 inches.

4.3.2 Weighted runoff coefficient

For a development of mixed land use, a watershed can be divided into various sub-areas based on the different land use. Using the area-based method, the weighted runoff coefficient for the entire watershed is calculated as:

$$C = \frac{\sum_{i=1}^{i=n} C_i A_i}{A} \tag{4.11}$$

where C = weighted runoff coefficient for the watershed, C_i = runoff coefficient for the i-th subarea, A_i = subarea in [L^2], and n = number of sub-areas.

Table 4.2 Runoff coefficients for type C/D soils (1 inch = 25.4 mm)

Soil Type	C/D		D_{vp} = 0.40	inch	
Infiltration F =	0.88	inch	Dvi = 0.10	inch	
Variable		Rainfall	Depth		
Return period	2 year	5 year	10 year	50 year	100 year
P (inch) =	0.95	1.35	1.60	2.20	2.60
Dvi/P =	0.11	0.07	0.06	0.05	0.04
Dvp/P =	0.42	0.30	0.25	0.18	0.15
F/P =	0.93	0.65	0.55	0.40	0.34
Imp I_a		Runoff	Coefficient		
0.05	0.04	0.10	0.24	0.45	0.53
0.10	0.09	0.14	0.27	0.47	0.55
0.20	0.18	0.23	0.35	0.53	0.60
0.30	0.27	0.31	0.42	0.58	0.64
0.40	0.36	0.40	0.50	0.63	0.69
0.50	0.45	0.49	0.57	0.69	0.73
0.60	0.54	0.58	0.64	0.74	0.78
0.70	0.63	0.66	0.72	0.79	0.83
0.80	0.72	0.75	0.79	0.85	0.87
0.90	0.81	0.84	0.86	0.90	0.92
0.99	0.89	0.92	0.93	0.95	0.96

Table 4.3 Runoff coefficients for Type B soils

Soil B Land Use	Imp I_a	2 year	5 year	10 year	50 year	100 year
Lawns, sandy soil	0.02	0.02	0.02	0.14	0.38	0.47
Parks/cemeteries	0.05	0.04	0.05	0.17	0.39	0.49
Playgrounds	0.10	0.09	0.09	0.21	0.42	0.51
Railroad yard areas	0.20	0.18	0.19	0.29	0.48	0.56
Gravel streets	0.30	0.27	0.28	0.37	0.54	0.61
Low-density residential	0.40	0.36	0.37	0.45	0.60	0.66
Schools	0.50	0.45	0.46	0.53	0.66	0.71
High-density apt	0.60	0.54	0.56	0.61	0.72	0.76
Business areas	0.70	0.63	0.65	0.69	0.78	0.81
Light industrial areas	0.80	0.72	0.74	0.78	0.84	0.86
Commercial areas	0.90	0.81	0.83	0.86	0.90	0.91
Roof, pavements	0.99	0.89	0.92	0.93	0.95	0.96

Example 4.5: A subdivision in Fig. 4.5 has an area of 350 ft by 500 ft. The local soil is Type B. The land use consists of residential and commercial areas. Determine the 5- and 100-year runoff coefficients for the entire area.

Solution: In this case, the residential area is $350 \times 350/43560 = 2.81$ acres and the commercial area is $150 \times 350/43560 = 1.21$ acres. The 5-yr and 100-yr runoff coefficients can be found in Table 4.5. Solutions are shown in Table 4.5.

Table 4.4 Runoff coefficients for Type A soils

Soil A Land Use	Imp I_a	2 year	5 year	10 year	50 year	100 year
Lawns, sandy soil	0.02	0.02	0.02	0.02	0.02	0.17
Parks/cemeteries	0.05	0.04	0.05	0.05	0.05	0.19
Playgrounds	0.10	0.09	0.09	0.09	0.10	0.23
Railroad yard areas	0.20	0.18	0.19	0.19	0.19	0.32
Gravel streets	0.30	0.27	0.28	0.28	0.29	0.40
Low-density residential areas	0.40	0.36	0.37	0.38	0.38	0.48
Schools	0.50	0.45	0.46	0.47	0.48	0.56
High-density apts	0.60	0.54	0.56	0.56	0.57	0.64
Business areas	0.70	0.63	0.65	0.66	0.67	0.72
Light industrial areas	0.80	0.72	0.74	0.75	0.76	0.80
Commercial areas	0.90	0.81	0.83	0.84	0.86	0.88
Roof, pavements	0.99	0.89	0.92	0.93	0.95	0.95

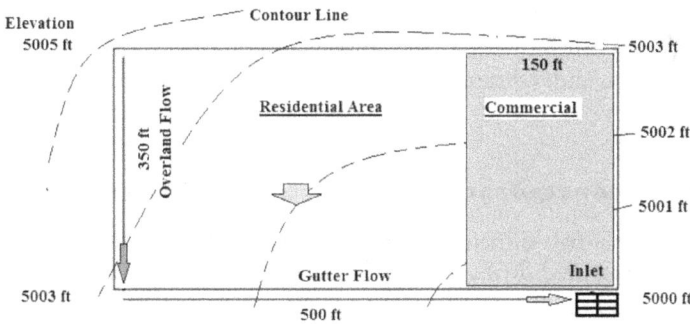

Figure 4.5 Example for mixed land uses

Table 4.5 Area-weighting method for runoff coefficients

For a 5-yr event

Variable	Land residential	Use commercial	Weighted C5
Area (acres)	2.810	1.210	
C5 (5-year)	0.37	0.83	0.51

For the design event

Variable	Land residential	Use commercial	Weighted C-design
Area (acres)	2.810	1.210	
C-design	0.66	0.91	0.74

4.4 TIME OF CONCENTRATION

By definition, the time of concentration of the watershed is the travel time required for stormwater to travel from the most upstream point along the waterway to the outlet. To estimate the time of concentration, it is recommended that the longest waterway be selected to represent the watershed. A waterway often begins with overland flows for a short distance and then becomes a gully or swale flow due to the concentration of flows. Farther downstream, a waterway is formed by well-defined cross-sections through reaches. The time of concentration along a waterway is the cumulative flow times through the reaches as:

$$T_c = T_o + \sum_{j=1}^{j=N} (T_f)_j \qquad (4.12)$$

where T_c = computed time of concentration in minutes, T_o = overland flow time in minutes, and T_f = gutter flow time in minutes, j = j-th reach, and N = number of reaches. Numerically, the flow time through each reach needs to be determined based on topography and hydraulic roughness. Because the development of a watershed is a continuous process through multiple stages, it is recommended that the time of concentration be first estimated under the existing conditions of the watershed and then compared with that under future conditions; whichever is shorter should be selected for drainage designs.

4.4.1 Time of concentration for existing conditions

An overland flow is a two-dimensional sheet flow. Overland flows occur over the areas upstream of concentrated flows. The maximum length of overland flows in an urban area is approximately 300 feet (90 meters) before the overland flow is intercepted by a street gutter or inlet. For a rural area, a maximum length of 500 feet (150 meters) is recommended for overland flows. Among many empirical formulas, the *airport formula* is recommended for urban drainage designs. The *airport formula* states (UDFCD 2010):

$$T_o = \frac{K_{To}(1.1 - C_5)\sqrt{L_o}}{S_o^{0.33}} \text{ for overland flow, where } L_o \leq L \qquad (4.13)$$

where T_o = overland flow in minutes, K_{To} = 0.395 for ft-second units or 0.715 for meter-second units, L_o = overland flow length in [L], C_5 = runoff coefficient for a 5-yr event, S_o = overland flow slope in [L/L], and L_* = maximum allowable distance in [L], such as 300 feet (90 meters) for an urban area or 500 feet (150 meters) for a rural area. Note that C_5 is recommended for Eq. 4.13, while the design runoff coefficient, C, should be used in Eq. 4.1.

  After the flow becomes concentrated, the *US Natural Resources Conservation Service* (NRCS 1976, 2013) recommends that *the upland method* be used to estimate the flow time through a swale as (NRCS 2013):

Table 4.6 Conveyance coefficients, K_f, for the upland method

Type ID	Type of Linings	K_f in ft/sec	K_f in m/sec
1	Forest or heavy meadow	1.5	0.46
2	Tillage or woodland	5.0	1.53
3	Short grain pasture	7.0	2.13
4	Bare soil	10.0	3.05
5	Grass swale	15.0	4.57
6	Paved gutter shallow flow	20.0	6.10

$$T_f = L_f / 60 V_f \text{ for shallow water flows} \tag{4.14}$$

$$V_f = K_f \sqrt{S_f} \tag{4.15}$$

where T_f = flow time in minutes, L_f = flow length in [L], V_f = flow velocity in [L/T], S_f = flow line slope and K_f = conveyance coefficient in [L/T] (NRCS 2013, McCuen 1982). *The NRCS's Soil Conservation Service (SCS) upland method* classifies the surface linings in shallow swales into six categories. Their conveyance coefficients are developed for various roughness surfaces, as shown in Table 4.6.

The *SCS upland method* in Eq. 4.14 was recommended for estimating flow velocities in shallow swales. For a well-defined stream or channel, Manning's formula should be used to estimate the flow velocity. However, when the design information is not readily available, the Types 5 and 6 conveyance coefficients in Table 4.6 can be used to estimate shallow water flow velocities in streams, channels, and street gutters.

It is important to understand that the SCS upland method was developed to estimate the average flow time through the entire flow length along the waterway. During a storm event, the upstream reach of a waterway carries shallow sheet flows, and the discharge in a waterway increases downstream. Eq. 4.15 estimates a *length-averaged velocity* through the entire waterway under an unsteady flow condition. In contrast, Manning's formula provides a *cross-sectional average velocity* for a steady flow. Therefore, these two flow velocity equations are not comparable. Eq. 4.15 gives much slower flow velocities than Manning's formula. Using Eqs 4.14 and 4.15, the time of concentration, T_{C1}, in minutes, for the existing condition is:

$$T_{C1} = T_o + T_f \text{ (minutes)} \tag{4.16}$$

4.4.2 Time of concentration for future conditions

In an urban area, waterways are often equipped with drop structures for grade controls on the stream bed or/and check dams for flow diversions. As a result, Eqs 5.17 and 5.18 do not reflect future hydraulic conditions along the waterway. To be conservative, a regional formula should be developed for *future time of concentration* under post-development conditions. For instance, the Cities of Denver and Las Vegas (UDFCD 2010, CCRFCD Manual 1999) recommend the regional time of concentration be computed based on the future watershed's imperviousness ratio as (Guo and McKenzie 2014):

$$T_{C2} = T_* + \frac{L}{60V_*} (\text{minutes}) \tag{4.17}$$

$$V_* = kK_*\sqrt{S_a} \tag{4.18}$$

$$T_* = 18 - 15I_a (3 \le T^* \le 18\,\text{min utes}, 0 \le I_a \le 1 \tag{4.19}$$

$$K_* = 24I_a + 12 (12 \le K^* \le 36, 0 \le I_a \le 1) \tag{4.20}$$

where T_{C2} = regional time of concentration in minutes, L = total length of waterway in [L], including all reaches along the waterway, T_* = initial overland flow time in minutes, V_* = post-development concentrated flow velocity in [L/T], K_* = conveyance factor in [L/T], k = 1 for foot-second units or 0.305 for meter-second units, and S_a = average slope along waterway. An *initial time* represents the overland flow time through upland areas. Eq. 4.19 implies that the initial time for urban overland flows is 18 minutes on a pervious surface with I_a = 0, and then reduced to 3 minutes for an impervious surface under I_a = 1.0. Eq. 4.19 reveals the fact that the higher the imperviousness in watershed, the shorter the length for the overland flow. The *conveyance parameter* of K_* is for shallow, concentrated flows. Eq. 4.20 reveals that K_* varies from 12 to 36 ft/sec (3.6 to 11 m/s), depending on the watershed's imperviousness ratio. For instance, on a slope of 1%, Eq. 4.20 sets the limits for the length-averaged flow velocity between 1.2 and 3.6 ft/sec, with an average velocity of 2.0 ft/sec. It agrees well with the recommended K_* = 20 ft/sec in Table 4.6 for a paved surface.

In practice, the design time of concentration is the smaller of the *computed time representing pre-development conditions* and the *regional times of concentration representing post-development conditions* as:

$$T_c = \min (T_{C1}, T_{C2}) \tag{4.21}$$

where T_c = design time of concentration in minutes.

4.4.3 Empirical formulas for time of concentration

There are two distinct approaches developed to estimate the time of concentration. One is the *velocity-based method* by which the time of concentration is defined as the flood wave travel time determined by the flow velocity and waterway parameters. Typical examples for the velocity-based method are Kirpitch's formula (Kirpitch 1940), and the upland method recommended by the Natural Resources Conservation Service (NRCS). The other approach is the *time-lag method*, by which the time of concentration is defined as the time difference between the mass center of rainfall excess and the inflection point on the recession limb of the runoff hydrograph. A typical example in this category is the NRCS lag-time method. There are many empirical formulas developed for estimating the time of concentration. These empirical formulas indicate that the time of concentration is a function of waterway length, slope, hydraulic roughness, and rainfall amount. Examples are:

$$T_c = 0.0078 \left(\frac{L}{\sqrt{S_0}} \right)^{0.77} \quad \text{(Kirpitch's formula in 1940)} \tag{4.22}$$

$$T_c = 0.93 n^{0.6} \frac{L^{0.6}}{\sqrt{S_0}} I^{-0.4} \quad \text{(Kinematic wave by Wooding 1965)} \tag{4.23}$$

where T_c = time of concentration in minutes, L = waterway length in feet, S_0 = waterway average slope in ft/ft, and I = average rainfall intensity in inch/hr. None of the above widely used formulas could provide either the true or reproducible values of the time of concentration. They are empirical for estimations only (McCuen et al. 1984). In fact, the time of concentration varies with respect to antecedent soil conditions and the distribution of rainfall. As a part of the hydrograph convolution process, the time of concentration cannot be directly measured by the time difference between the hyetograph and the hydrograph (Singh and Cruise in 1992). However, it can be indirectly derived by minimizing the least-square errors between the predicted and observed hydrographs using the runoff coefficient and time of concentration as the system parameters (Guo 2001a). Based on 44 observed events, the regression equation for estimating the time of concentration was derived as (Guo 2001b):

$$T_c = M \left(\frac{L}{\sqrt{S_0}} \right)^{0.66} \quad (M = 0.054 \text{ for metric units or } 0.025 \text{ for English units}) \tag{4.24}$$

$$V = \frac{1}{N} L^{0.66} S_0^{0.33} \quad (N = 3.48 \text{ for metric units or } 1.50 \text{ for English units}) \tag{4.25}$$

where T_c = time of concentration in minutes, L = waterway length in meters or feet, V = length-averaged flow velocity in mps or fps, and N = conversion factor for units.

Eq. 4.25 represents the length-averaged velocity over the entire waterway. Such an average value represents a spatial and temporal unsteady flow process. As expected, Eq. 4.25 gives smaller velocities than those from Manning's formula. For instance, Eq. 4.25 results in a length-averaged flow velocity of 3.25 fps along a 3000 feet waterway on a slope of 5.0%, while the cross-sectional average flow velocity is estimated to be 7.0 to 9.0 fps by Manning's formula with a roughness coefficient of 0.030. Applying the SCS upland method to the same example, the predicted flow velocity is 3.35 ft/sec in a grass waterway. Therefore, Eq. 4.25 is in the same nature as, and numerically equivalent to, the SCS upland method (NRCS 2013).

Example 4.6: Estimate the 100-yr peak flow for the subdivision in Fig. 4.5.

(A) Calculation of runoff coefficient

From Table 4.5, the weighted $C_5 = 0.51$ and $C_{100} = 0.74$.

(B) Calculation of T_{C1} under existing conditions:

B.1 *Overland flow time*

The slope for the overland flow is:

$$S_0 = \frac{5005 - 5003}{350} = 0.0057$$

The overland flow time is calculated up to 300 feet as:

$$T_0 = \frac{0.393(1.1 - 0.54)\sqrt{300}}{0.0057^{0.33}} = 20.9 \, \text{minutes}$$

B.2 *Swale flow time from 300 to 350 feet:*

Let $K = 20$ and $S_0 = 0.0057$. The swale flow velocity is:

$$V_2 = 20 \times \sqrt{0.0057} = 1.50 \, fps$$

The swale flow time is:

$$T_2 = \frac{50}{60 \times 1.51} = 0.55 \, \text{minutes}$$

B.3 *Gutter flow time*

The slope for the gutter flow is:

$$S_3 = \frac{5003 - 5000}{500} = 0.006$$

Let $K = 20$ and $S_3 = 0.006$. The gutter flow velocity is:

$$V_3 = 20 \times \sqrt{0.0060} = 1.55 \, fps$$

The gutter flow time is:

$$T_3 = \frac{500}{60 \times 1.55} = 5.38 \, \text{minutes}$$

B.4 *Time of concentration, T_{C1}, under existing conditions:*

$$T_{C1} = 20.9 + 0.55 + 5.38 = 26.83 \, \text{minutes}$$

(C) Calculation of T_{C2} under post-development conditions

With $I_a = 0.55$, $T_* = 18 - 15I_a = 9.75$ minutes and $K_* = 24I_a + 12 = 25.2 \, fps$

$$S_a = \frac{5005 - 5000}{(350 + 500)} = 0.006$$

$$T_{C2} = 9.75 + \frac{(350 + 500)}{60 \times 25.2\sqrt{0.006}} = 17.1 \text{ minutes}$$

(D) Prediction of the peak flow as:

$$T_c = \min(T_{C1}, T_{C2}) = 17.1 \text{ minutes}$$

The 100-yr one-hr rainfall depth at the site is 2.62 inches and the IDF curve is given as:

$$I = \frac{28.5 \times 2.62}{(10 + 17.1)^{0.789}} = 5.50 \text{ inch / hr}$$

$$Q = 0.74 \times 5.50 \times \frac{(350 \times 500)}{43560} = 15.70 \text{ cfs}$$

Example 4.7: The catchment in Fig. 4.6 has an area imperviousness percentage of 90% on soils C and D. Estimate the 5-year peak runoff for the street inlet in Fig. 4.6.
   The overland flow time is predicted with C_s = 0.92 (see Table 4.2) as:

$$T_0 = \frac{0.395(1.1 - 0.92)\sqrt{40}}{0.02^{0.33}} = 1.64 \text{ min utes}$$

The gutter flow time is predicted with K_f = 20 as:

$$T_s = 400 / \left(60 \times 20.0\sqrt{0.01}\right) = 3.33 \text{ min utes}$$

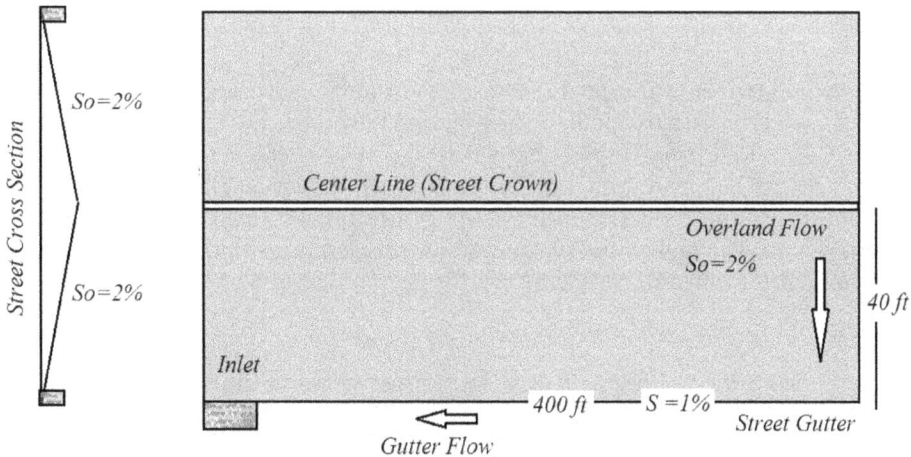

Figure 4.6 Catchment for street drainage

$$T_{C1} = 1.64 + 3.33 = 4.97 \, \text{min utes}$$

The future T_{C2} is calculated with $I_a = 1$, $T^* = 3.0$ minutes, $K^* = 36$ fps, $V^* = 2.79 \, fps$:

$$S_a = \frac{S_o L_o + S_f L_f}{L_o + L_f} = \frac{0.02 \times 40 + 0.01 \times 400}{(40 + 400)} = 0.011$$

$$T_{C2} = 3.00 + \frac{(40 + 400)}{60 \times 36\sqrt{0.011}} = 4.95 \, \text{minutes}$$

Use $T_c = 5$ minutes for this case. Therefore, the 5-minute rainfall intensity is applied to this case. The 5-year 1-hr precipitation at the site is 1.35 inch. The design rainfall intensity and peak discharge are:

$$I = \frac{28.5 \times 1.35}{(10 + 5)^{0.789}} = 4.55 \, \text{inch/hr}$$

$$Q_P = 0.92 \times 4.55 \times \left(\frac{40 \times 400}{43560} \right) = 1.54 \, \text{cfs}$$

Example 4.8: A residential subdivision in Fig. 4.2 has a tributary area of 40 acres and an imperviousness ratio of 0.60. The soil type in this watershed is Type C/D. As shown in Fig. 4.2, the waterway is marked from Point 1 to Point 2 as overland flow; from Point 2 to Point 3 as street-gutter flow; and from Point 3 to Point 4 as grass swale flow. The lengths and slopes for the flow segments are summarized in Table 4.7.

Predict the 100-yr peak flow using the 100-year IDF formula as:

$$I \, (\text{inch/hr}) = \frac{74.5}{(10 + T_c)^{0.789}}, \text{ where } T_c = \text{time of concentration in minutes}$$

With $I_a = 0.60$ for soil type C/D, the 5-yr runoff coefficient is $C_5 = 0.58$ (from Table 4.2), which is used to calculate the overland flow time, and the 100-yr runoff coefficient is $C_{100} = 0.78$ (from Table 4.2), which is used to calculate the 100-yr peak flow.

The times of concentration are determined as shown in Table 4.8. The post-development condition offers a faster flow. As a result, for this case, the time of concentration is determined to be 16.42 minutes for design conditions, and the 100-yr peak flow is determined to be 138.52 cfs.

Table 4.7 Flow lengths and slopes for example watershed

Reach	Length (ft)	Type	Slope (%)
1 to 2	250.00	Overland flow	2.0
2 to 3	1100.00	Street gutter flow ($K_f = 20.0$)	1.0
3 to 4	600.00	Grass swale flow ($K_f = 15.0$)	2.0

Table 4.8 Example of flow prediction using the rational method

Basin ID ID	Area acres A	Runoff Coef 5-yr C5	Runoff Coeff Design C-design	Impervious Area Ratio Ia	Effective Area acres CA
Watershed	40.00	0.58	0.78	0.60	31.20

Slope %	Overland Length ft	Flow Time min	Street Slope %	Gutter Length ft	Flow SCS K fps	Time min	Grass Slope %	Swale Length ft	Flow SCS K fps	Time min
So 2.00	Lo 250.00	To 11.81	S2 1.00	L2 1100.00	K2 15.00	T2 12.22	S3 2.00	L3 600.00	K3 15.00	T3 4.71

Avg Slope %	Exist Tc min	Future Tc min	Storm Duration min	Rainfall Intensity in/hr	Peak Flow cfs
Sa 1.44	Tc 28.75	Tc 19.27	Td 19.27	I 2.68	Qp 83.62

4.5 RATIONAL HYDROGRAPH METHOD

The ideal conditions for producing triangular or trapezoidal hydrographs do not exist in engineering practice. Therefore, we shall expand the Rational method from the ideal case under a uniform rainfall distribution to the real case under a non-uniform hyetograph.

Runoff generation from a watershed is a response to the loading of rainfall on the watershed. Hydrologic systems are continuous in time and are also causal, because the output cannot precede its corresponding inputs over a length of time in the past. Such a time period is termed the *system memory*, over which the historical input affects the present system behavior (Singh 1982). When introducing the Rational method to stormwater drainage design, Kuichling (1889) stated that the peak rate of runoff at a design point is a direct function of tributary area and the tributary rainfall amount up to the time of concentration of the watershed. As illustrated in Fig. 4.7, a hyetograph consists of a series of rain blocks. The loading of rain blocks onto a watershed is similar to the weight of a train onto a bridge. No matter how long the train is, the loading on the bridge depends on the bridge length.

Applying the same analogy to the rainfall loading on the watershed, the waterway length is converted to the time of concentration as the system memory, T_c, which defines the rainfall amount as the input in the past for producing a flow rate as the output at the present. The flow rate, $Q(T)$ at time T, on the hydrograph in Fig. 4.7 depends on the contributing rainfall amount from $(T - T_c)$ to T. Both can be computed as (Guo and Urbonas 2014):

$$I(T) = \frac{1}{T_c} \sum_{t=T-T_c}^{t=T} \Delta P(t) \text{ where } T_c \leq T \leq T_d \tag{4.26}$$

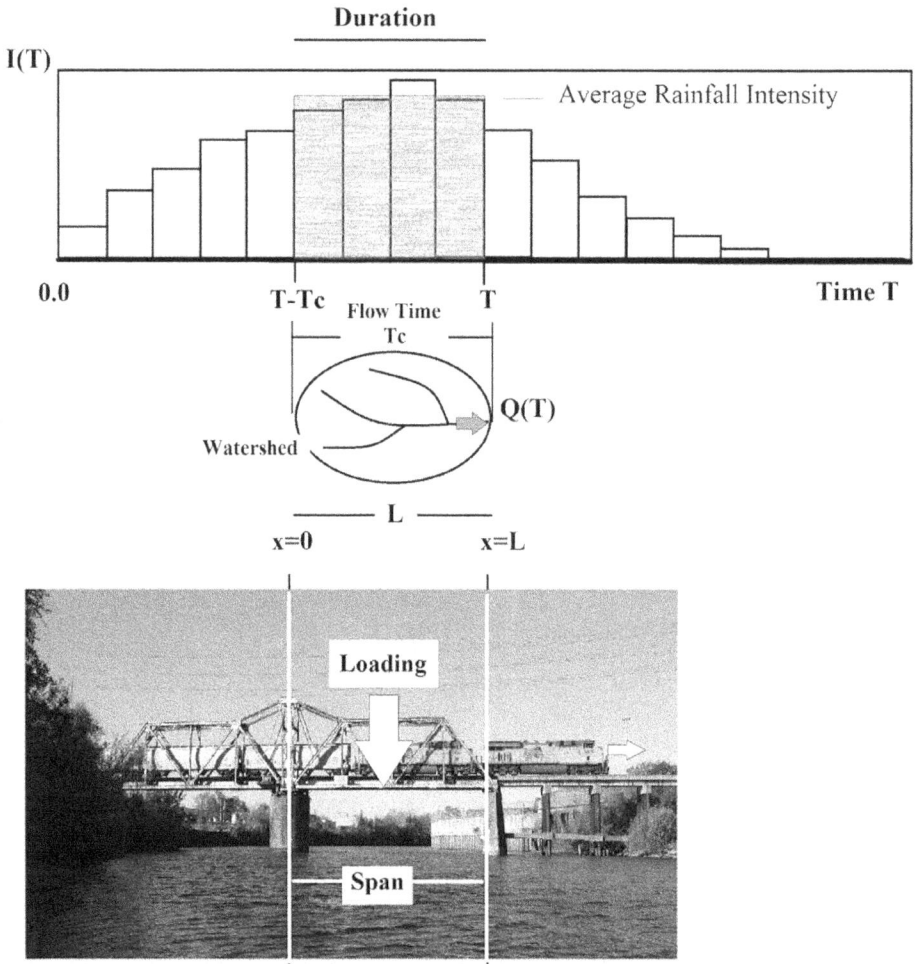

Figure 4.7 Illustration of the rational method with a given rainfall event

$$Q(T) = KCAI(T) \quad \text{where } T_c \leq T \leq T_d \tag{4.27}$$

where $I(T)$ = moving average rainfall intensity at time T for a period of T_c prior to T, $Q(T)$ = runoff rate at time T, t = time variable, T_d = event duration, A = tributary area, and $\Delta P(t)$ = incremental rainfall depth at time t. Using Eqs 4.26 and 4.27, the Rational method is expanded into the rainfall-runoff convolution process to convert a non-uniform rainfall distribution into the corresponding runoff hydrograph (Guo 2001a, 2001b).

Eqs 5.35 and 5.36 are applicable to the peaking hydrograph from T_c to T_d. Before the time of concentration, the rising hydrograph represents the runoff flow from the tributary area, which is on an increasing rate through each time step until the entire watershed becomes a tributary at $T = T_c$. Assuming a linear increasing rate, the tributary area is approximated as (Guo 2000):

$$A_e = A\frac{T}{T_c}, \quad \text{where } T \le T_c \tag{4.28}$$

where A_e = tributary area to runoff flow at time T. Aided by Eq. 4.28, the corresponding runoff flow on the rising hydrograph is estimated as:

$$Q(T) = KCA_e I(T) = KC\frac{A}{T_c}I(T)\text{ for } T_c \ge T \ge 0 \tag{4.29}$$

$$I(T) = \frac{1}{T}\sum_{t=0}^{t=T}P(t)\text{ for } T_c \ge T \ge 0 \tag{4.30}$$

The recession hydrograph begins as soon as the rain ceases. Runoff flows on the recession hydrograph form a decay curve in nature. For a small watershed, a linear approximation is developed as:

$$Q(T) = Q(T_d)\left(1 - \frac{T - T_d}{T_c}\right)\text{ for } T_d \le T \le (T_d + T_c) \tag{4.31}$$

$$Q(T_d) = KCAI(T_d) \tag{4.32}$$

where $Q(T_d)$ = runoff flow at T_d determined by Eq. 4.29, and $I(T_d)$ = average rainfall from $(T_d - T_c)$ to T_d determined by Eq. 4.31.

Example 4.9: A rainfall-runoff event was recorded on November 6, 1977 at USGS Gauge Station 06714300 located at the Concourse D Storm Drain at Stapleton Airport, Denver, Colorado (USGS Open File 82-873). The watershed area is 96.75 acres with an imperviousness of 38%. The waterway has a length of 2,530 feet on a slope of 0.012 ft/ft. The total precipitation for this event was 0.28 inch, with duration of 80 minutes. The observed peak runoff rate was 12.0 cfs.

Solution: The time of concentration of this watershed is estimated to be 20 minutes, and the runoff coefficient is approximately 0.32. Therefore, the runoff flows were predicted by Eqs 4.29 and 4.30 before 20.0 minutes, Eqs 4.26 and 4.27 after 20.0 minutes, and Eqs 4.31 and 4.32 after 80 minutes. Table 4.9 presents a comparison between the observed and predicted hydrographs by the Rational method. As shown in Fig. 4.8, the predicted hydrograph by the Rational method reflects the temporal variations on the hyetograph, and results in good agreement with the observations.

   In this case, the peaking rainfall blocks are changed from a non-uniform distribution to a uniform distribution, as were the runoff flows. The recession began when the rainfall ceased.

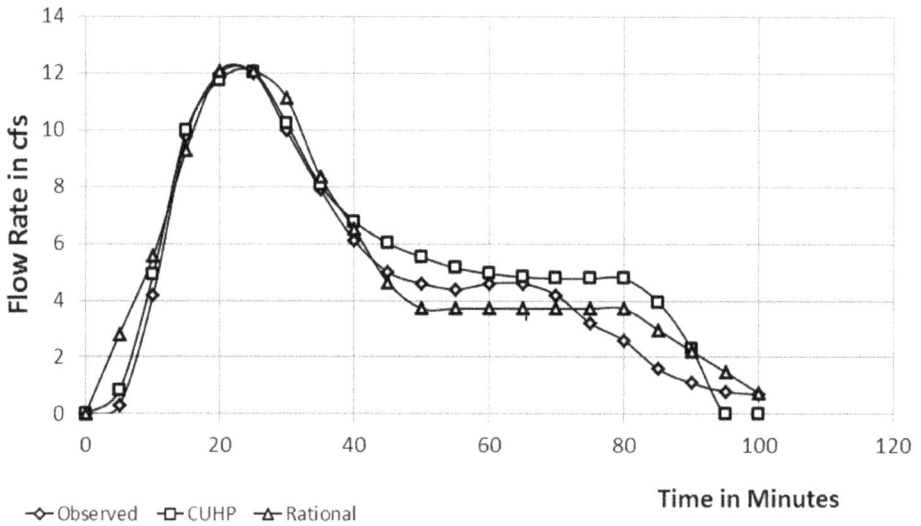

Figure 4.8 Case study for Stapleton Airport watershed, Denver, Colorado

Table 4.9 Predicted hydrograph for Stapleton Airport watershed in Denver, Colorado. A = 96.75 acres, T_c = 20 minutes, C = 0.32 for 50% imperviousness

Time (min)	Incremental Precipitation (inch)	Observed Hydrograph (cfs)	Rational Moving Average Intensity (inch/hr)	Hydrograph Predicted Runoff Rate (cfs)	Remarks
0.00	0.00	0.00	0.00	0.00	rising
5.00	0.03	0.30	0.36	2.79	
10.00	0.03	4.20	0.36	5.57	
15.00	0.04	9.80	0.40	9.29	
20.00	0.03	**12.00**	0.39	**12.07**	peaking
25.00	0.03	12.00	0.39	12.07	
30.00	0.02	10.00	0.36	11.15	
35.00	0.01	7.90	0.27	8.36	
40.00	0.01	6.10	0.21	6.50	
45.00	0.01	5.00	0.15	4.64	
50.00	0.01	4.60	0.12	3.72	
55.00	0.01	4.40	0.12	3.72	
60.00	0.01	4.60	0.12	3.72	
65.00	0.01	2.60	0.12	3.72	$Q(T_d)$ = 3.72
70.00	0.00	1.60	0.09	2.97	recession
75.00	0.00	1.10	0.06	2.23	
80.00	0.00	0.80	0.00	1.49	
85.00	0.00	0.70	0.00	0.74	

Table 4.10 Predicted 50-year runoff hydrograph

Duration D min	IDF Curve I (D) inch/hr	Precip Depth P(D)=DxI(D) inch	Incremental Depth dP(D) inch	Clock Time t minutes	Rainfall Distribtuion dp(t) inch	Moving Average I(t) inch/hr	Runoff Hygraph Q(t) cfs
0	0	0	0	0	0	0.00	0
5	5.42	0.45	0.45	5	0.05	0.65	15.3
10	4.32	0.72	0.27	10	0.07	0.74	34.5
15	3.62	0.90	0.19	15	0.09	0.86	60.4
20	3.13	1.04	0.14	20	0.14	1.20	84.3
25	2.78	1.16	0.11	25	0.27	2.00	140.2
30	2.50	1.25	0.09	30	0.45	3.44	240.6
35	2.28	1.33	0.08	35	0.19	3.62	253.4
40	2.09	1.40	0.07	40	0.11	2.99	209.6
45	1.94	1.46	0.06	45	0.08	1.50	105.3
50	1.81	1.51	0.05	50	0.06	1.01	70.4
55	1.70	1.56	0.05	55	0.05	0.76	52.9
60	1.61	1.61	0.05	60	0.05	0.62	43.5

Example 4.10: Predict the 50-yr runoff hydrograph generated from an urban watershed with $A = 100$ acres and $T_c = 15$ minutes. The rainfall IDF curve is defined with $P_1 = 1.61$ inch, $C_1 = 28.5$, $C_2 = 10$, and $C_3 = 0.789$.

Solution: The predicted hydrograph is presented in Table 4.10.

4.6 APPLICABILITY LIMIT

The basic assumptions for the application of the Rational method are summarized as:

1. The runoff flow rate in the Rational method is varies linearly with rainfall depth. Over a single storm event, the maximum rainfall amount over a period of time of concentration is the contributing rainfall depth to the peak runoff flow rate. Statistically, it means that the 100-yr rainfall depth produces a 100-yr peak flow, and so on.
2. The hydrologic losses in the watershed are homogenous and uniform. The runoff coefficient varies with respect to the type of soils, imperviousness ratio, and rainfall frequency. The runoff coefficient used in design represents average soil antecedent moisture conditions.
3. The time of concentration is assumed to be equivalent to the time of equilibrium when the entire watershed becomes the tributary area to the peak flow. For composite soils and land use, an area-weighted method is recommended to derive the average hydrologic parameters.
4. This method does not involve any hydrograph routing; as a result, it is not applicable to watersheds with a significant depression area or storage capacity, such as ponds and lakes.
5. This method tends to slightly overestimate the combined peak flow at a design point where several upstream flows come together, because the accumulation

of time of concentrations is not adequate to compensate for flow attenuation through hydrograph routing.

The applicable limit of the *Rational Hydrograph Method* (RHM) was examined by the SCS unitgraph method in the HEC-1 Flood Prediction Package using a series of hypothetical square watersheds ranging from 0.01 to 1.0 square miles. The watershed slope was assumed to be 0.01 ft/ft for all test watersheds and the runoff coefficient was assigned to be 0.75, equivalent to a SCS curve number of 85. An SCS 6-hour rainfall distribution curve was adopted as the design rainfall distribution with a total precipitation of 2.77 inches. Fig. 4.9 presents a comparison of the predicted peak flows by HEC-1 and RHM models. It can be seen that the predicted peak runoff rates are comparable until the watershed area exceeds 150 acres.

As a linear model, the major assumptions in the RHM are that the surface storage effect in the watershed is negligible and the present runoff flow is linearly related to the cumulative rainfall depth within a period of the time of concentration. In general, the RHM tends to overestimate the rising hydrograph. After the entire watershed becomes a tributary to the runoff, the RHM does accurately reflect the temporal changes in rainfall distribution. When applying a uniform design rainfall distribution to a small watershed, the RHM produces a triangular hydrograph when the rainfall duration is equal to the time of concentration, or a trapezoidal hydrograph when the rainfall duration is longer or shorter than the time of concentration. Hydrographs predicted by the RHM are comparable to sophisticated models such as CUHP, SWMM, and HEC-1. Considering the variations of natural depression in watersheds, it is suggested that the RHM be applicable up to 150 acre (AGU Committee 2015, CCRFCD Manual 1999).

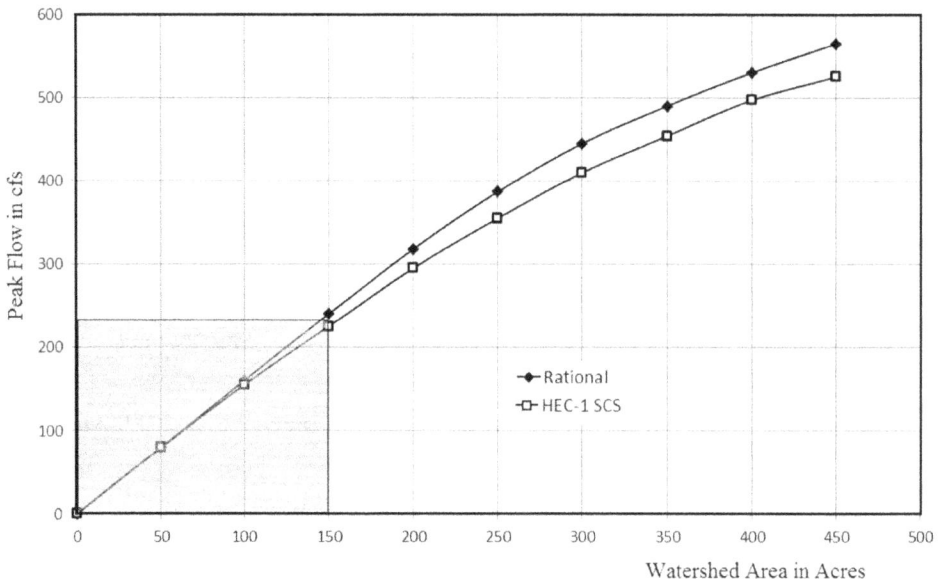

Figure 4.9 Applicability limit for the rational hydrograph method

HOMEWORK

Q4-1   The project site in Figure Q4-1 has a tributary area of 1000-ft by 500-ft. The runoff flows from the roof are directly drained into concrete chapter 5ditches and grass swales. The 5-year rainfall IDF formula at the site is described as:

$$I\left(\frac{inch}{hr}\right) = \frac{28.5 \times 1.35}{(10 + T_c)^{0.789}}, \text{ where } T_c = \text{time of concentration in minutes.}$$

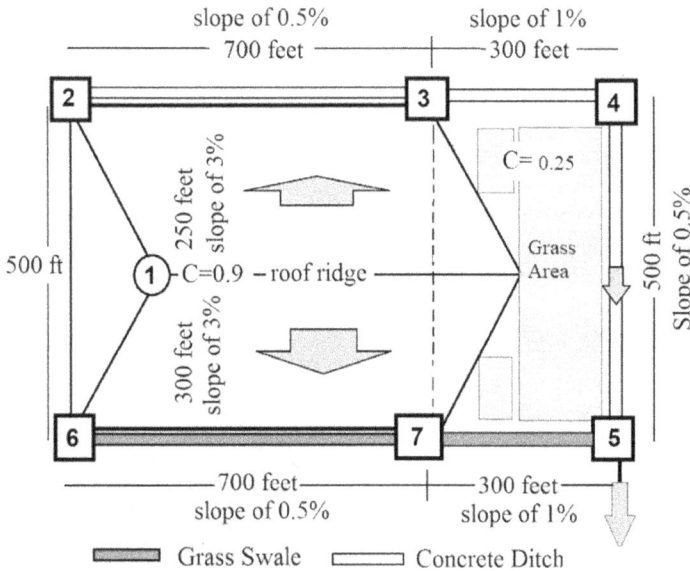

Figure Q4-1 Roof drainage problem

Your tasks are:

1. Determine the area-weighted runoff coefficients.
2. Considering that the airport formula for the overland flow and the SCS upland method for the ditch and swale flows, calculate the flow times through the two flow paths: (1) path one: 12345 and (2) path two: 1675. Determine the time of concentration at the project site.
3. Estimate the peak flow.
4. Comment on the assumption that the 5-yr rainfall depth produces the 5-yr peak flow.

Q4-2 1. Referring to Fig. Q4-2, use the IDF formula in Q4-1 to estimate the peak flow at Point C.
2. Use the IDF formula in Q4-1 to generate the design temporal distribution, and then apply Eqs 4.29 to 4.32 to predict the storm hydrograph at Point C.
3. Comment on the difference, if any, between the two predicted peak flows and also the difference, if any, between the two times to peak flow.

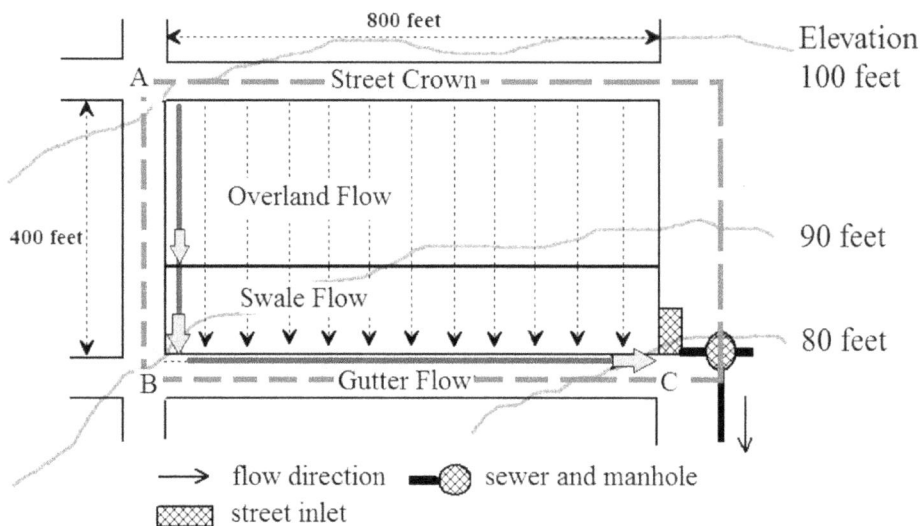

Figure Q4.2 Street drainage problem

REFERENCES

AGU Committee (2015). American Geophysical Union, *Committee on Runoff*, http://sites. agu.org/

CCRFCD Manual (1999). Clark County Regional Flood Control District, Design Manual, http://gustfront.ccrfcd.org/pdf_arch1/hcddm/Current%20Manual%20by%20Section/Section%20100.pdf

Guo, J.C.Y. (1998). "Overland Flow on a Pervious Surface", *International Journal of Water*, vol. 23, no. 2, June, pp. 91–95

Guo, J.C.Y. (2000). "A Semi Virtual Watershed Model by Neural Networks", *Journal of Computer-Aided Civil and Infrastructure Engineering*, vol. 15, pp. 439–444.

Guo, J.C.Y. (2001a). "Rational Hydrograph Method", *ASCE of Hydrologic Engineering*, vol 6, no. 4, July/August, pp. 352–357

Guo, J.C.Y. (2001b). "Storm Hydrographs from Small Urban Catchment", *Journal of International Water, American Water Resources Association*, vol. 25, no. 3, September, pp. 481–487.

Guo, J.C.Y. (2003). "Response to Discussion on Rational Hydrograph Method for Small Urban Catchments", *ASCE Journal of Hydrologic Engineering*, vol. 2, no. 1, May/June.

Guo, J.C.Y. and MacKenzie, K. (2014). "Modeling Consistency for small to large watershed studies", *ASCE Journal of Hydrologic Engineering*, vol. 19, no. 8, August, 04014009-1–7.

Guo, J.C.Y. and Urbonas B. (2014). "Volume-based Runoff Coefficient", *ASCE Journal of Irrigation and Drainage Engineering*, vol. 140, no. 2, February, 04013013-1–5.

HMS (2010). *Hydrologic Modeling Simulation Package*, Corp of Engineers, Hydrologic Engineering Center, Davis, CA.

Kirpitch, Z.P. (1941). "Time of Concentration for Small Agricultural Watersheds", *Civil Engineering, ASCE*, vol. 10, no. 6, June, p. 362.

Kuichling, E. (1889). "The Relation between Rainfall and the Discharge of Sewers in Populous Districts", *Transactions of the American Society of Civil Engineers*, vol. 20, pp. 1–56.

McCuen, R. (1982). *A Guide to Hydrologic Analysis Using SCS Methods*, Prentice Hall, Englewood Cliffs, NJ.

McCuen, R.H., Wong, S.L., and Rawls W.J. (1984). "Estimating Urban Time of Concentration", *Journal of Hydraulic Engineering*, vol. 110, no. 7, July.

Morgali, J.R. (1970). "Laminar and Turbulent Overland Flow Hydrographs", *Journal of Hydraulic Engineering*, HY 2, pp. 441–360.

Natural Resource Conservation Service (NRCS 1976). "Rainfall-runoff for Small Watersheds", Technical Release no. 55, US Government Printing Office, Washington DC.

NRCS (2013). SCS Upland method, www.nrcs.usda.gov/wps/portal/nrcs/site/national/home/

Rossman, L. (2005). EPA SWMM5, www2.epa.gov/water-research/storm-water-management-model-swmm

Singh, V.P. (1982). *Hydrologic Systems: Rainfall-Runoff Modeling*, vol. 1, Prentice Hall, Englewood Cliffs, NJ.

Singh, V.P. and Cruise, J.F. (1992). "Analysis of the Rational Formula Using a System Approach", in *Catchment Runoff and Rational Formula*, edited by B.C. Yen, Water Resources Publication, Colorado, pp. 39–51.

UDFCD (2010). "Rainfall and Runoff", Volume 1, Urban Storm Water Design Criteria Manual (USWDCM) published by Urban Drainage and Flood Control District, Denver, CO.

Wooding, R.A. (1965). "A Hydraulic Model for a Catchment-Stream Problem", *Journal of Hydrology*, vol. 3, pp. 254–267

Yu, Y.S. and McNown, R.K. (1965). "Runoff from Impervious Surfaces", *Journal of Hydraulic Research*, vol. 2, no. 1, pp. 3–24.

Chapter 5

Unit hydrograph

5.1 AGRICULTURAL SYNTHETIC UNITGRAPH

The concept of the unit hydrograph was originally introduced to analyze the relationship between observed rainfall amounts and runoff flows (Sherman 1932). Later on, the approach of the unit hydrograph was expanded into the synthetic unit hydrograph method for flood predictions (Snyder 1955). A synthetic unit hydrograph is constructed using the peaking factors and flow times. The coefficients used in Snyder's formulas are correlated to watershed topographic parameters including drainage area, length of waterway, length to the centroid of the watershed, and slope of watershed. For instance, the *Soil Conservation Service Unit Hydrograph* (SCSUH) method utilizes the time to peak and peak discharge as the parameters to predict the entire storm hydrographs. For convenience, the SCSUH's were further normalized to simplify the convolution process when generating storm hydrographs (NRCS SCS 1986). The SCSUH method was derived from a large number of agricultural unit hydrographs observed from the Appalachian Mountain region in the USA though the application of the SCSUH has been extended to metropolitan areas, acceptance of SCSUH is often justified by the consistence of the method, rather than the accuracy.

The *US Natural Resources Conservation Service* (NRCS or the former US Soil Conservation Service, SCS) suggests that the curve runoff hydrograph from a rural watershed be approximated by a triangular shape, as shown in Fig. 5.1. The SCSUH method is an empirical approach developed from the east mountainous forest areas in the continental US.

The key factor in the SCSUH method is the *lag time*, T_{lag}, which is defined as a time span between the mass centers of the excess rainfall hyetograph and runoff hydrograph. The *base time*, T_B, for this storm hydrograph is the sum of the *time to peak*, T_p, and the *time of recession*, T_R. The empirical formulas used in the SCS UH were derived for large forest areas using English units. For instance, the tributary area, A, is measured in square miles and the waterway length, L, is expressed in feet. Care must be taken when using such a unit-sensitive method. The empirical formula developed for the lag time is:

$$T_{lag} = \frac{L^{0.8}(S+1)^{0.7}}{1900\sqrt{S_0}} \quad \text{(for large rural and forest areas)} \tag{5.1}$$

DOI: 10.1201/9781003284239-5

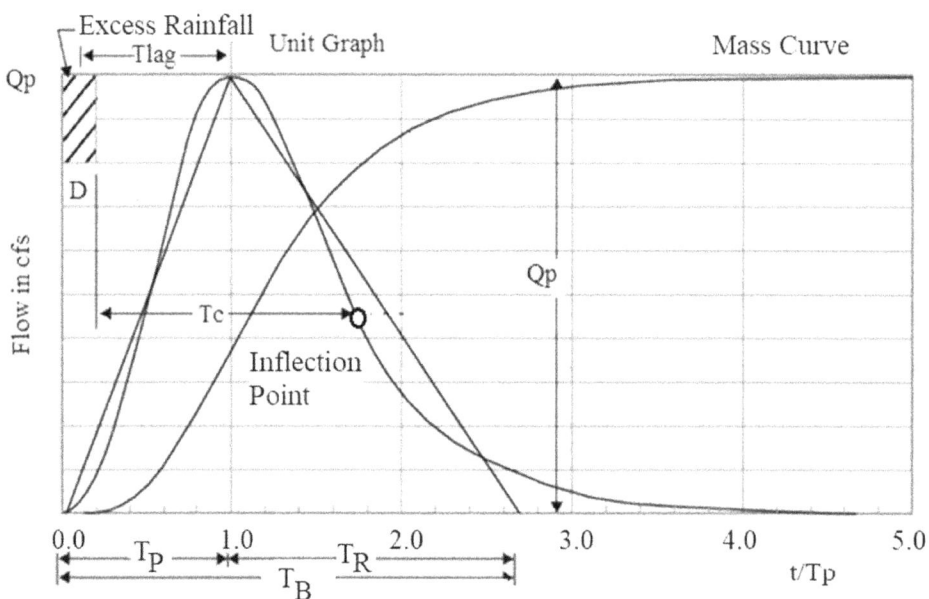

Figure 5.1 SCS triangular unit hydrograph

where T_{lag} = lag time in hours, S_o = waterway slope as a percentage, L = waterway length in feet and S = maximum soil retention volume (maximal soil moisture deficit), which is defined by the curve number (CN) as:

$$S = \frac{1000}{CN} - 10 \tag{5.2}$$

The CN in Table 5.1 is a special index system that was developed to describe the soil infiltration loss used in the SCSUH method. The CN varies between 30 and 100; the more impervious the watershed, the higher the CN. Because the CN was developed from agricultural watersheds, it does not adequately represent the hydrologic response from impervious surfaces in urban areas. Impervious surfaces are more hydraulically efficient and often result in a shorter time to peak and a higher runoff volume. Therefore, it is necessary to modify the SCS lag time for urban catchments as:

$$T_{lag} = 0.5 \text{ to } 0.6 \, T_c \text{ (for urbanized small areas)} \tag{5.3}$$

$$T_c = \frac{1}{60} M \left(\frac{L}{\sqrt{S}} \right)^{0.66} \quad (M = 0.054 \text{ for metric units or } 0.025 \text{ for English units}) \tag{5.4}$$

where T_c = time of concentration in hours for the tributary area and S = waterway slope in ft/ft. As illustrated in Fig. 5.1, the length of T_c is determined by the point of inflection on the recession hydrograph. In practice, the value of T_c is estimated using a valid empirical formula (see Chapter 4 Rational Method). According to Fig. 5.1, the time to peak is then calculated as:

$$T_P = \frac{D}{2} + T_{lag} \tag{5.5}$$

where T_p = time to peak flow in hours and D = user defined rainfall duration in hours. As a rule of thumb, the rainfall duration must be 1.5 to 2 times T_c to make sure the entire tributary area is covered under the design storm. As shown in Fig. 5.1, the time to peak and the time for recession hydrograph are estimated as:

$$T_R = \frac{5}{3}T_P \tag{5.6}$$

$$T_B = T_P + T_R \tag{5.7}$$

where T_B = base time in hours and T_R = time for recession in hours. The volume under the SCS UH is represented by the area of the triangular hydrograph, which must be equal to the unit volume of one-unit depth covering the tributary area as:

$$V = \frac{Q_p T_B}{2} \times 3600 = \frac{Q_p}{2}(T_P + T_R) \times \frac{3600 \text{ seconds}}{\text{hr}} \tag{5.8}$$

$$Q_p = \frac{2V}{3600(T_P + T_R)} = \frac{0.75}{3600\,T_P}V = \frac{0.75}{3600\,T_P} \times \left[\frac{1}{12} \times A \times \left(\frac{5280}{1}\frac{\text{ft}}{\text{mile}}\right)^2\right] = 484\frac{A}{T_P} \tag{5.9}$$

where V = runoff volume in cubic ft, Q_p = peak flow in cfs, A = tributary area in square miles, and T_p = time to peak in hours. Care must be taken when using a unit-sensitive formula such as Eq. 5.9.

Example 5.1: Derive an SCS triangular unit graph for a desert land parcel covered with Type B soils. Use the following parameters: CN = 77, D = 0.3 hr, A = 50.0 sq mile, L = 9.5 mile, S_o = 3 %.

Solution:

$$S = \frac{1000}{77} - 10 = 2.99$$

$$T_{lag} = \frac{(9.5 \times 5280)^{0.8}(2.99 + 1)^{0.7}}{1900\sqrt{3.0}} = 4.61 \text{ hr}$$

$$T_P = \frac{D}{2} + T_{lag} = \frac{3.0}{2} + 4.61 = 6.11 \text{ hr}$$

$$T_R = \frac{5}{3}T_P = 10.18 \text{ hr}$$

Table 5.1 SCS curve number for various land uses

Cover description		Curve numbers for hydrologic soil group			
Cover type and hydrologic condition	Average percent impervious area[2]	A	B	C	D
Fully developed urban areas (vegetation established)					
Open space (lawns, parks, golf courses, cemeteries, etc.)[3/]:					
Poor condition (grass cover < 50%)		68	79	86	89
Fair condition (grass cover 50% to 75%)		49	69	79	84
Good condition (grass cover > 75%)		39	61	74	80
Impervious areas:					
Paved parking lots, roofs, driveways, etc. (excluding right-of-way)		98	98	98	98
Streets and roads:					
Paved; curbs and storm sewers (excluding right-of-way)		98	98	98	98
Paved; open ditches (including right-of-way)		83	89	92	93
Gravel (including right-of-way)		76	85	89	91
Dirt (including right-of-way)		72	82	87	89
Western desert urban areas:					
Natural desert landscaping (pervious areas only)[4/]		63	77	85	88
Artificial desert landscaping (impervious weed barrier, desert shrub with 1- to 2-inch sand or gravel mulch and basin borders)		96	96	96	96
Urban districts:					
Commercial and business	85	89	92	94	95
Industrial	72	81	88	91	93
Residential districts by average lot size:					
1/8 acre or less (town houses)	65	77	85	90	92
1/4 acre	33	61	75	83	87
1/3 acre	30	57	72	81	86
1/2	25	54	70	80	85
1 acre	20	51	68	79	84
2 acres	12	46	65	77	82
Developing urban areas					
Newly graded areas (pervious areas only, no vegetation)[5/]		77	86	91	94
Idle lands (UN'S are determined using cover types similar to those in table 2-2c).					

[1] Average runoff condition, and $I_a = 0.2S$.

[2] The average percent impervious area shown was used to develop the composite CN's. Other assumptions are as follows: impervious areas are directly connected to the drainage system, impervious areas have a CN of 98, and pervious areas are considered equivalent to open space in good hydrologic condition. CN's for other combinations of conditions may be computed using figure 2-3 or 2-4.

[3] CN's shown are equivalent to those of pasture. Composite CN's may be computed for other combinations of open space cover type.

[4] Composite CN's for natural desert landscaping should be computed using figures 2-3 or 2-4 based on the impervious area percentage (CN = 98) and the pervious area CN. The pervious area CN's are assumed equivalent to desert shrub in poor hydrologic condition.

[5] Composite CN's to use for the design of temporary measures during grading and construction should be computed using figure 2-3 or 2-4 based on the degree of development (impervious area percentage) and the CN's for the newly graded pervious areas.

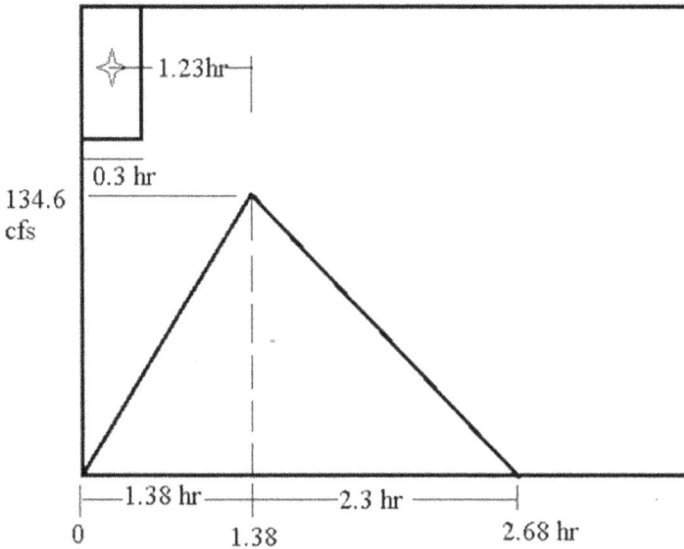

Figure 5.2 SCS triangular unit graph for Example 5.1

$T_B = 6.1 + 10.18 = 16.28$ hr

$$Q_p = \frac{484 \times 50.0}{6.11} = 3962.67 \text{ cfs}$$

$$V = \frac{3962.67 \times 16.28}{2} \times \frac{3600}{43560} = 2665.8 \text{ ac} - \text{ft volume under the UH}$$

$$V = 1 \text{ inch} \times \frac{\text{ft}}{12 \text{ inch}} \times 50 \text{ mile} \times 645 \frac{\text{ac}}{\text{mile}^2} = 2665.8 \text{ ac-ft volume for excess rainfall}$$

The hydrograph parameters for this case are plotted in Fig. 5.2.

Example 5.2: An industrial park is covered with Type A/B soils. Derive the SCS UH for $D = 0.25$ hr, $A = 0.15$ sq mile, $L = 0.43$ miles, CN = 85 and $S_o = 2.0\%$.

Solution: Notice that the tributary area is less than one square mile. The lag time is set to be 50% of the time of concentration, which is calculated as: $T_{lag} = 0.5$, $T_c = 0.13$ hr. The SCSUH for this case is predicted in Table 5.2 and Fig. 5.3. In practice, it is advisable to make sure that $D \approx T_c$.

The SCS method originated from the field data collected from the Apalachee Mountain areas and rural watersheds in the east coast areas of the US. Urbanization induces changes in a watershed's response to precipitation. An increase in impervious areas and decrease in time of concentration can jointly increase the peak flows generated from an urbanized watershed. The SCS empirical formulas ignored the impacts of natural reservoirs and man-made flood storage works on peak-flow reduction. Care must

Table 5.2 Example of SCS unit graph

SCS CN (50 to 100)	CN=	85.00	
Watershed Area	A =	0.15	Sq miles
Length of Main Stream	L=	0.43	Mile
Waterway Slope	So=	2.00	percent
Time of Concentration	Tc=	0.25	hours
User Defined Duration	D=	0.25	hours
Max Soil Retention	S=	1.76	
Lag Time	**Tlag=**	**0.13**	hours
Time to Peak Flow	Tp=	0.25	hours
Time for Recession Limb	TR=	0.42	hours
Based Time	TB=	0.67	hours
Peak Flow	Qp=	290.05	cfs
Volume under UH	UH Vol=	8.00	ac-ft
Unit-depth Volume	1-inch Vol=	8.00	ac-ft
Volume Difference	Check dV=	0.00	ac-ft

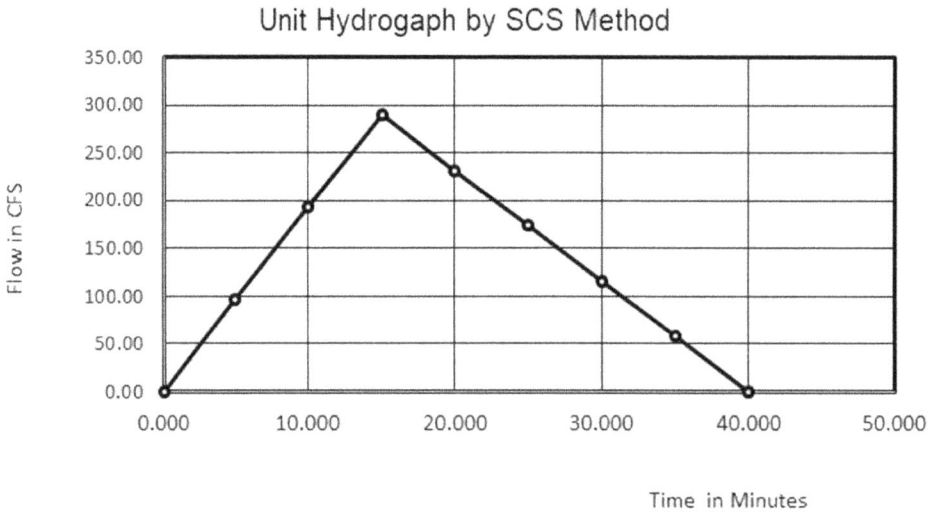

Figure 5.3 SCS triangular unit graph for Example 5.2

be taken when selecting CN and T_c for an urban watershed. More information about how to apply the SCS method can be found in *Technical Releases 20 and 55* (NRCS SCS 1986).

5.2 RATIONAL UNIT GRAPH

Considering the symmetric triangular hydrograph, the Rational method is expanded into the Rational unit hydrograph (RUH) method. By definition, the rainfall duration

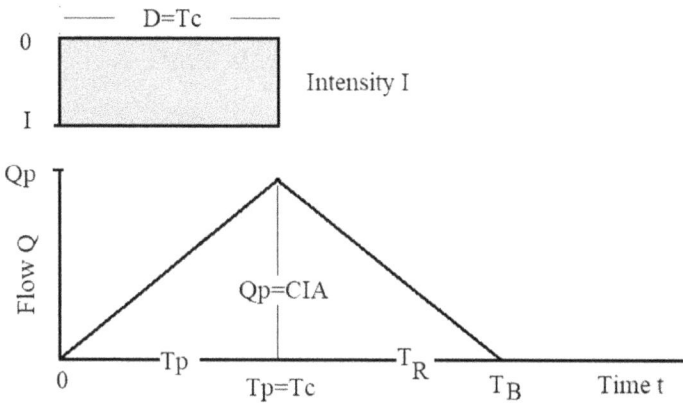

Figure 5.4 Rational unit hydrograph

is set to be the time of concentration to make sure the entire watershed is the tributary, and the excess rainfall intensity, I_e, is calculated as:

$$D = T_c \tag{5.10}$$

$$I_e = \frac{1}{D} \tag{5.11}$$

Referring to Fig. 5.4, the peak flow is centered with the time to peak and the time for the recession hydrograph equal to the time of concentration, T_c.

Both the rising and recession limbs on the RUH are linear and equal to T_c. The triangular unit graph is defined with the following parameters:

$$Q_p = kCIA = kI_eA \tag{5.12}$$

where $k = 1$ and Q_p = peak flow in cfs if I = rainfall intensity in inch/hr and A = tributary area in acres or $k = 1/360$ and Q_p = peak flow in cms if I is given in mm/hr and A in hectares, and C = runoff coefficient (Guo and MacKenzie 2014)

$$T_p = T_R = T_c \tag{5.13}$$

$$T_B = 2T_c \tag{5.14}$$

$$V = CIAT_c = \text{unit depth} \times \text{watershed area} \tag{5.15}$$

Example 5.3: An industrial park is covered with Type B soils. Derive the SCS UH for $D = 0.25$ hr, $A = 0.155$ sq mile, $L = 0.43$ miles, $CN = 85$, $S_o = 2.0\%$, and $T_c = 0.25$ hr (see Example 5.2).

Solution: The drainage area is: $A = 0.155$ sq mile $= 0.155 \times 645$ acre/sq mile $= 100$ acres.
In this case, the time of concentration is estimated using the empirical formula (Chapter 4, Guo 2001b) as:

$$T_c = \frac{1}{60} M \left(\frac{L}{\sqrt{S}} \right)^{0.66} = \frac{1}{60} \times 0.025 \left(\frac{5280 \times 0.43}{\sqrt{0.02}} \right)^{0.66} = 0.25 \, hr$$

where T_c = time of concentration in hours, L = waterway length in feet or meters, S = waterway slope in ft/ft or m/m, and M = 0.054 for metric units or 0.025 for English units.
The hydrograph parameters for this case are plotted in Fig. 5.5.

Table 5.3 Solution for Example 5.3

under I inch of Excess Rain			Under I mm of Excess Rain	
Watershed Area	A =	100.00 acre	40.00	hectares
Waterway Length	L =	0.43 mile	0.70	kilometer
Waterway Slope	So =	0.02 ft/ft	0.02	m/m
Time of Concentration	Tc =	0.25 hr	0.25	hr
Rainfall Duration	D =	0.25 hr	0.25	hr
Excess Rainfall Depth	Pe =	1.00 inch	1.00	mm
Rainfall Intensity	Ie =	4.00 inch/hr	4.06	mm/hr
Peak Flow	Qp =	400.00 cfs	0.45	cms
Time to Peak	Tp =	0.25 hr	0.25	hr
Base Time	TB =	0.50 hr	0.49	hr
Unit-depth Rain Volume	V-rain =	8.33 acre-ft	40.00	ha-mm
Runoff Volume	V-runoff =	8.26 acre-ft	40.00	ha-mm
dV=Vrain−Vrunof= close to zero	dV =	0.07 acre-ft	0.00	ha-mm

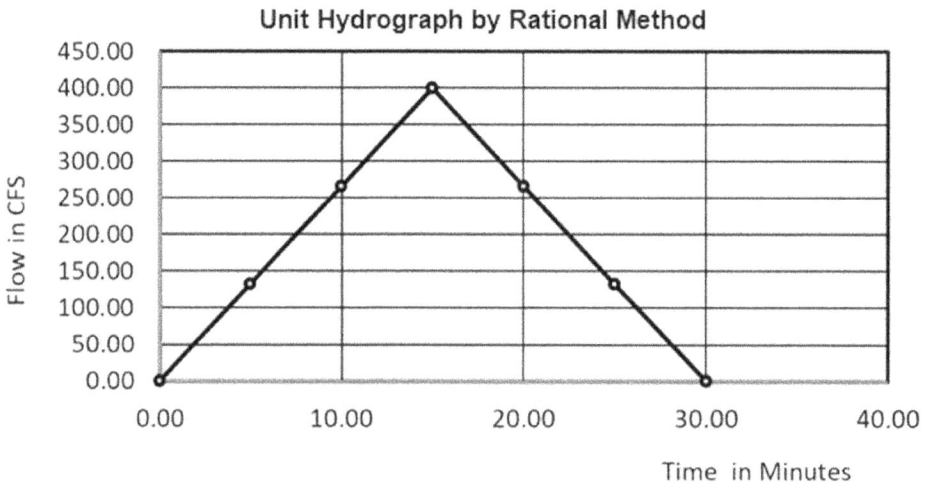

Figure 5.5 Rational unit graph for Example 5.3

5.3 URBAN SYNTHETIC UNIT GRAPH

In the State of Colorado, the *Colorado unit hydrograph procedure* (CUHP) was developed for metropolitan areas (CUHP 2005). The major parameters used in the CUHP were calibrated using the watershed's area imperviousness ratio using urbanized watershed data. The CUHP has been widely accepted for urban flood predictions in the front range of the Rocky Mountains (UDFCD 2010). Both the SCSUH and CUHP are similar to Snyder's approach. Both the CUHP and SCSUH have been revised for small, urbanized catchments. For instance, SCSTR55 was published to introduce the peaking factor as an adjustment to the SCSUH when applied to urban areas (SCS 1986). The time to peak used in the CUHP was also replaced with the time of concentration when the CUHP was applied to urban catchments of less than 150 acres (Guo and Urbonas 2014).

The CUHP was developed to predict the storm runoff generated from urbanized areas. From the aspect of stormwater drainage, street curbs, and gutters represent the level of urbanization. The CUHP was calibrated with the rainfall-runoff data collected from several selected urban watersheds in the metro Denver area, Colorado. The CUHP is recommended for urban hydrologic planning. The CUHP applies Snyder's synthetic unit hydrograph (Snyder 1938, 1955)procedure to determine the unit hydrograph for urbanized catchments. The empirical formulas are summarized as follows:

$$C_p = PC_t A^{0.15} \tag{5.9}$$

$$t_p = C_t \left(\frac{LL_c}{\sqrt{S_0}} \right)^{0.48} \quad \text{(hour)} \tag{5.10}$$

$$q_p = 640 \frac{C_p}{t_p} \quad \text{(cfs/sq mile)} \tag{5.11}$$

$$T_p = 60t_p + 0.5t_u \quad \text{(hour)} \tag{5.12}$$

$$Q_p = q_p A \quad \text{(cfs)} \tag{5.13}$$

$$t_u = \max\left(\frac{1}{3}t_p \times 60, 5 \right) \text{(minutes)} \tag{5.14}$$

where A = drainage area (sq mile), L = waterway length (mile), L_a = waterway length to the centroid of watershed (mile), S_o = watershed slope (ft/ft), P = coefficient for peak flow (Fig. 5.6), C_t = coefficient for time to peak (Fig. 5.7), t_u = rainfall duration (minute), t_p = time to peak flow (hr) from one square mile, q_p = peak flow (cfs/sq mile) from one square mile, T_p = time to peak (minutes) on the unit hydrograph, and Q_p = peak flow (cfs) on the unit hydrograph.

Although Eq. 5.14 is recommended for determining the rainfall duration, in practice, the selection is one of 5, 10, or 15 minutes. For convenience, the computational time increment is kept the same as the rainfall duration. The construction of a synthetic unit graph (Snyder 1938) requires time widths at 50 and 75% of the peak flow, as shown in Fig. 5.8. Empirical equations for these time widths are:

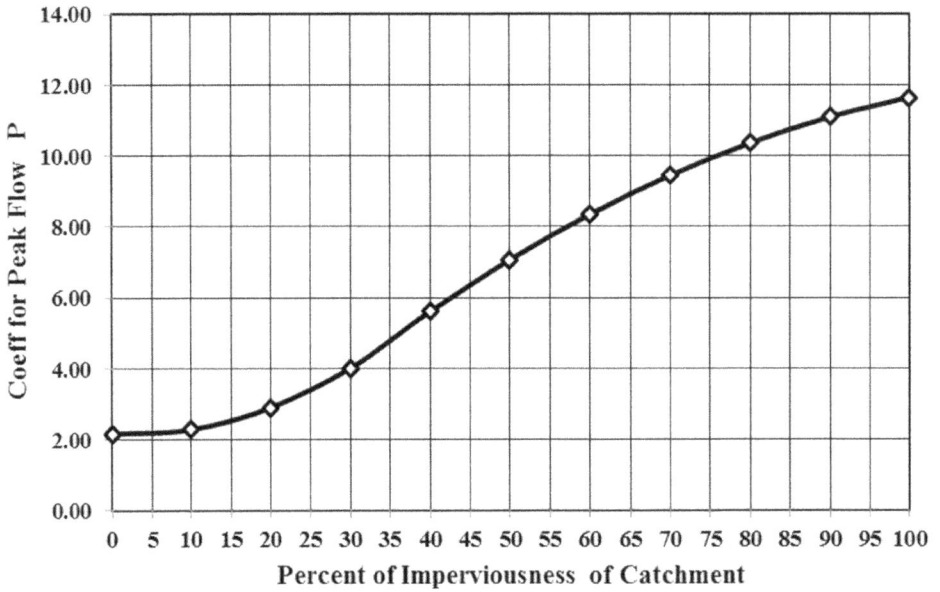

Figure 5.6 Coefficient, P, of peak flow for the CUHP method (CUHP2005)

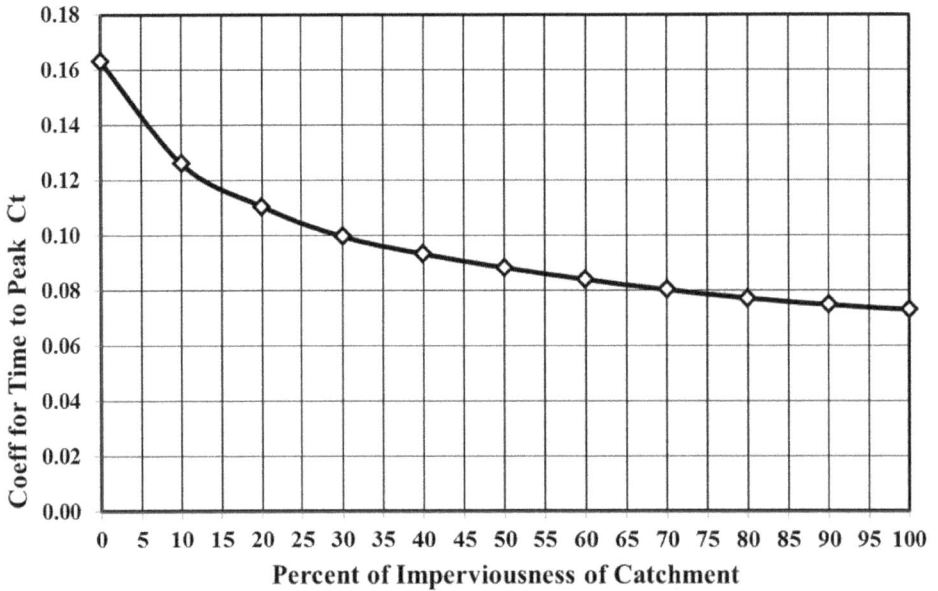

Figure 5.7 Coefficient, C_t, for time to peak in the CUHP method (CUHP2005)

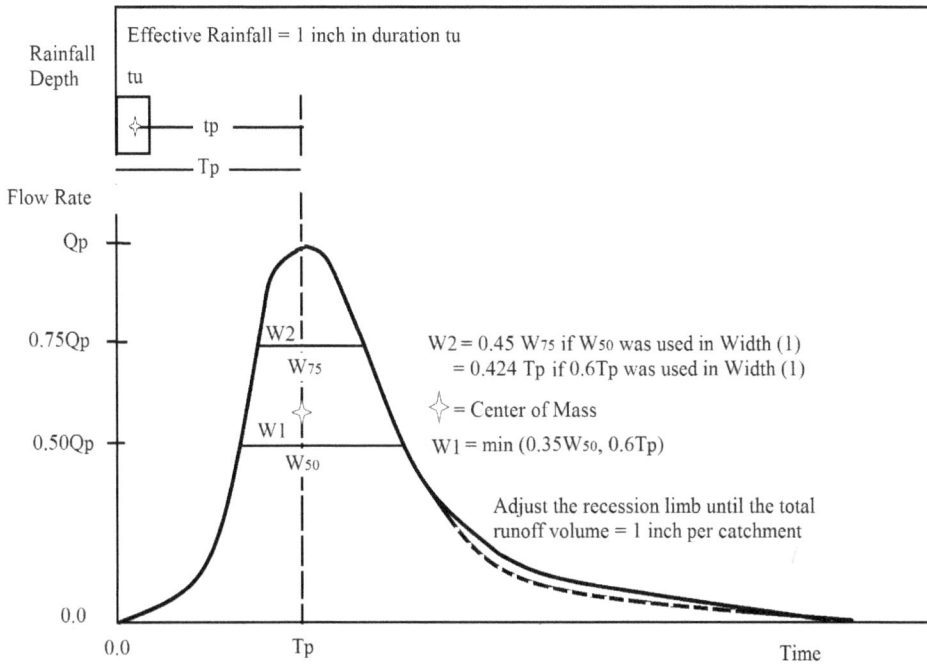

Figure 5.8 Shape of Colorado urban unit graph (CUHP 2005)

$$W_{50} = \frac{500}{q_p} \quad (\text{hour}) \tag{5.15}$$

$$W_{75} = \frac{260}{q_p} \quad (\text{hour}) \tag{5.16}$$

where W_{50} = time width at 50% of peak discharge (hour), and W_{75} = time width at 75% of peak discharge (hour). Fig. 5.8 illustrates how an urban UH is defined by six points and the recession tail is adjusted to satisfy the volume of one inch per watershed. The applicable range of the CUHP is for urban watersheds between 5 and 3,000 acres, and the waterway slope should not be steeper than 6%.

Example 5.4: Derive the 5-minute unit graph using the CUHP for a watershed with the following parameters: A = 0.38 sq mile, L = 1.28 mile, L_c = 0.52 mile, I_a = 44%, S_o = 0.0102 ft/ft.

Solution:
    Step 1: C_t = 0.091 based on 44% imperviousness from Fig. 5.7.
    Step 2: Because the watershed is larger than 90 acres, the time to peak should be calculated as:

$$t_p = 0.091\left(\frac{1.28 \times 0.52}{\sqrt{0.0102}}\right)^{0.48} = 0.225\,\text{hour}$$

Step 3: $t_u = \max\left(\frac{1}{3}t_p \times 60,\, 5.0\right) = (4.5,\, 5.0) = 5.0\,\text{minutes}$

Step 4: $P = 6.2$ based on 44% imperviousness from Fig. 5.6.

Step 5: $q_p = 640 \times \dfrac{0.49}{0.225} = 1394$ cfs/sq mile

Step 6: $Q_p = 1394 \times 0.38 = 530$ cfs

Step 7: $T_p = 60t_p + 0.5t_u = 60 \times 2.25 + 0.5 \times 5.0 = 16.0$ minutes

Step 8: $W_{50} = 0.369$ hour = 21.0 minutes (7.4 minutes ahead of Q_p)
$\phantom{\text{Step 8: }}W_{75} = 0.186$ hour = 11.2 minutes (5.0 minutes ahead of Q_p)

Step 9: The unit volume from the watershed $= \dfrac{1}{12} \times 0.38 \times 645 = 20.3$ acre-ft used
to adjust the recession hydrograph for water volume balance.

Step 10: The unit graph is constructed with the six points and a recession tail adjusted for one-inch volume as shown in Fig. 5.9.

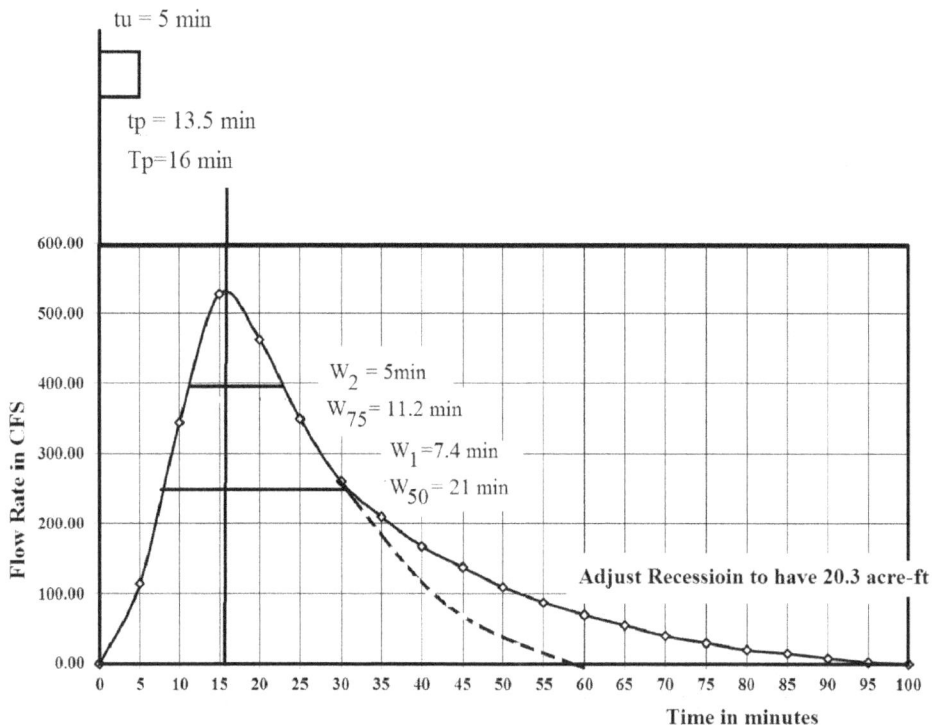

Figure 5.9 Urban UH for example watershed using CUHP2005

Example 5.5: The watershed has a drainage area of 0.38 square miles. The soil infiltration parameters are f_o = 3.0 inch/hr, f_c = 0.60 inch/hr, and k = 5.50 1/hr. The interception loss is 0.10 inch and the depression loss is 0.10 inch. The tasks are: (1) derive the excess rainfall amount from the given storm hyetograph in Table 5.4, and (2) predict the hydrograph using the CUHP.

    (a) Calculation of the rainfall excess as shown in Table 5.4.
    (b) Prediction of the storm hydrograph by CUHP as shown in Table 5.5.

    The individual and total hydrographs are presented in Fig. 5.10.
    In this example, the 15-min excess rainfall is the sum of three 5-min rainfall blocks as:

$$P_e = 0.1 + 0.28 + 0.01 = 0.39 \text{ inch}$$

Applying the above 15-min excess rainfall to the SCS UH in Example 5.2 and the Rational UH in Example 5-3, the peak flows are respectively predicted to be:

$$Q_p = 400 \times 0.39 = 156.0 \text{ cfs (the predicted peak flow on the Rational}$$
$$\text{hydrograph)}$$

$$Q_p = 290.05 \times 0.39 = 113.1 \text{ cfs (the predicted peak flow on the SCS hydrograph)}$$

In this case, the watershed area is 0.38 square miles (approximately 245 acres or 100 hectares), which is beyond the application limits (<150 acres) for the Rational UH. The depression loss for this watershed is 0.1 inch, which is far less than the depression loss (0.5 inch) on pervious areas. It is noticed that the SCSUH is applicable to rural watersheds, not urbanized watersheds as in Example 5.5.

Table 5.4 Calculation of rainfall excess for Example 5.5

Time t (minutes) (1)	$p(t)/P_t$ (percent)	Incremental Precipitation dP(t) (inch) (2)	Initial Loss $I_a(t)$ (inch) (3)	(2)–(3) (inch) (4)	Incremental infiltration dF(t) (inch) (5)	(4)–(5) (inch) (6)	Depression Loss D(t) (inch) (7)	Rainfall Excess (7)–(8) (inch) (8)
0.00	0.00	0.00	0.00	0.00				0.00
5.00	2.00	0.03	0.03	0.00				0.00
10.00	3.70	0.06	0.06	0.00		0.00*		0.00
15.00	8.20	0.13	0.01	0.12	0.21	0.00*	0.00*	0.00
20.00	15.00	0.24		0.24	0.15	0.09	0.09	0.00
25.00	25.00	0.40		0.40	0.11	0.29	0.01	0.28
30.00	12.00	0.19		0.19	0.09	0.10	0.00	0.10
35.00	5.60	0.09		0.09	0.08	0.01	0.00	0.01
40.00	4.30	0.07		0.07	0.07	0.00*		0.00*
45.00	3.80	0.06		0.06	0.06	0.00*		0.00*
total	115.70	1.86	0.10	1.76	0.05	0.50	0.10	0.39

Note: 0.00* means that the rainfall excess is not enough to produce runoff.

Table 5.5 Convolution of hydrographs for Example 5.5

Time (minutes)	Rainfall Excess (inch)	Unit Graph (cfs)	DRH-1 (cfs)	DRH-2 (cfs)	DRH-3 (cfs)	Storm DRH (cfs)
0.00	0.00	0.00				0.00
5.00	0.28	115.00	0.00			0.00
10.00	0.10	345.00	32.20	0.00		32.20
15.00	0.01	528.00	96.60	11.50	0.00	108.10
20.00		463.00	147.84	34.50	1.15	183.49
25.00		350.00	129.64	52.80	3.45	185.89
30.00		260.00	98.00	46.30	5.28	149.58
35.00		210.00	72.80	35.00	4.63	112.43
40.00		168.00	58.80	26.00	3.50	88.30
45.00		138.00	47.04	21.00	2.60	70.64
50.00		110.00	38.64	16.80	2.10	57.54
55.00		88.00	30.80	13.80	1.68	46.28
60.00		70.00	24.64	11.00	1.38	37.02
65.00		55.00	19.60	8.80	1.10	29.50
70.00		40.00	15.40	7.00	0.88	23.28
75.00		30.00	11.20	5.50	0.70	17.40
80.00		20.00	8.40	4.00	0.55	12.95
85.00		15.00	5.60	3.00	0.40	9.00
90.00		8.00	4.20	2.00	0.30	6.50
95.00		2.00	2.24	1.50	0.20	3.94
100.00		0.00	0.56	0.80	0.15	1.51
105.00			0.00	0.20	0.08	0.28
110.00				0.00	0.02	0.02
115.00					0.00	0.00

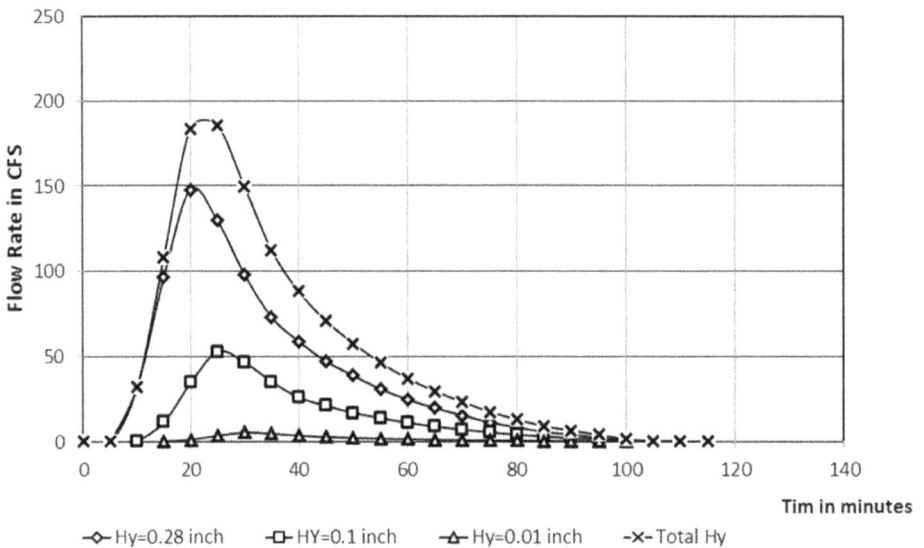

Figure 5.10 Convolution of unit graphs for Example 5.5

5.4 CONCLUSIONS

The SCS triangular unit hydrograph is characterized by the drainage nature of rural, forest, undeveloped, large watersheds. Time parameters are the major factors for shaping the SCSUH. In general, the lag time in Eq. 5.1 tends to give a long period of time because it was inherited the nature of large watersheds under storage effects associated with depressed areas. An overestimated lag time tends to decrease the peak flow. Therefore, it is important to replace Eq. 5.1 with Eqs 5.3 and 5.4 when the watershed is urbanized with street curbs and gutters. Urban watershed hydrology is sensitive to the area imperviousness percentage. As indicated in the CUHP method, all major parameters are related to the watershed imperviousness. Although CUHP was derived from the Rocky Mountain foothills, it is also applicable to metro urban areas. It is noted that CUHP is numerically sensitive to the waterway slope. When generalizing the CUHP for other urban areas, the slope should not exceed 6%, so 6% should be set as the upper limit for applications of CUHP.

HOMEWORK

Q5-1   A highway in Fig. Q5-1 runs across a natural waterway. Knowing that $A = 1.25$ sq miles, $I_a = = 0.70$, $L = 1.60$ miles, $L_c = 0.75$ mile, and $S_o = 0.014$, apply the CUHP to construct the unit hydrograph at the highway bridge in Fig. Q5-1.

Q5-2   Repeat Q5-1. With a soil Type C, apply the SCSUH method with CN = 90 ($I_a = 70\%$) to construct the unit hydrograph at the highway bridge in Fig. Q5-1.

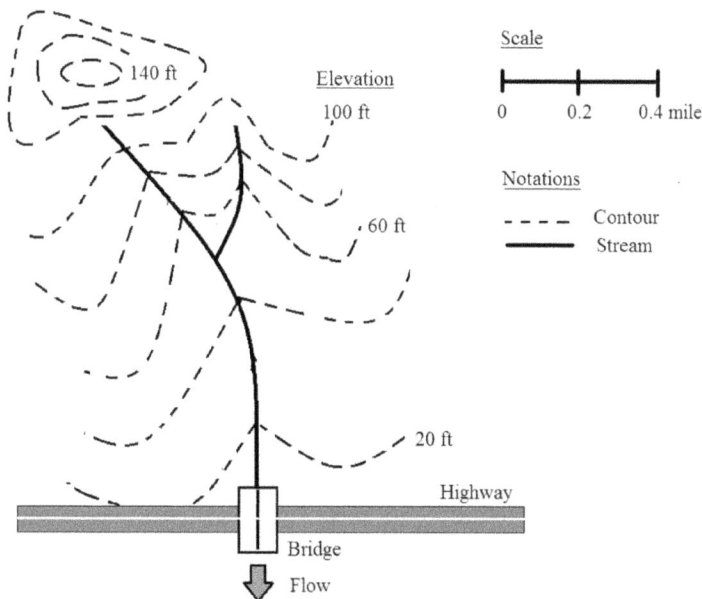

Figure Q5-1 Watershed for unit hydrograph

Q5-3 Of the CUHP and SCSUH, which one is more suitable in this case? Explain why.

Solution:

Watershed parameters		SCSCN(50 to 100)	CN=	90 00
Imp=	70.00 %	Watershed Area	A =	1.25 sq miles
Ct=	0.080	Length of Main Stream	L=	1.60 mile
P=	9.50	Waterway Slope	So=	1.40 percent
A=	1 25 sq miles	Times of Concentration	Tc=	0.67 hours
L=	1 60 miles	User Defined Duration	D=	0.67 hours
Lc=	0.76 miles	Max Soil Retention	s=	1.11
S=	0.0140 ft/ft	Lag Time	Tlag=	1.04 hours
Unit graph parameters		Time to Peak Flow	Tp=	1.37 hours
UNIT AREA Cp=	0.79	Time for Recession Limb	TR=	2 29 hours
tp=	0.24 hrs	Based Time	TB=	3.66 hours
qp=	2067.78 cfs	Peak Flow	Qp=	440.22 cfs
BASIN AREA Tu=	5.00 min	Volume under UH	UHVol=	66.67 ac-ft
Tp=	17.09 min	Unit-depth Volume	1-inch Vol=	66.67 ac-ft
Q=	2584.72 cfs	Volume Difference	Check dV=	0.00 ac-ft
W75=	7.54 min			
W50=	14.51 min			
Vol=	67.1875 ac-ft			

Q5-1: Solution: CUHP Unit Graph   *Q5-2: Solution: SCS Unit Graph*

REFERENCES

CUHP (2005). "Colorado Urban Unit hydrograph Procedure", in *Runoff, Urban Storm Water Design Criteria Manual*, UDFCD, Denver, CO

Dooge, J.C.I. (1959). "A General Theory of the Unit Hydrograph", *Journal of Geophysical Research*, vol. 64, no. 2, pp. 241–256.

Guo, J.C.Y. (1988). "Colorado Unit Hydrograph Procedures – Its Synthetic Unit Hydrograph Characteristics", Proceeding of ASCE International Conference on Hydraulic Engineering held at Colorado Springs, Colorado, Aug.

Guo, James C.Y. (1998). "Overland Flow on a Pervious Surface", *International Journal of Water*, vol. 23, no. 2, June, pp. 91–95.

Guo, J.C.Y. (2000) "A Semi Virtual Watershed Model by Neural Networks", *Journal of Computer-Aided Civil and Infrastructure Engineering*, vol. 15, pp. 439–444.

Guo, J.C.Y. (2001a) "Rational Hydrograph Method", *ASCE of Hydrologic Engineering*, vol. 6, no. 4, July/August, pp. 352–357

Guo, J.C.Y. (2001b). "Storm Hydrographs from Small Urban Catchment", *Journal of International Water, American Water Resources Association*, vol. 25, no. 3, September, pp. 481–487.

Guo, J.C.Y. and MacKenzie, K. (2014). "Modeling Consistency for small to large watershed studies", *ASCE Journal of Hydrologic Engineering*, vol. 19, no. 8, August, 04014009-1–7.

Guo, J.C.Y. and Urbonas B. (2014) "Volume-based Runoff Coefficient", *ASCE Journal of Irrigation and Drainage Engineering*, vol. 140, no. 2, February, 04013013-1–5.

NRCS SCS (1986). *Urban Hydrology for Small Watersheds*, 2nd ed., Tech. Release no. 55 (NTIS PB87-101580), US Department of Agriculture, Washington, DC.

Sherman, L.K. (1932). "Stream Flow from Rainfall by Unit-Graph Method", *Engineering News-Record*, vol. 108, April 7, pp. 501–505

Snyder, F.F. (1938). "Synthetic Unit hydrographs", *Transactions of the American Geophysical Union*, vol. 19, pp. 447–454.

Snyder, W.M. (1955). "Hydrograph Analysis by the Method of Least Squares", *Journal of the Hydraulics Division*, vol. 81, pp. 1–25.

Tauxe, G.W. (1978). "S-Hydrographs and Change of Unit Hydrograph Duration", *Journal of the Hydraulics Division*, vol. 104, no. HY3, pp. 439–444.

Taylor, A.B. and H.E. Schwartz (1952). "Unit Hydrograph Lag and Peak Flow Related to Basin Characteristics", *Transactions of the American Geophysical Union*, vol. 33.

UDFCD (2010). *Urban Stormwater Design Criteria Manual*, Urban Drainage and Flood Control District, Denver, CO.

Chapter 6

Kinematic wave method

6.1 KINEMATIC WAVE APPROACH

The KW method is applicable to a rectangular KW plane using the unit-width flow approach (Bedient, Huber, and Vieux 2008). To convert an irregular natural watershed into its equivalent rectangular KW plane depends on how to select the plane width (Rossman 2010, EPA SWMM 2005). Parking lots in Fig. 6.1 are the best example of a KW rectangular plane. The overland flow is then calculated by the unit-width KW procedure using the rectangular (x,y) coordinates.

For modeling convenience, the watershed is further divided into a left-impervious plane and a right-pervious plane, as shown in Fig. 6.2, according to the impervious area ratio, Ia, in the watershed. The two flows from the left and right planes drain into the central collector channel. The total flow, Q, collected through the central channel is the sum of the unit-width flows from the left and right planes. The *unit-width kinematic wave model* for overland flow consists of the continuity and simplified momentum principles as (Blackler and Guo 2012).:

$$\frac{\partial y}{\partial t} + \frac{\partial q}{\partial x} = I_e \tag{6.1}$$

and

$$S_f = S_w \tag{6.2}$$

where y = overland flow depth in [L], q = flow rate in [L²/T], x = distance from the upstream boundary in [L], I_e = excess rainfall intensity in [L/T], S_f = friction loss, and S_w = ground slope.

Eq. 6.2 implies that the momentum principle for the KW flow is reduced to the Manning formula, which described the relationship between q and y. As a result, at a time step, the KW flow system has two equations for two unknowns, q and y. Before going into the numerical detail, let us discuss how to convert a watershed into its equivalent rectangle.

DOI: 10.1201/9781003284239-6

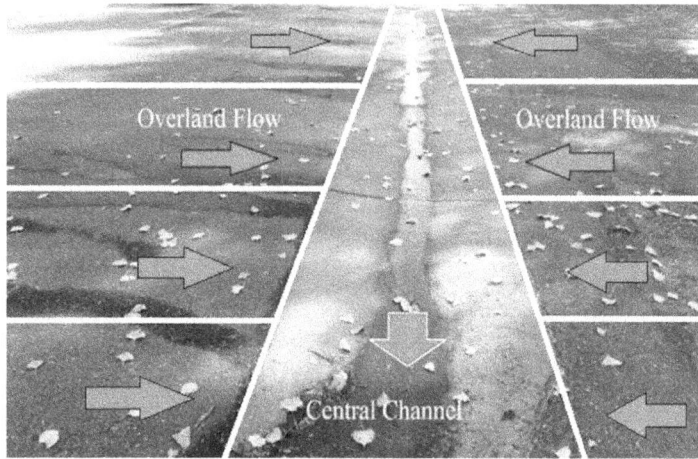

Figure 6.1 Kinematic wave rectangular planes

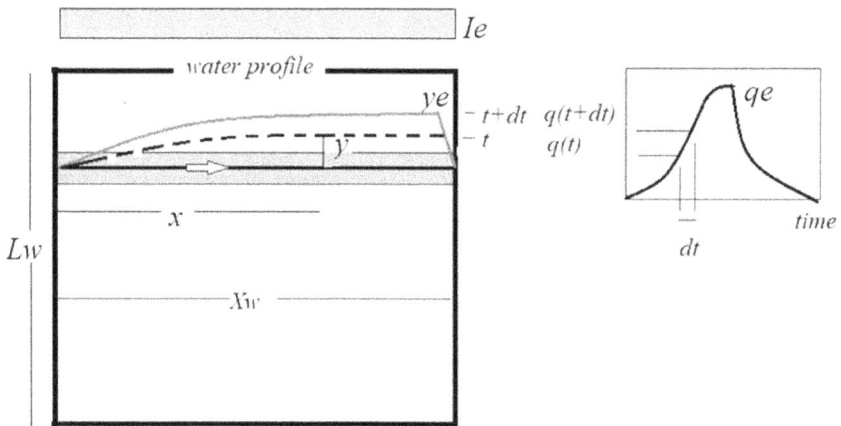

Figure 6.2 Unit-width approach used on kinematic wave rectangular plane

6.2 CONVERSION OF A WATERSHED INTO A RECTANGULAR PLANE

The KW procedure requires the conversion of a real watershed into its virtual rectangle on the KW plane. Fig. 6.3 illustrates the major parameters between the real and virtual flow systems. According to the principle of continuity, the total tributary area must satisfy:

$$A = X_w L_w \tag{6.3}$$

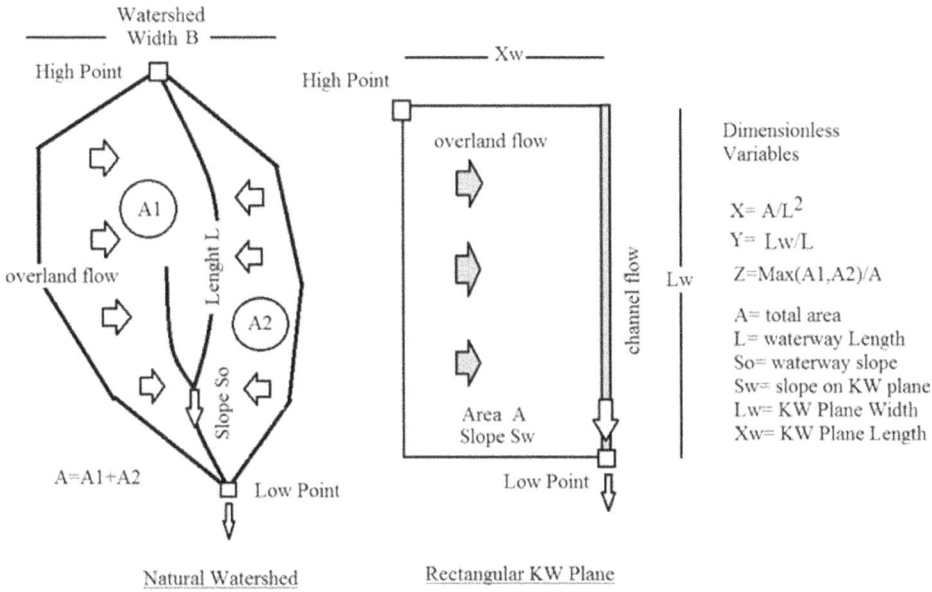

Figure 6.3 Natural watershed and the KW rectangular wave plane

where A = watershed area in $[L^2]$, X_w = rectangular length for overland flow on the KW plane in [L], and L_w = rectangular width of the KW plane in [L]. The fall over the waterway is the elevation difference from the high point on the upstream boundary to the outlet. Between these two flow systems, the potential energy in terms of the vertical fall along the waterway must be preserved as:

$$S_o L = S_w \left(X_w + L_w \right)$$ (6.4)

where S_o = longitudinal slope along the waterway through the watershed, S_w = slope on the KW plane, and L = length of waterway in [L].

   Using the waterway length, L, to normalize the parameters, Eqs 6.3 and 6.4 are converted into:

$$\frac{A}{L^2} = \frac{X_w}{L} \frac{L_w}{L}$$ (6.5)

$$\frac{S_o}{S_w} = \frac{X_w}{L} + \frac{L_w}{L}$$ (6.6)

Eq. 6.5 implies that the watershed shape must be preserved between these two flow systems. Watershed shape factor represents how the overland flows are collected into the waterway. Referring to Fig. 6.3, the shape factors for the real watershed and virtual KW plane are defined as (Guo and Urbanos 2009)

$$X = \frac{A}{L^2} \cong \frac{B}{L}(\le K) \tag{6.7}$$

$$Y = \frac{L_w}{L} \tag{6.8}$$

where X = watershed shape factor, B = average width of watershed in [L], Y = KW shape factor for the KW plane, and K = upper limit of shape factor. In practice, it is advisable that a large watershed be divided into smaller sub-areas and each sub-area should have a shape factor not to exceed the limit, $K \le 4$ (UDFCD 2005); otherwise, the peak runoff may be overestimated because the sub-area is too wide in shape. Aided by Eqs 6.5 and 6.8, Eq. 6.6 becomes:

$$\frac{S_o}{S_w} = \frac{X}{Y} + Y(X \le K) \tag{6.9}$$

The relationship between X and Y was derived using the parabolic equation as (Guo and Urbonas 2009):

$$Y = (1.5 - Z)\left[\frac{2}{1 - 2K}X^2 - \frac{4K}{1 - 2K}X\right] \tag{6.10}$$

$$Z = \frac{A_m}{A} = \frac{\max(A_1, A_2)}{A} \tag{6.11}$$

where Z = area skewness coefficient between 0.5 and 1.0, and A_m = the larger of A_1 and A_2, which are the two sub-areas divided by the waterway in Fig. 6.3. As illustrated in Fig. 6.4, $Z = 0.5$ for a symmetric watershed. $Z = 1.0$ for a side channel along the watershed boundary.

It is noted that Eq. 6.10 is reduced to $Y = 2$ for a square watershed with a central channel, and $Y = 1$ for a square watershed with a side channel. As indicated in Eq. 6.10, the relationship between X and Y depends on the application limit of the watershed shape factor, X. For instance, the width to length ratio should not exceed 4. Substituting $K = 4$ into Eq. 6.10 yields:

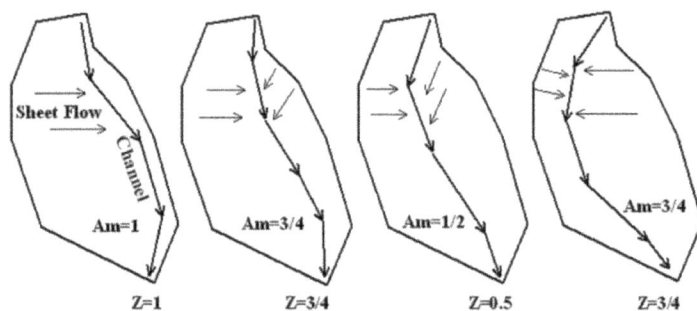

Figure 6.4 Area skewness

$$Y = (1.5 - Z)(2.286X - 0.286X^2) \text{ for all watersheds with } K \leq 4 \tag{6.12}$$

Example 6.1: Determine the KW plane width for the watershed given in Fig. 6.5. The watershed parameters are A = = 40 acs, L = = 1986 ft, So = = 1.39%, Z = = 0.55 (1 acre = 43560 sq ft), Impervious percent Ia = 85%.

Solution: Convert the natural watershed into KW plane as:

$$X = \frac{A}{L^2} = \frac{40 * 43560}{1986^2} = 0.44$$

Z = 0.55 (estimate from Fig. 6.5)

$$Y = (1.5 - Z)(2.286X - 0.286X^2) = (1.5 - 0.55)(2.286 \times 0.44 - 0.286 \times 0.44^2) = 0.91$$

$$Y = \frac{L_w}{L} = \frac{L_w}{1986} = 0.91 \quad \text{So, we have } L_w = 1800 \text{ ft.}$$

$$X_w = \frac{A}{L_w} = \frac{40 \times 43560}{1800} = 968 \text{ft}$$

$$\frac{S_o}{S_w} = \frac{X}{Y} + Y = \frac{0.44}{0.91} + 0.91 = 1.39 \text{ So, we have } S_w = 1.0\%.$$

It is critically important to understand that the overland flow length, X_w, is derived for the virtual KW plane which is not a real or rough surface. This KW flow length is results from a conformal mapping approach that projects the actual flow motion onto a virtual surface. As a result, the overland flow length, X_w, on the KW plane

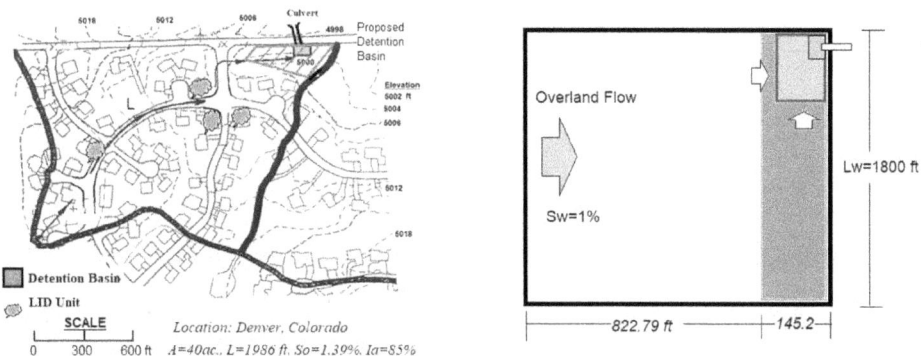

Area	L	So	Z=Am/A	X=A/L^2	Y=Lw/L	So/Sw	Sw	Lw	Xw	IaXw	(1-Ia)Xw
acre	ft	%					%	ft	ft	ft	ft
40.00	1986.00	1.39	0.55	0.44	0.91	1.39	1.00	1800.02	967.99	822.79	145.20

Figure 6.5 Watershed to be converted into the KW plane

is not subject to the maximum allowable overland flow length of 300 to 500 feet as recommended for real watersheds.

6.3 OVERLAND KW FLOW

On an impervious surface, both infiltration and depression losses are negligible. An overland flow is often portrayed as a wide, shallow, one-dimensional open channel flow. The rating curve for an overland flow is expressed as:

$$q = \frac{k_n}{n} y^{\frac{5}{3}} \sqrt{S_w} = \alpha y^m \qquad (6.13)$$

where α and m are constants. Eq. 6.13 is Manning's formula derived for wide and shallow overland flows. The constants in Eq. 6.13 are defined as (Guo 1988, 2000):

$$a = \frac{k_n}{n} \sqrt{S_w} \quad \text{and} \quad m = \frac{5}{3} \text{ for Manning's formula} \qquad (6.14)$$

where n = Manning's roughness, k_n = 1.0 for meter-second units or 1.486 for foot-second units and S_w = slope in [L/L] of the ground surface. Taking the first derivative of Eq. 6.13 with respect to x yields:

$$\frac{\partial q}{\partial x} = \alpha m y^{m-1} \frac{\partial y}{\partial x} \qquad (6.15)$$

Substituting Eq. 6.15 into Eq. 6.1 yields:

$$\frac{\partial y}{\partial t} + \alpha m y^{m-1} \frac{\partial y}{\partial x} \approx \frac{\partial y}{\partial t} + u \frac{\partial y}{\partial x} = I_e \qquad (6.16)$$

where u = overland flow velocity in [L/T]. Eq. 6.16 is the total derivative of the flow depth. Solutions for Eq. 6.16 are composed of two characteristic curves as:

$$\frac{dy}{dt} = I_e \qquad (6.17)$$

$$\frac{dx}{dt} = u = \alpha m y^{m-1} \qquad (6.18)$$

The initial condition for the overland flow in Fig. 6.2 is a dry bed everywhere as:

$$y(t,x) = y(0,x) = 0.0 \qquad (6.19)$$

The upstream boundary does not have any inflow. As a result, the boundary condition is:

$$y(t,x) = y(t,0) = 0.0 \qquad (6.20)$$

Aided by the initial and boundary conditions, integrating Eq. 6.17 yields the flow depth at $x = L$ as:

$$y = I_e t \text{ at x} = L \tag{6.21}$$

Integrating Eq. 6.18 yields:

$$x = \alpha I_e^{m-1} t^m \tag{6.22}$$

Substituting Eq. 6.21 into Eq. 6.22 yields the water surface profile, (x,y), as:

$$x = \alpha \frac{y^m}{I_e} \tag{6.23}$$

When the kinematic wave reaches the outlet at $x = L$, the flow time, t, is called the time of equilibrium of the watershed, T_e. Therefore Eq. 6.23 is converted into an equation for the time of equilibrium as:

$$T_e = \left(\frac{L}{\alpha I_e^{m-1}} \right)^{\frac{1}{m}} \tag{6.24}$$

When Manning's formula is used, $m = 5/3$, the time of equilibrium in Eq. 6.24 becomes

$$T_e = \left(\frac{nL}{k_n \sqrt{S_o} I_e^{0.67}} \right)^{0.60} \tag{6.25}$$

The equilibrium flow depth, y_e, and discharge, q_e, at the outlet are (Wooding 1965):

$$y_e = I_e T_e \tag{6.26}$$

$$q_e = \alpha y_e^m \tag{6.27}$$

The time of equilibrium is similar to the time of concentration, except that the rainfall excess must be uniform in time and space. After the time of equilibrium, the entire unit-width area becomes a tributary to the flow at the outlet. In comparison, a long rainfall event that has a duration longer than the time of equilibrium is more critical to the design. Therefore, further discussions of the overland hydrograph are based on the assumption of a long event. To apply Eqs 6.21, 6.25, and 6.26 to the unit-width area, the overland runoff hydrograph can be predicted as shown in Fig. 6.6. An overland runoff hydrograph consists of three segments: *the rising limb* before the time of equilibrium, *the peaking portion* between the time of equilibrium and the end of rainfall, and *the recession* after the rain ceases.

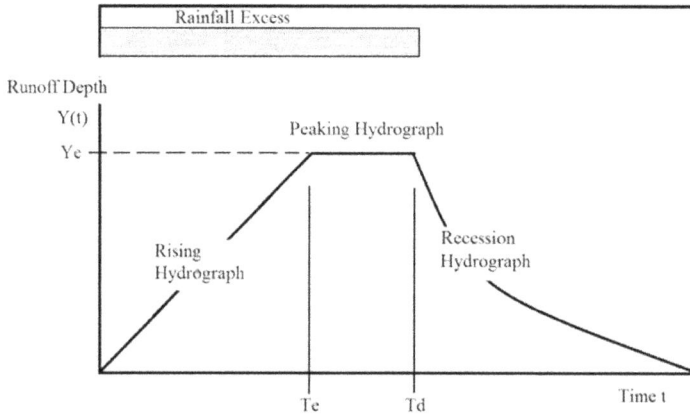

Figure 6.6 Runoff hydrograph

(1) *Rising hydrograph* $0 \le t \le T_e$.

Eq. 6.21 is converted to

$$y_e = I_e t \tag{6.28}$$

where $y(t)$ = runoff depth in [L] at time t

(b) *Peaking hydrograph* $T_e \le t \le T_d$.

Eq. 6.26 is applied to the peak depth as:

$$y(t) = I_e T_e \tag{6.29}$$

where T_d = rainfall duration in [T].

I   *Recession* $t \ge T_d$

After the rain ceases, the equilibrium water profile begins to recede. As shown in Fig. 6.7, the water depth y at the location, x, has to travel through the distance of $(L - x)$ to reach the outlet. Aided by Eq. 6.18, the recession wave movement is described as:

$$\frac{dx}{dt} = \frac{L - x}{t - T_d} = u \tag{6.30}$$

Substituting Eqs 6.23 for x and 6.18 for u into Eq. 6.30 yields:

$$L = \alpha \frac{y^m}{I_e} + \alpha m y^{m-1} \left(t - T_d \right) \tag{6.31}$$

For a specified t, $t \ge T_d$, the outlet depth, $y(t) = y$ in Eq. 6.31, can be solved iteratively.

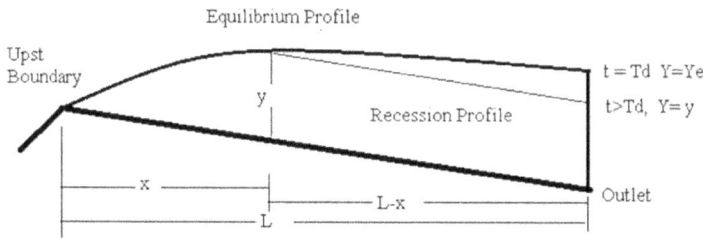

Figure 6.7 Recession of equilibrium water profile for overland flow

Example 6.2: Continued from Example 6.1. Predict the overland flow hydrograph for I_e = 3.7 inch/hr applied to the impervious surface in the watershed in Example 6.1 with A = 40 acres, L_w = 1800 ft, X_w = 968 ft, S_w = 1%, N = 0.025, and T_d = 20 minutes (1200 seconds).

$$\alpha = \frac{k}{n}\sqrt{S_w} = \frac{1.486}{0.025}\sqrt{0.01} = 5.94 \, \text{and} \, m = 5/3 = 1.67$$

I_e = 3.7 inch/hr = 0.000086 ft/second,

$L = X_w = A/L_w$ = (40 × 43560)/1800 = 968 feet is the length of the unit-width overland flow.

$$(2) \quad T_e = \left(\frac{NL}{k_n\sqrt{S_o}I_e^{0.67}}\right)^{0.60} = \left(\frac{0.025 \times 968}{1.486 \times \sqrt{0.01} \times 0.0000868^{0.67}}\right)^{0.60} = 900 \, \text{seconds}$$

(A) Rising hydrograph for t < 900 seconds:

$$q(t) = \alpha y(t)^m = 5.94 \times (0.000086 \, t)^{1.67} \, \text{cfs/ft}$$

$$Q(t) = q(t)L_w = \left[5.94 \times (0.000086 \, t)^{1.67}\right] \times 1800 \, \text{cfs}$$

(B) Peaking hydrograph for 900 ≤ t ≤ 1200 seconds:

$$Q(t) = q_e L_w = 0.0829 \times 968 = 149.2 \, \text{cfs}$$

(C) Recession hydrograph for t > 1200 seconds:

The flow depth on the recession hydrograph is determined by an iterative procedure, guessing the water depth Y for a given time, t, in Eq. 6.31 as:

$$968 = 5.96\frac{Y^{1.67}}{0.000086} + 5.96 \times 1.67 \times Y^{0.67} \times (t - 1200) \, \text{for} \, t > 1200 \, \text{seconds} \quad (6.32)$$

$$y(t) = Y$$

$$Q(t) = q(t)L_w = 5.96Y^{0.67} \times 1800 \text{ cfs}$$

The above process is repeated until the flow depth in Eq. 6.32 is close to zero. The predicted overland runoff hydrograph is summarized in Table 6.1 and plotted in Fig. 6.8.

Table 6.1 Overland flow hydrograph from impervious surface

t second	$y(t)$ ft	$q(t)$ cfs/ft	$Q(t)$ cfs
0	0.0000	0.0000	0.0
150	0.0128	0.0042	7.5
300	0.0257	0.0133	23.9
450	0.0385	0.0261	47.1
600	0.0514	0.0422	76.0
750	0.0642	0.0612	110.2
900	0.0770	0.0829	149.2
1050	0.0770	0.0829	149.2
1200	0.0770	0.0829	149.2
1350	0.0649	0.0623	112.2
1500	0.0544	0.0464	83.5
1650	0.0454	0.0343	61.8
1800	0.0379	0.0254	45.8
1950	0.0318	0.0190	34.2
2100	0.0269	0.0144	25.8
2250	0.0229	0.0110	19.8
2400	0.0197	0.0086	15.4
2550	0.0171	0.0068	12.2
2700	0.0150	0.0054	9.8

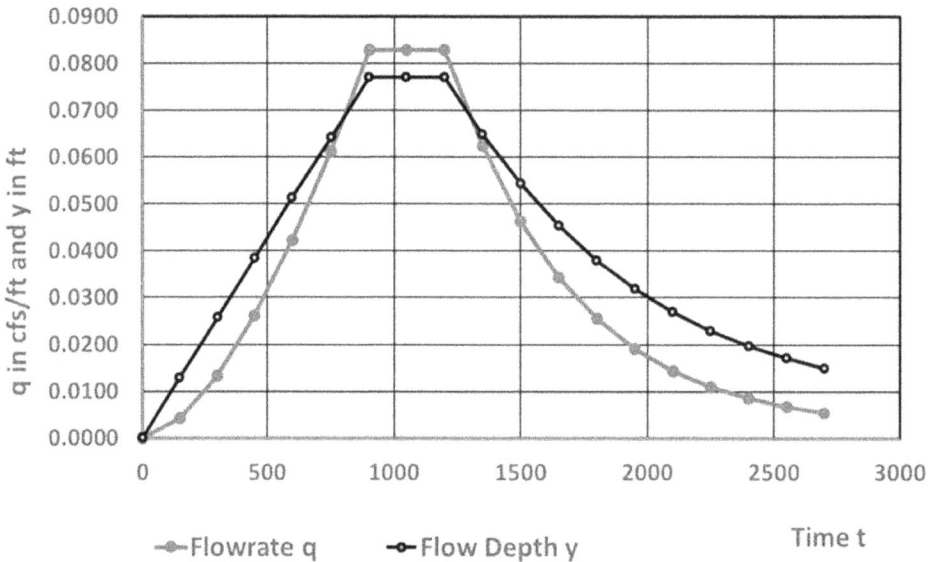

Figure 6.8 Predicted KW overland flow from an impervious area

6.4 KW DIMENSIONLESS UNIT GRAPH

When predicting a design flood flow, it is essential that the design rainfall duration be longer than the equilibrium time of the catchment so that the entire catchment becomes tributary to the peak flow. Under equilibrium conditions, the discharge from the tributary area is equal to the rate of rainfall amount fallen on the tributary area. After the rain ceases, the runoff begins to taper off accordingly. To derive the non-dimensional KW UH, the three parameters, Y_e, T_e, and L, are employed to normalize to the variableles (Guo 2006)

(1) Rising limb $(0 \leq t \leq T_e \leq T_d)$ or $(0 \leq t^* \leq 1 \leq T_d^*)$

At an elapsed time t, the flow depth, $y(t)$, and its location, x, can be determined by solving Eqs 6.21 and 6.23 simultaneously. At $x = L$, the flow depth at an elapsed time, t, on the rising hydrograph is normalized as:

$$y^* \left(t^* \right) = \frac{y(t)}{y_e} = \frac{t}{T_e} = t^* \tag{6.33}$$

Aided by Eq. 6.27, the normalized flow rate per unit width, q^*, is expressed as:

$$q^* \left(t^* \right) = \frac{q(t)}{q_e} = \alpha t^{*m} \tag{6.34}$$

Eqs 6.33 and 6.34 agree with previous studies (Eagleson 1970 and Wooding 1965). They represent the rising hydrograph for the flow depth as a 45-degree line between zero and one.

(2) Peaking portion $(T_e \leq t \leq T_d)$ or $(1 \leq t^* \leq T_d^*)$

During the peaking portion, the inflow volume is equal to the outflow volume. Therefore, the normalized peak flow depth and rate are defined as:

$$y^* \left(t^* \right) = 1.0 \tag{6.35}$$

$$q^* \left(t^* \right) = 1.0 \tag{6.36}$$

(3) Recession Limb $(t \geq T_d$ or $t^* \geq T_d^*)$

After the time of equilibrium, the equilibrium water profile is defined by the pairs of (x, y) using Eqs 6.21 and 6.23. For instance, at the exit, the pair is $(x = L, y = y_e)$. Using Eq. 6.23, we have:

$$L = \frac{\alpha y_e^m}{I_e} \tag{6.37}$$

As soon as the rain ceases, the flow depth at location, x, under the equilibrium water surface profile in Fig. 6.7 begins to propagate toward the outfall point at $x = L$. Re-arranging Eq. 6.30 yields:

$$t = \frac{L - x}{u} + T_d \quad (t \geq T_d \text{ for the recesssion hydrograph}) \tag{6.38}$$

Substituting Eqs 6.37, 6.23, and 6.18 into Eq. 6.38, the normalized kinematic wave movement during the recession period is derived as:

$$t^* = \frac{1}{m \, y^{*m-1}} - \frac{1}{m} y^* + \frac{T_d}{T_e} (T_d > T_e) \tag{6.39}$$

With $m = 5/3$, and $T_d = T_e$, Eq. 6.39 becomes

$$t^* = \frac{1}{1.67 \, Y^{*0.667}} - \frac{3}{5} y^* + 1 \tag{6.40}$$

For a given elapsed time, t, on the recession hydrograph at the outlet, the flow depth, y, which reaches the outlet at time t, is calculated by solving Eq. 6.40. The corresponding normalized flow depth and flow rate are:

$$y^* (t^*) = y^* \tag{6.41}$$

$$q^* (t^*) = \alpha y^{*m} \tag{6.42}$$

The recession hydrograph ends when the elapsed time, t^*, becomes so long that the flow depth in Eq. 6.40 vanishes. The above procedure produces a generalized KW hydrograph for the special case $T_d = T_e$. It is noted that the excess rainfall for this generalized KW hydrograph is not one unit depth. Fig. 6.9 presents the dimensionless KW hydrograph.

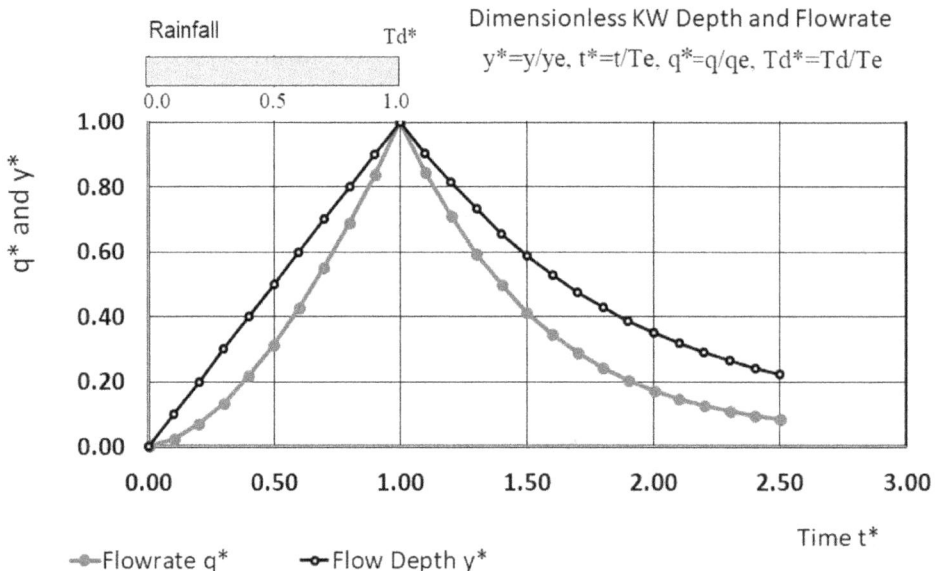

Figure 6.9 Dimensionless KW hydrograph

Table 6.2 Unit hydrograph derived from KE dimensionless hydrograph

t*	y*(t*)	q*(t*)	t min	Q(t) cfs
0.000	0.000	0.000	0.00	0.00
0.333	0.333	0.160	5.00	25.96
0.667	0.667	0.509	10.00	82.42
1.000	1.000	1.000	15.00	162.00
1.333	0.706	0.560	20.00	90.65
1.667	0.492	0.307	25.00	49.75
2.000	0.349	0.173	30.00	28.07
2.333	0.256	0.103	35.00	16.75

Example 6.3: Continued from Example 6.2. Predict the unit hydrograph for the watershed in Example 6.1, which has A = 40 acres, L_w = 1800 ft, X_w = 968 ft, S_w = 1%, and N = 0.025. In this case, T_d = T_e and excess rainfall P_e = 1.0 inch.

Solution: From Example 6.2, with I_e = 3.7 inch/hr, we have T_e = 900 seconds (15 minutes), q_e = 0.083 cfs/ft and y_e = 0.077 ft. In this case, the excess rainfall is calculated as:

$$P_e = T_e \times I_e = 900 \text{ seconds} \frac{1 \text{ hr}}{3600 \text{ sec}} \times 3.7 \frac{\text{inch}}{\text{hr}} = 0.925 \text{ inch}$$

Scale = 1/0.925 = 1.08. It implies that the peak discharge should be multiplied with the scale factor as:
  q_e = 1.08 × 0.083 = 0.09 cfs/ft unit-width peak discharge for the unit-depth excess rainfall. For this case, the peak flow using the Rational method is calculated as:
  $Q_e = q_e L_w$ = 0.09 × 1800 = 162 cfs peak flow on the KW unit graph.
  Aided by q_e = 0.09 cfs/ft and T_e = 15 minutes, the KW unit graph is derived in Fig. 6.9 and summarized in Table 6.2.

6.5 CONCLUSIONS

The main purpose of stormwater modeling is to predict the future flood potential under proposed development conditions. The KW solution provides a unit-width hydrograph that is suitable for urban areas of less than 150 to 200 acres without any significant storage effects (Guo 2000). It is a useful method, except that the tributary area has to be converted into its equivalent rectangular sloping plane. Care must be taken when selecting the value for the Manning's roughness, which will determine the surface detention effect. As a rule of thumb, the higher the Manning's roughness, the more the surface detention and the less the peak flow. The Rational method is a special case of the KW flow when the rainfall distribution is uniform. To expand the KW overland flow predictions under a non-uniform rainfall distribution, a proper numerical method should be employed to directly solve the differential equations for continuity and momentum principles. With a finite difference numerical algorithm, the KW approach can further provide numerical solution for cascading flows.

HOMEWORK

Q6-1   A concrete parking lot in Fig. Q6-1 is under a uniform rainfall intensity of 10 inch/hr for duration of 30 minutes. The overland flow has a slope of 2% for a length of 250 feet. The Manning's roughness coefficient is 0.025.

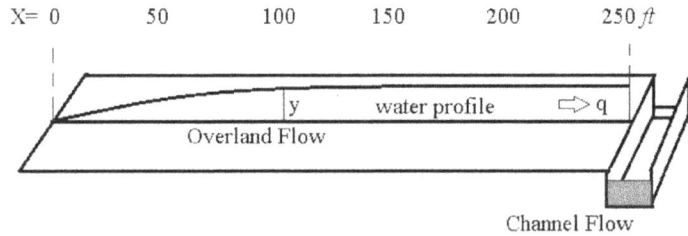

Figure Q6-1 Overland flow in a parking lot

1. Determine the time of equilibrium at the outlet.
2. Construct the rising overland flow hydrograph in cfs/ft at the outlet
3. Construct the peaking overland flow hydrograph in cfs/ft at the outlet
4. Construct the recession hydrograph at the outlet.

Q6-2   Determine the KW plane for the given watershed in Fig. Q6-2 if the watershed area A = 40.0 acres, waterway length, L_w = 1500 feet, waterway slope of 0.010, and imperviousness of 60%. The Manning's N = 0.015 for the impervious area and 0.035 for the pervious area. Your tasks are:

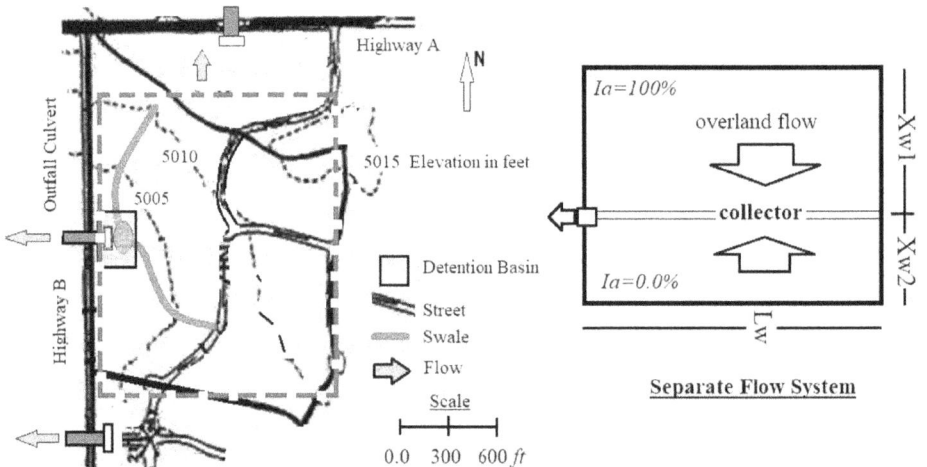

Figure Q6-2 Watershed for developing KW plane

1. Develop the KW planes for pervious and impervious areas.
2. Under a 10-min uniform rainfall distribution of $I = 10$ inch/hr and $f = 2$ inch/hr, calculate the overland hydrographs from the impervious and pervious planes.
3. Calculate the total hydrograph from a two-flow drainage system
4. Determine the runoff coefficient using $C = Q_p/(IA)$

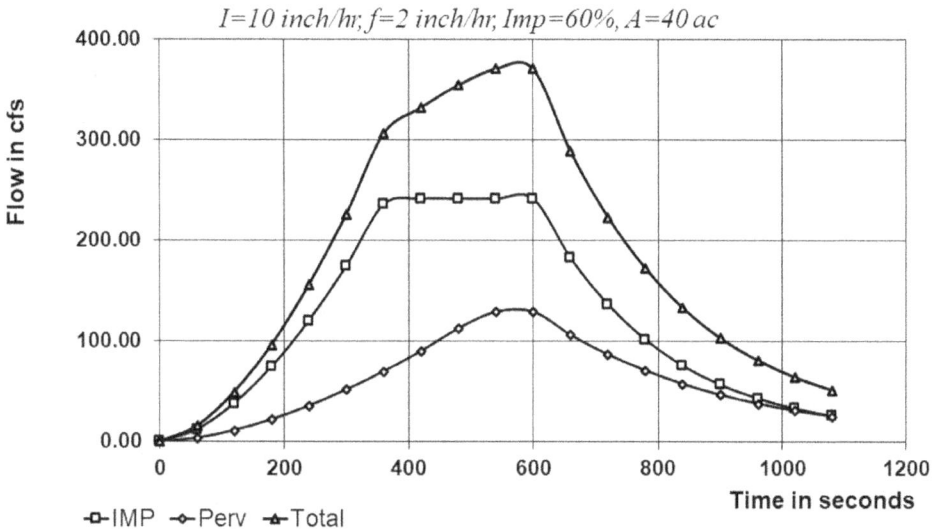

Figure Q6-2.2 Solution for overland hydrographs from a two-flow system

REFERENCES

Bedient, P.B, Huber, W.C. and Vieux B.E. (2008). *Hydrology and Floodplain Analysis*, 4th Edition, Prentice-Hall, NJ.

Blackler, G. and Guo, James C.Y. (2012). "Field Test of Paved Area Reduction Factors using a Storm Water Management Model and Water Quality Test Site", *ASCE Journal of Irrigation and Drainage Engineering*, vol. 17, no. 8, August.

Eagleson, P.S. (1970). *Dynamic Hydrology*, McGraw Hill, New York.

EPA SWMM (2005). Stormwater Management Model, US EPA, www2.epa.gov/water-research/storm-water-management-model-swmm.

Guo, J.C.Y. (1984). "Effects of Infiltration on Hydrograph", Proceedings of ASCE International Conference on Irrigation and Drainage Engineering, Flagstaff, AZ, July.

Guo, J.C.Y. (1988). "Dynamics and Kinematics of Overland Flow", Proceedings of ASCE International Conference on Hydraulic Engineering held at Colorado Springs, Colorado, Aug.

Guo, J.C.Y. (1998). "Overland Flow on a Pervious Surface", *International Journal of Water*, vol. 23, no. 2, June.

Guo, J.C.Y. (2000). "Storm Hydrographs for Small Catchments", *International Journal of Water*, vol. 23, no. 2, September.

Guo, J.C.Y. (2001). "Rational Hydrograph Method for Small Urban Catchments", *ASCE Journal of Hydrologic Engineering*, vol. 6, no. 4, July.

Guo, J.C.Y. (2006). "Dimensionless Kinematic Wave Unit Hydrograph for Storm Water Predictions", ASCE Journal of Irrigation and Drainage Engineering, vol. 132, no. 4, July.

Guo, J.C.Y. and Urbanos, B. (2009). "Conversion of Natural Watershed to Kinematic Wave Cascading Plane", *Journal of Hydrologic Engineering*, vol. 14, no. 8, August, pp. 839–846.

Guo, J.C.Y. (2012). "Storm Centering Approach for Flood Predictions from Large Watersheds", *ASCE Journal of Hydrologic Engineering*, vol. 17, no. 9, September 1.

Guo, J.C.Y. (2014). "Closure on Storm Centering Approach for Flood Predictions from Large Watersheds", *ASCE Journal of Hydrologic Engineering*, vol. 19, no. 1, January, pp. 272–274.

Guo, J.C.Y. Cheng, J., and Wright, L. (2012). "Field Test on Conversion of Natural Watershed into Kinematic Wave Rectangular Planes", *ASCE Journal of Hydrologic Engineering*, Vol. 17, no. 8, August.

Chapter 7

Kinematic wave watershed modeling

7.1 KW CASCADING FLOWS

Many innovative concepts suggest that drainage systems should be laid out to take advantage of soil infiltration benefits. For instance, roof runoff flows should drain onto vegetal beds and parking lots should drain onto a grass buffer to slow down the flow and to remove solids from the stormwater for water quality enhancement. A traditional drainage plan is *a separate flow system*, as shown in Fig. 7.1a, where an *impervious-area flow path* is from the rooftop, downspout, driveway, paved ditch, and to the street gutter, and a *pervious-area flow path* starts from the grass yard to the swale and then drains into the street. These two flow paths are independent and produce two hydrographs that are merged at the street inlet. A green drainage plan for stormwater management prefers a *cascading flow system* to a *separate flow system* because of its increased benefits from filtering and infiltration processes. As shown in Fig. 7.2b, the impervious area drains onto the pervious area before the runoff flows reach the street. In practice, the percentage of flow interception is approximately 60 to 80% or approximately 70% of impervious areas are tied into the downstream porous areas. As shown in Fig. 7.2c, a portion of the impervious area is directly connected to the street.

Example 7.1: The parking lot in Fig. 7.3 is laid out with a grass buffer to intercept the major portion of the runoff flows generated from the impervious area. In this case, the overland flow uniformly moves towards the downstream property boundary. (1) What is the impervious area percentage? (2) Estimate the area-interception percentage.

Solution:

The total drainage area	$A_t = 300 \times 500 = 150{,}000$ sq ft
The impervious area	$A_I = 200 \times 300 + 180 \times 300 = 114{,}000$ sq ft
Impervious percentage	$I_a = 114{,}000/150{,}000 = 76\%$
Intercepted impervious area	$A_c = 300 \times 180 = 54{,}000$ sq ft
Area Interception percentage	$A_c/A_I = 47\%$

In this case, 47% of runoff flows from the impervious area will drain onto the grass area

DOI: 10.1201/9781003284239-7

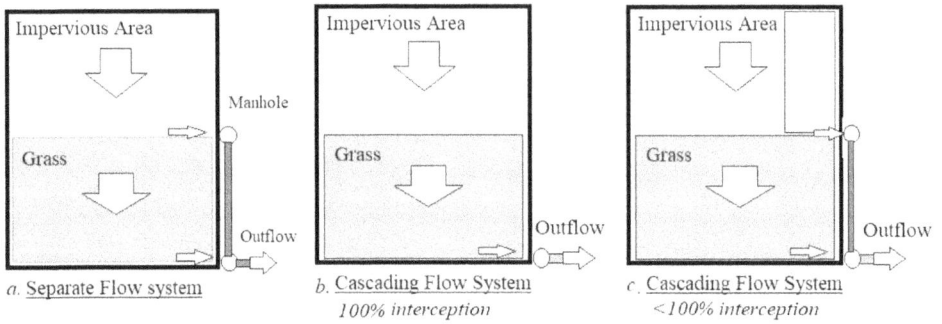

Figure 7.1 Flow paths in the drainage system

(a) Separate Flow System   (b) Cascading Flow System

Figure 7.2 Examples for urban stormwater drainage systems

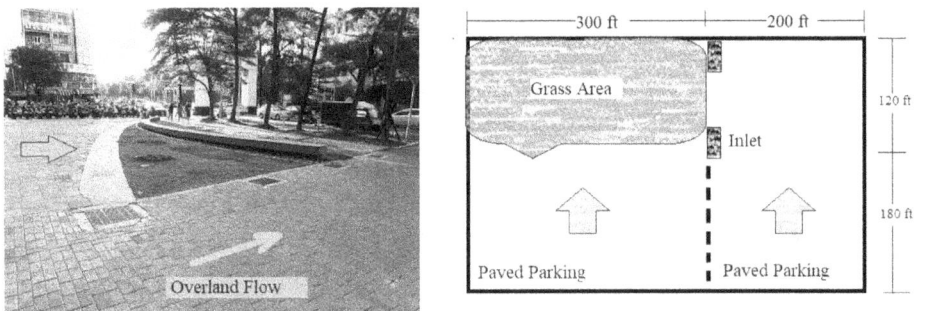

Figure 7.3 Percentage of impervious area intercepted by a grass area

7.2 KW OVERLAND FLOW

The overland flow shown in Fig. 7.4 *is* shallow and wide, and is often described as a one-dimensional sheet flow. Overland flows are often modeled using the unit-width KW method. The *Manning's formula* for a sheet flow is reduced to a unit-width approach in which all flow parameters are only related to the flow depth as:

$$u = \frac{k}{n} y^{\frac{2}{3}} \sqrt{S_w}$$ (7.1)

$$q = uy$$ (7.2)

$$F_r = \frac{u}{v_w} = \frac{\text{flow velocity}}{\text{wave celerity}} = \frac{u}{\sqrt{gy}}$$ (7.3)

where u = unit-width flow velocity in [L/T], y = flow depth in [L], k = 1.486 for using foot-second units and 1.0 if using meter-second units, S_w = ground slope in [L/L], q = flow rate in [L²/T], v_w = wave celerity in [L/T], and F_r = Froude number.

When $F_r > 1$, it is a supercritical flow in which the wave speed is faster than the flow velocity. When $F_r = 1$, it is a critical flow in which $v_w = u$. When $F_r < 1$, it is a subcritical flow in which the flow velocity is slower than the wave movement. As a shallow water flow, an overland flow is sensitive to the surface roughness. Care must be taken when selecting the Manning's n. Table 7.1 gives the recommended values for various surface textures.

Using Eq. 7.1, the unit-width flow rate in Eq. 7.2 is calculated based on the flow depth. For convenience, a lumped parameter, α, is often used in computations.

$$q = \frac{k}{n} y^{5/3} \sqrt{S_w} = \alpha y^m$$ (7.4)

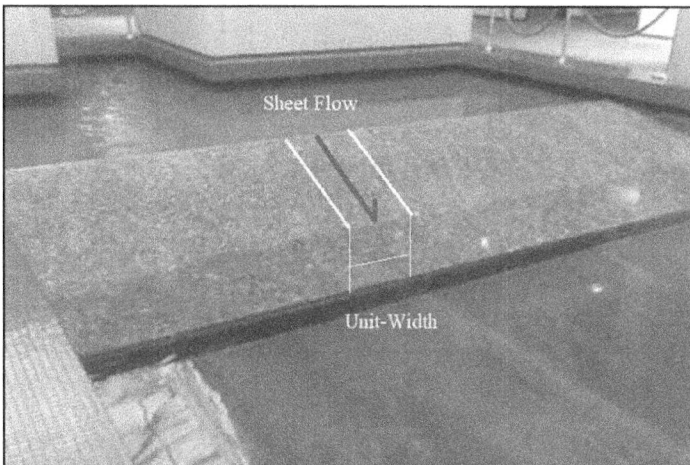

Figure 7.4 Overland sheet flow

Table 7.1 Manning's n for overland flows

Surface Texture	Manning's n for Overland Flows
Dense growth	0.40–0.50
Pasture	0.30–0.40
Lawn	0.20–0.30
Bluegrass sod	0.20–0.50
Short grass	0.10–0.20
Sparse vegetation	0.05–0.13
Bare clay-loam soil	0.01–0.03
Concrete/Asphalt (depth <1/4 inch)	0.10–0.15
Concrete/Asphalt (depth >1/4 inch)	0.05–0.10

where α = lumped factor, and m = empirical factor between 1.5 and 2.0. A value of 5/3 is recommended for Eq. 7.5 based on Manning's formula. It is noted that the values of α and m vary with respect to the empirical formula selected to model the overland flow.

Example 7.2: The sheet flow in Fig. 7.4 has a depth of 1.0 inch on a slope of 1.0%. Determine the value for α and the unit-width flow rate.

Solution: Referring to Table 7.1, N = 0.10.

$$\alpha = \frac{k}{n}\sqrt{S_w} = \frac{1.486}{0.10}\sqrt{0.01} = 1.486$$

$$q = \alpha y^{5/3} = 1.486 \times \left(\frac{1}{12}\right)^{5/3} = 0.023 \text{ cfs/ft}$$

7.3 NUMERICAL SCHEME FOR KW OVERLAND FLOW

The EPA Storm Water Management Model (EPA SWMM) suggests that the KW *routing method* be applied to both overland and channel flows for numerical simulations. The basic concept is to consider each reach as a reservoir. For each time step, the change in the flow depth represents the change in the storage volume due to the balance between the inflows and outflow through the reach. A unit-width overland flow in Fig. 7.5 is depicted as:

$$I - f - q = \frac{\Delta Vol}{\Delta t} \tag{7.5}$$

where I = rainfall intensity in [L/T], f = soil infiltration rate in [L/T], q = overland flow in [L^3/T], Δvol = change of water volume in [L^3/T], and Δt = time step in [T].
   Convert Eq. 7.6 into its finite difference format as:

$$(I - f)\Delta t X_w - q\Delta t = \Delta y X_w \tag{7.6}$$

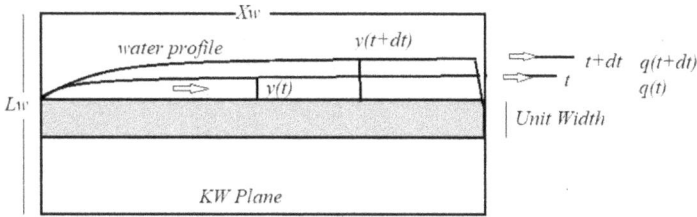

Figure 7.5 Overland flow routing scheme

Where Δy = change in water depth in [L], and X_w = length of overland flow in [L]. As an unsteady flow, both the flow rate and flow depth are varied with respect to time. Considering that the average values over a time step are instantaneously uniform and constant, the finite difference form for Eq. 7.7 is divided into two volumes: *flow volume* and *storage volume*. These two water volumes are respectively calculated as:

$$DV = X_w\Delta t\left[\frac{I(t)+I(t+\Delta t)}{2} - \frac{f(t)+f(t+\Delta t)}{2}\right] - \left[\frac{q(t+\Delta t)+q(t)}{2}\right]\Delta t \qquad (7.7)$$

$$DS = X_w\left[y(t+\Delta t) - y(t)\right] \qquad (7.8)$$

where DV = change in flow volumes in [L³], and DS = change in storage volume in [L3]. Using Eq. 7.5, the flow rates, $q(t)$ at time t and $q(t + \Delta t)$ at time $t + \Delta t$, are calculated as:

$$q(t) = \alpha[y(t)]^{\frac{5}{3}} \qquad (7.9)$$

$$q(t+\Delta t) = \alpha[y(t+\Delta t)]^{\frac{5}{3}} \qquad (7.10)$$

$$DV = DS \qquad (7.11)$$

Eqs 7.10 and 7.11 are simultaneously solved for the two unknowns: $q(t + \Delta t)$ and $y(t + \Delta t)$. The *boundary condition* for Eq. 7.7 includes (1) the given inflow flow at the upstream entrance if this is a cascading plane, (2) the excess rainfall distribution to the entire length of overland flow, and (3) negligible tailwater effect at the downstream exit. *The initial condition* is a dry bed, i.e. $y = 0$ and $q = 0$ at $t = 0$. The solution is achieved by balancing the two volumes in Eqs 7.8 and 7.9. During the iterative process, the error tolerance used in the trial-and-error procedure is recommended as:

$$\eta = \left|\frac{DV - DS}{DV}\right| < 0.5\% \qquad (7.12)$$

where η = error tolerance. In general, the kinematic wave numerical algorithm is stable and reliable as long as the flow length, X_w, is long enough.

$$Q(t) = q(t)L_w \tag{7.13}$$

where Q = channel flow in $[L^3/T]$. For simplicity, the runoff flow in Eq. 7.13 from the entire KW rectangular plane is the linear sum of the unit-width flow. Of course, we may use the channel flow routing method to collect the unit-width flow through the reach.

Example 7.3: Referring to Fig. 7.4, the rectangular 10-acre parking lot is under a 60-min uform rainfall with I = 4 inch/hr. The overland flow length is 726 ft on a slope of 1.5%. Manning's n is 0.05 for the concrete pavements. Determine the storm hydrograph.

Solution: At t = 0, the dry bed condition leads to y = 0. Let Δt = 15 minutes. At t = 900 sec, we have:

$$DV = 726 \times \frac{900 \left[\frac{4+4}{2} - 0 \right]}{12 \times 3600} - \left[q(t + \Delta t) - 0 \right] \times 900$$

$$Q(t+\Delta t) = L_w \, q(t + \Delta t) = 600 \times q(t + \Delta t)$$

By trial and error, $y(t + \Delta t)$ = 0.0618 ft, $q(t + \Delta t)$ = 0.035 cfs/ft, and $Q(t + \Delta t)$ = 20.88 cfs

Table 7.2 Solution for Example 7.3

Watershed Area	A=	10.00	Coeff m on Rating Curve	m=	1.67	
Imperviousness	Ia =	1.00	Coeff α for Rating Curve	α=	3.640	
Watershed Slope	Sw=	0.015	Soil Initial Infiltration Rate	fi =	0.00	inch/hr
Overland flow width	Lw=	600.00	Soil Final Infiltration Rate	fc =	0.00	inch/hr
Overland flow Length	Xw=	726.00	Infiltration Decay Coef	k=	0.00	1/hr
Surface Roughness	n=	0.05	Computation dt =	dt=	900.0	sec

Time t t+dt Seconds	Rainfall Intensity I(t) inch/hr	Infiltration Rate f(t) inch/hr	Guess Depth y(t) ft	Flow q(t) cfs/ft	DV= [(I−f)−q]dt cu ft/ft	DS= dY Xw cu ft/ft	Check Balance Dv−Ds = 0 cu ft/ft	Hydrograph Q(t) cfs
0.0	4.000	0.000	0.000	0.000	0.000	0.000	0.000	0.00
900.0	4.000	0.000	0.0618	0.035	44.839	44.839	0.000	20.88
1800.0	4.000	0.000	0.0860	0.061	17.609	17.609	0.000	36.31
2700.0	4.000	0.000	0.0908	0.066	3.467	3.467	0.000	39.74
3600.0	4.000	0.000	0.0915	0.067	0.510	0.510	0.000	40.25
4500.0	0.000	0.000	0.0669	0.040	−17.843	−17.843	0.000	23.87
5400.0	0.000	0.000	0.0342	0.013	−23.744	−23.744	0.000	7.79
6300.0	0.000	0.000	0.0222	0.006	−8.686	−8.686	0.000	3.80
7200.0	0.000	0.000	0.0161	0.004	−4.497	−4.497	0.000	2.20
8100.0	0.000	0.000	0.0123	0.002	−2.711	−2.711	0.000	1.41
9000.0	0.000	0.000	0.0099	0.002	−1.791	−1.791	0.000	0.97
9900.0	0.000	0.000	0.0081	0.001	−1.259	−1.259	0.000	0.70

Repeat the above procedure until the flow depth is close to zero. In this case, the peak flow is 40.25 cfs, which agrees with the Rational method:

$Q_p = CIA = 1.0 \times 4 \times 10 = 40$ cfs where C = runoff coefficient and A = tributary area in acres.

7.4 NUMERICAL MODELING FOR KW OVERLAND FLOW

To apply the KW method, an irregular watershed has to be converted to its equilibrium rectangular KW plane. As illustrated in Fig. 7.5, the key factors are: drainage area, A, waterway length and slope, L and S_o, and area skewness, Z.

The watershed's area skewness, Z, is defined by the area ratio of the larger of A_1 and A_2 to the total area as:

$$Z = \frac{A_m}{A} = \frac{\max(A_1, A_2)}{A} \tag{7.14}$$

where Z = area skewness factor between zero and unity, and A = watershed area in [L_2], which is divided by the waterway into the left half, A_1, and right half, A_2 in Fig. 7.6. Eq. 7.14 is determined by the larger half. In practice, the value of Z can be approximated by an eye-ball measurement. The watershed shape factor, X, is defined as the watershed width to length ratio as:

$$X = \frac{B}{L} \approx \frac{A}{L^2} \tag{7.15}$$

where X = watershed shape factor, B = watershed average width in [L], and L = length of waterway in [L]. Next, we need to convert the watershed shape factor, X, into the KW shape factor, Y, as:

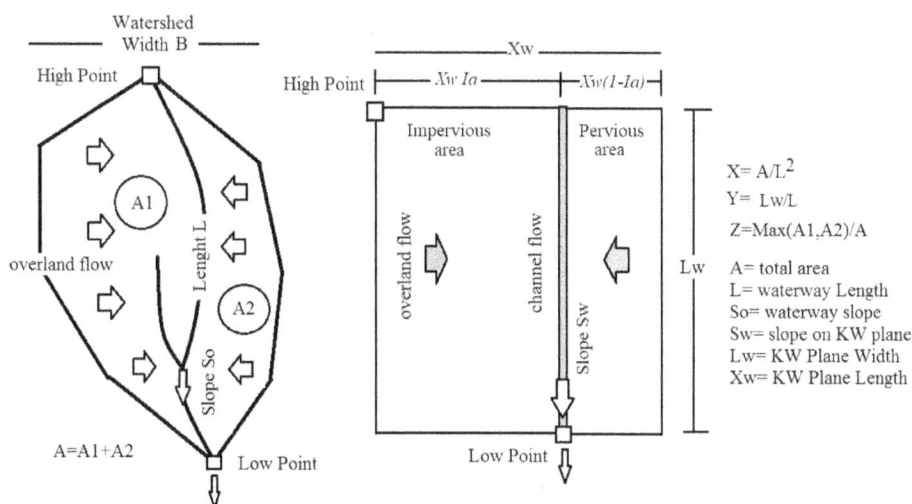

Figure 7.6 Conversion of the natural watershed into a rectangular KW plane

$$Y = \frac{L_w}{L} = (1.5 - Z) \left[2.286 \frac{A}{L^2} - 0.286 \left(\frac{A}{L^2} \right)^2 \right] \text{ for } A / L^2 \leq 4 \qquad (7.16)$$

where Y = KW shape factor representing the rectangular KW plane. The overland flow length, X_w, on the rectangular KW plan is calculated as:

$$X_w = \frac{A}{L_w} \qquad (7.17)$$

Based on the energy principle, the vertical fall along the two flow paths should be balanced as:

$$\frac{S_o}{S_w} = \frac{X_w}{L} + \frac{L_w}{L} \qquad (7.18)$$

Using Eq. 7.18, the slope, S_w, on the rectangular KW plane is directly related to the waterway slope, S_o. As a distributed model, the overland flow length, X_w, is further divided into two segments, according to the watershed's impervious percent, I_a, as:

$$X_1 = I_a X_w \qquad (7.19)$$

$$X_2 = (1 - I_a) X_w \qquad (7.20)$$

where X_1 = flow length on impervious area, and X_2 = flow length on pervious area. Depending on the layout of the drainage system, we may have a *separate flow system* when the impervious and pervious areas are laid in parallel to the central collector channel, or a *cascading flow system* when the impervious area drains onto the pervious area, and then drains into the side collector channel.

Example 7.4: A watershed in Fig. 7.7 is located in Southwest Denver, Colorado. The tributary area is 67.91 acres, which is to be developed with $I_a = 0.65$. The watershed has a length of $L = 2323$ ft on a slope of $S_o = 0.02$ ft/dt. Convert the watershed to its equilibrium KW rectangular plane and layout (1) side channel drainage system, and (2) central channel drainage system.

Solution:

$$Z = \frac{A_m}{A} = \frac{\max(A_1, A_2)}{A} = 0.53 \text{ (eye-ball estimate)}$$

$$X = \frac{A}{L^2} = \frac{67.9 \times 43560}{2323^2} = 0.58$$

$$Y = \frac{L_w}{L} = (1.5 - Z) \left[2.286 \frac{A}{L^2} - 0.286 \left(\frac{A}{L^2} \right)^2 \right] = 1.20. \text{ So, } L_w = 2709 \text{ ft}$$

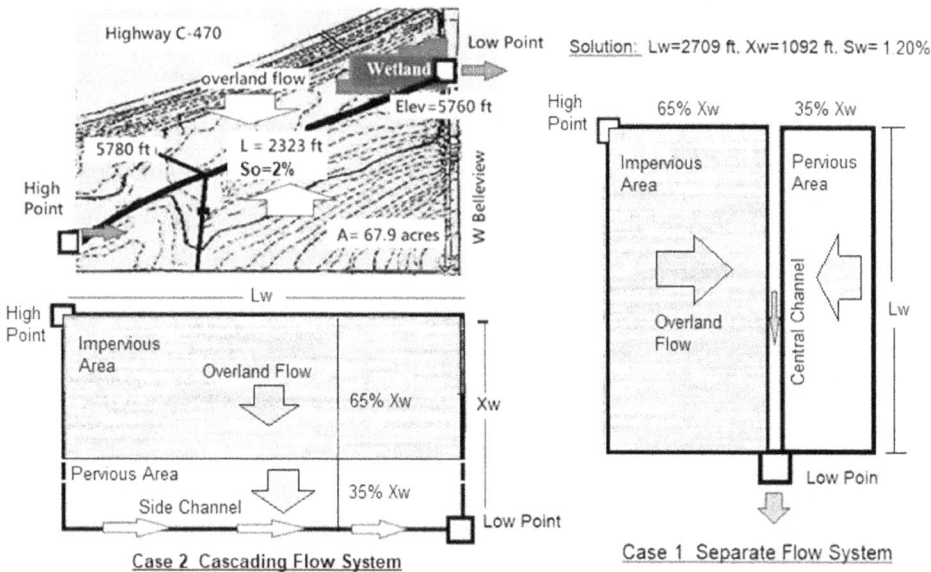

Figure 7.7 Side channel and central channel drainage plans

Table 7.3 Conversion of watershed into KW rectangular plane

Subarea	Area	L	So	Z = Am/A	X = A/L^2	Y=Lw/L	So/Sw	Sw	Lw	Xw
ID	acre	ft	%					%	ft	ft
I	67.91	2252.50	2.00	0.53	0.58	1.20	1.69	1.1852	2709.00	1092.0

$$X_w = \frac{A}{L_w} = 1092 \ ft, X_1 = 0.65 \times 1092 = 709.8 \ ft \ \text{and} \ X_2 = 382.8 \ ft$$

$$\frac{S_o}{S_w} = \frac{X_w}{L} + \frac{L_w}{L} = 1.20. So, S_w = 0.012$$

7.4.1 Separate flow system

A separate flow system in Fig. 7.9 has the impervious and pervious planes placed in parallel to the central collector channel. Both the overland flows are independently generated and then merged into the collector channel.

The combined hydrograph is the sum of the two flows as:

$$Q(t) = [q_1(t) + q_2(t)]L_w \tag{7.21}$$

Table 7.4 Overland flow generated on impervious area (Fig. 7.8a)

Watershed Area	A=	67.9 acres		Surface Roughness	n=	0.15		
Imperviousness	Ia=	0.65		Coeff m on Rating Curve	m=	1.67		
Waterway slope	Sw=	0.012 ft/ft		Coeff α on Rating Curve	α=	1.085		
Overland flow Length	Xw=	1092.0 ft		Initial Infiltration Rate	fi =	0.00 inch/hr		
Overland flow Width	Lw=	2709.0 ft		Final Infiltration Rate	fc =	0.00 inch/hr		
Overland Flow Length Xw(1−Ia)=		709.8 ft		Infiltration Decay Coef	k=	0.00 1/hr		
				Time Interval dt=Δt=		300.0 sec		

Time t t+dt Seconds	Rainfall Intensity I(t) inch/hr	Infiltration Rate f(t) inch/hr	Guess Depth y(t) ft	Unit-width Flow q(t) cfs/ft	Flow Volume DV = [(i−f−q]dt cu ft/ft	Storage Vol DS = dy Xwla cu ft/ft	Volume Balance Dv−Ds = 0 cu ft/ft	Hydrograph Imp Area Q(t) cfs
0.0	0.654	0.000	0.000	0.0000	0.000	0.000	0.000	0.00
300.0	0.739	0.000	0.0048	0.0001	3.411	3.411	0.000	0.40
600.0	0.863	0.000	0.0102	0.0005	3.848	3.848	0.000	1.40
900.0	1.205	0.000	0.0170	0.0012	4.837	4.837	0.000	3.27
1200.0	2.002	0.000	0.0274	0.0027	7.323	7.323	0.000	7.22
1500.0	3.438	0.000	0.0444	0.0060	12.111	12.111	0.000	16.21
1800.0	3.620	0.000	0.0653	0.0114	14.790	14.790	0.000	30.82
2100.0	2.994	0.000	0.0823	0.0168	12.082	12.082	0.000	45.38
2400.0	1.504	0.000	0.0902	0.0195	5.643	5.643	0.000	52.94
2700.0	1.005	0.000	0.0907	0.0197	0.299	0.299	0.000	53.35
3000.0	0.756	0.000	0.0886	0.0190	−1.457	−1.456	0.000	51.35
3300.0	0.621	0.000	0.0856	0.0179	−2.132	−2.132	0.000	48.48
3600.0	0.000	0.000	0.0806	0.0162	−3.578	−3.577	0.000	43.80

where Q = flow in the channel in $[L^3/T]$, q_1 = flow from impervious plane, q_2 = flow from pervious area, and t = elapsed time after it rains.

Example 7.5: Determine the storm hydrograph from the separate flow system for the watershed in Fig. 7.6.

Solution: In this case, at each time step, the flows in Table 7.4 from the impervious area and Table 7.5 from the pervious area are summed together as shown in Table 7.6. The peak flow is determined to be 76.80 cfs in Fig. 7.10.

7.4.2 Cascading flow system

A cascading flow system in Fig. 7.11 is laid out to have the upper impervious area drained onto the lower pervious area. The unit-width flow from the upper area serves as an input to the lower area. As a result, the overland flow equation is revised to:

$$I - f + q_i - q = \frac{\Delta Vol}{\Delta t}$$
(7.22)

Table 7.5 Overland flow generated on pervious area (Fig. 7.8b)

Watershed Area	A=	67.9 acres	Surface Roughness	n=	0.25
Imperviousness	Ia=	0.65	Coeff m on Rating Curve	m=	1.67
Waterway Slope	Sw=	0.012 ft/ft	Coeff α on Rating Curve	α=	0.651
Overland flow Length	Xw=	1092.0 ft	Initial Infiltration Rate	fi =	1.50 inch/hr
Overland flow Width	Lw=	2709.0 ft	Final Infiltration Rate	fc =	0.50 inch/hr
Overland Flow Length Xw(1-Ia)=		382.2 ft	Infiltration Decay Coef	k=	6.25 1/hr
			Time Interval dt=Δt=		300.0 sec

Time t t+dt Seconds	Rainfall Intensity I(t) inch/hr	Infiltration Rate f(t) inch/hr	Guess Depth y(t) ft	Flow Volume Unit-width Flow q(t) cfs/ft	Flow Volume DV = [(i−f) −q]dt cu ft/ft	Storage Vol DS = dy Xw(1−Ia) cu ft/ft	Volume Balance Dv − Ds = 0 cu ft/ft	Pervious Area Q(t) cfs
0.0	0.654	1.500	0.000	0.000	0.000	0.000	0.000	0.00
300.0	0.739	1.094	−0.0042	0.000	−1.594	−1.594	0.000	0.00
600.0	0.863	0.853	−0.0054	0.000	−0.458	−0.458	0.000	0.00
900.0	1.205	0.710	−0.0036	0.000	0.670	0.670	0.000	0.00
1200.0	2.002	0.625	0.0029	0.000	2.476	2.476	0.000	0.17
1500.0	3.438	0.574	0.0171	0.001	5.438	5.438	0.000	3.29
1800.0	3.620	0.544	0.0356	0.004	7.080	7.080	0.000	11.21
2100.0	2.994	0.526	0.0503	0.007	5.631	5.631	0.000	19.98
2400.0	1.504	0.516	0.0560	0.009	2.160	2.160	−0.001	23.87
2700.0	1.005	0.509	0.0544	0.008	−0.611	−0.611	0.000	22.74
3000.0	0.756	0.505	0.0508	0.007	−1.390	−1.390	0.000	20.26
3300.0	0.621	0.503	0.0466	0.006	−1.603	−1.603	0.000	17.54
3600.0	0.000	0.502	0.0407	0.005	−2.255	−2.255	0.000	13.99

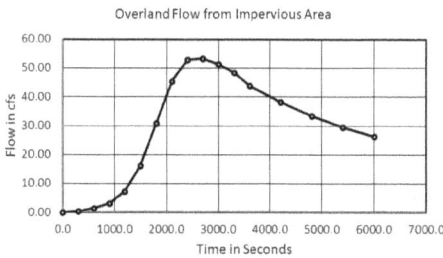

(a) Hydrograph from Impervious Area

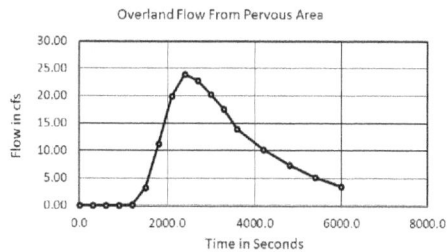

(b) Hydrograph from Pervious Area

Figure 7.8 Examples of predicted hydrographs

where q_i = unit-width inflow in $[L^2/T]$ from the upper impervious area. Convert Eq. 7.22 into its finite difference form as:

$$(I - f)\Delta t X_w + q_i \Delta t - q\Delta t = \Delta y X_w \qquad (7.23)$$

Using the average values over a time step, Eq. 7.22 is converted into the inflow volume changes and storage volume changes as:

Rainfall *I=12 in/hr f=2 in/hr   Td= 30 minutes*

Info

Ia= 50%
Lw=500 ft
Xw=1000 ft
Sw=0.02
N= 0.15

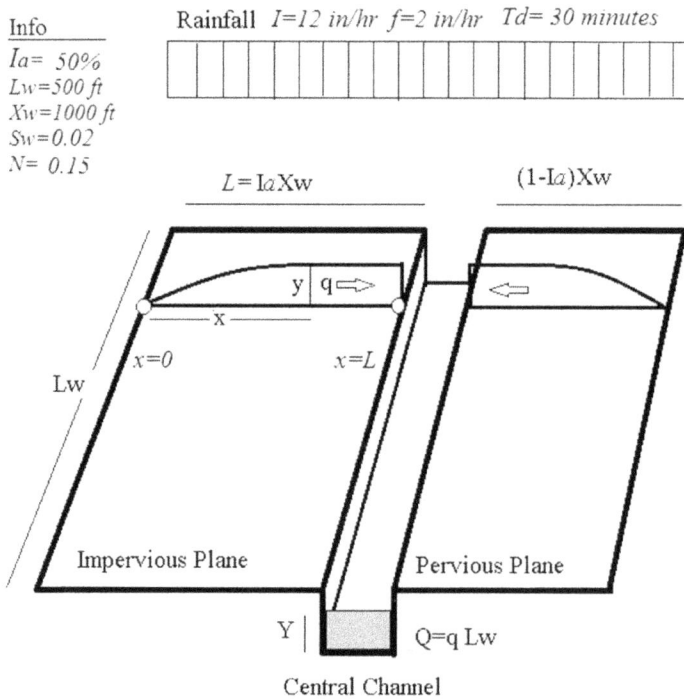

Figure 7.9 KW plane for a central channel system

Table 7.6 Storm hydrograph for separate flow system

Elapsed Time t Seconds	Pervious Area Q(t) cfs	Impervious Area Q(t) cfs	Watershed Storm Q(t) cfs
0.0	0.00	0.00	0.00
300.0	0.00	0.40	0.40
600.0	0.00	1.40	1.40
900.0	0.00	3.27	3.27
1200.0	0.17	7.22	7.38
1500.0	3.29	16.21	19.50
1800.0	11.21	30.82	42.02
2100.0	19.98	45.38	65.36
2400.0	23.87	52.94	76.80
2700.0	22.74	53.35	76.09
3000.0	20.26	51.35	71.61
3300.0	17.54	48.48	66.02
3600.0	13.99	43.80	57.79
4200.0	10.21	38.15	48.36
4800.0	7.35	33.48	40.83
5400.0	5.16	29.56	34.72
6000.0	3.47	26.27	29.74

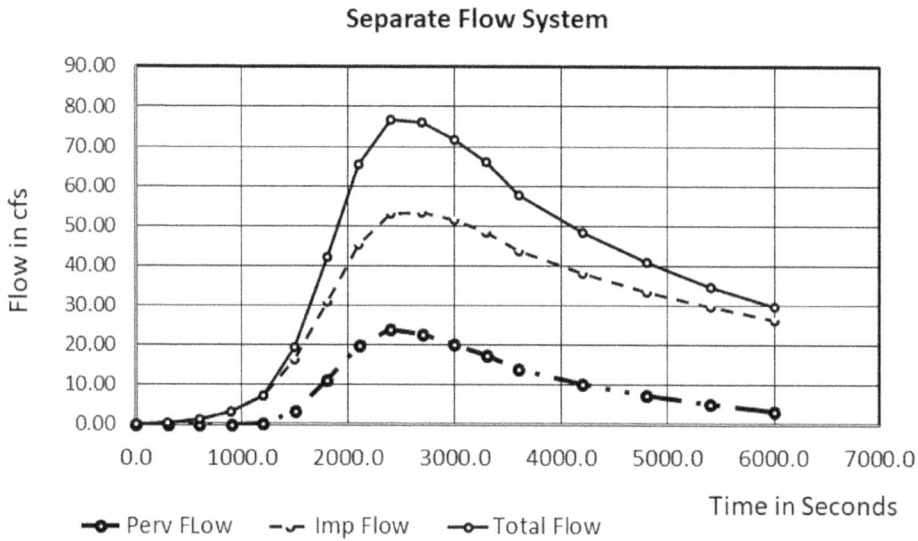

Figure 7.10 Hydrographs predicted for a separate flow system

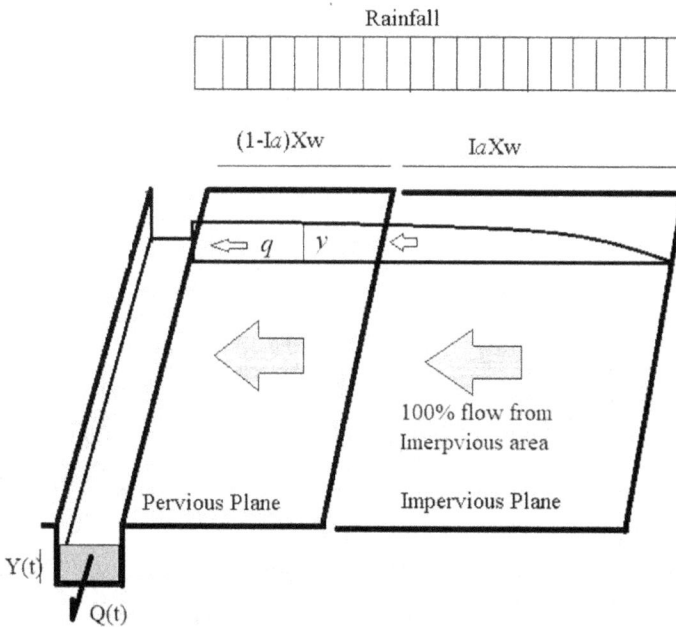

Figure 7.11 KW plane for a cascading flow system

$$DV = X_w \Delta t \left[\frac{I(t)+I(t+\Delta t)}{2} - \frac{f(t)+f(t+\Delta t)}{2} \right]$$
$$- \frac{[q(t)+q(t+\Delta t)]}{2} \Delta t + \frac{[qi(t)+qi(t+\Delta t)]}{2} \Delta t$$
(7.24)

$$DS = X_w \left[y(t+\Delta t) - y(t) \right]$$
(7.25)

$$DV = DS$$
(7.26)

where DV = flow volume changes in [L³/T] due to inflows and outflows, and DS = storage volume changes in [L³/T] due to the flow depth over the flow length under the water profile. As mentioned above, Eq. 7.4 describes the relationship between flow rate and flow depth. The solution for each time step is to determine $y(t+\Delta t)$ to satisfy the principle of continuity. During the iterative process, the error tolerance is the same as Eq. 7.12.

Example 7.6: Determine the storm hydrograph from the cascading flow system for the watershed in Fig. 7.10.

Table 7.7 Overland flow from cascading flow system with 100% area interception

Watershed Area	A=	67.9 acres	Surface Roughness	n=	0.15
Imperviousness	Ia=	0.65	Coeff m on Rating Curve	m=	1.67
Waterway slope	Sw=	0.012 ft/ft	Coeff α on Rating Curve	α=	1.085
Overland flow Length	Xw=	1092.0 ft	Initial Infiltration Rate	fi =	1.50 inch/hr
Overland flow Width	Lw=	2709.0 ft	Final Infiltration Rate	fc =	0.50 inch/hr
Overland Flow Length	Xw(1−Ia)=	382.2 ft	Infiltration Decay Coef	k=	6.25 1/hr
			Time Interval	dt=Δt=	300.0 sec

Time t t+dt Seconds	Upper Inflow qi(t) cfs/ft	Rainfall Intensity I(t) inch/hr	Infiltration Rate f(t) inch/hr	Guess Depth Y(t) ft	Unit-width Flow q(t) cfs/ft	Flow Volume DV = [(I−f) −q+qi]dt cu ft/ft	Storage Vol DS = dyXw(1−Ia) = 0 cu ft/ft	Volume Balance Dv − Ds ft/ft	Watershed Hydrograph Q(t) cfs
0.0	0.0000	0.6539	1.500	0.000	0.0000	0.000	0.000	0.000	0.00
300.0	0.0001	0.7389	1.094	−0.0041	0.0000	−1.572	−1.572	0.000	0.00
600.0	0.0005	0.8627	0.853	−0.0051	0.0000	−0.359	−0.359	0.000	0.00
900.0	0.0012	1.2047	0.710	−0.0026	0.0000	0.929	0.929	0.000	0.00
1200.0	0.0027	2.0024	0.625	0.0053	0.0002	3.040	3.040	0.000	0.47
1500.0	0.0060	3.4378	0.574	0.0226	0.0019	6.609	6.609	0.000	5.25
1800.0	0.0114	3.6199	0.544	0.0467	0.0065	9.218	9.219	0.000	17.65
2100.0	0.0168	2.9942	0.526	0.0695	0.0126	8.703	8.703	0.000	34.25
2400.0	0.0195	1.5041	0.516	0.0840	0.0173	5.534	5.535	0.000	46.97
2700.0	0.0197	1.0051	0.509	0.0901	0.0195	2.331	2.331	0.000	52.81
3000.0	0.0190	0.7561	0.505	0.0922	0.0203	0.823	0.823	0.000	54.93
3300.0	0.0179	0.6215	0.503	0.0921	0.0202	−0.058	−0.058	0.000	54.78
3600.0	0.0162	0.0000	0.502	0.0887	0.0190	−1.284	−1.284	0.000	51.48

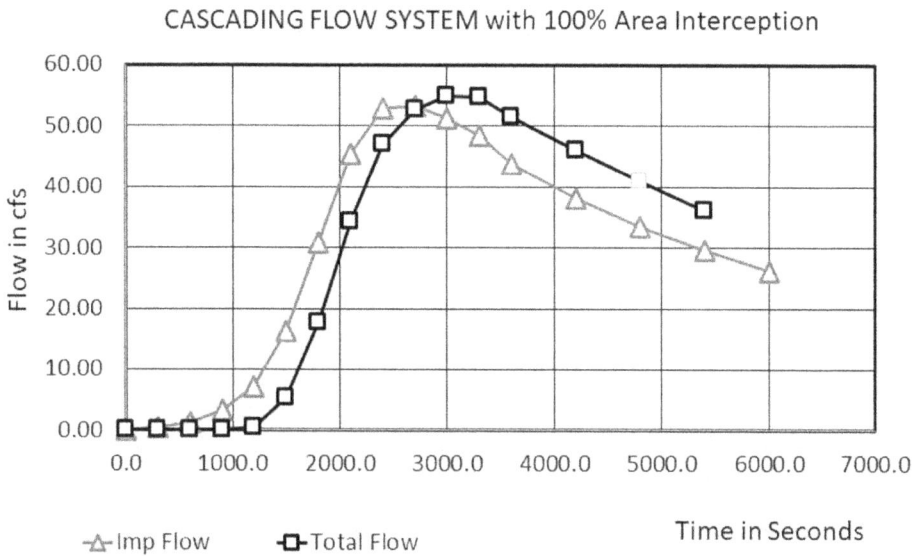

Figure 7.12 Hydrographs from a cascading system with 100% area interception

Solution: In this case, the upper inflow is prescribed in Table 7.4 which is added to Table 7.5 based on the continuity principle. Details are presented in Table 7.7. The peak flow in Fig. 7.12 is 54.93 cfs. Comparing to the *separate flow system*, the *cascading flow system* reduces the peak flow from 76.8 to 54.93 cfs due to the additional infiltration benefits.

Example 7.6 presents an ideal case in which the upper impervious area is completely drained onto the lower pervious area. In practice, the flow interception is approximately 60 to 80 %. Without any specific on-site information, an area-interception percentage of 75% is recommended.

Example 7.7: Continued from Example 7.6. Revise the numerical procedure used in Example 7.6 when an area interception percentage of 75% as shown in Fig. 7.13.

Solution: Consider 25% of the runoff flows, $q_i(t)$, from the impervious area as the bypass flow. Load 75% of $q_i(t)$ to the lower pervious area. Repeat Table 7.7 to calculate the cascading flow. The total flow is the sum of the cascading flow and bypass flow. As shown in Fig. 7.14, the peak flow is slightly higher than that in Example 7.6.

7.5 KINEMATIC WAVE CHANNEL FLOW

For simplicity, the channel flows in Examples 7.2 to 7.7 may be calculated as the linear sums of overland flows. In fact, all channels have a storage effect to reduce the peak flow. A linear sum of overland flows can be improved with channel flow routing

Rainfall

$(1-Ia)Xw$ $IaXw$

q y

75% Impervious area

25% Impervious area

Pervious area

Y(t)   25% Flow from Impervious Area

Q(t) Cascading Flow

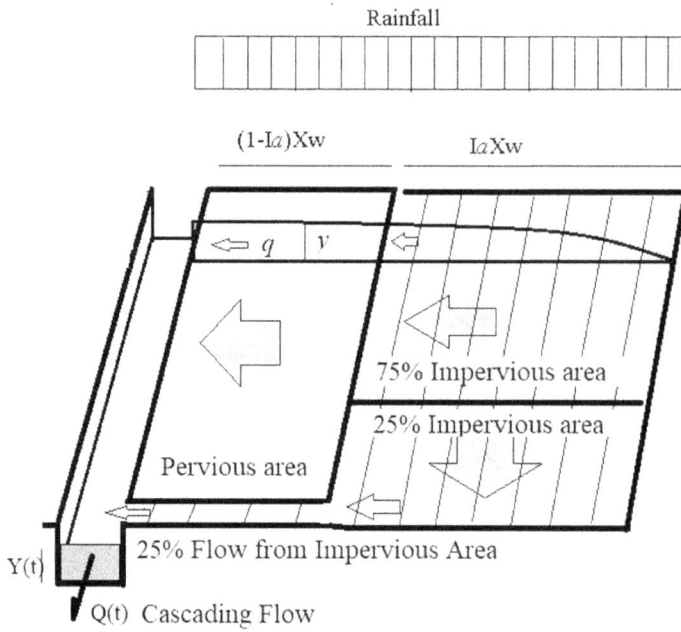

Figure 7.13 Cascading plan with a 75% area interception

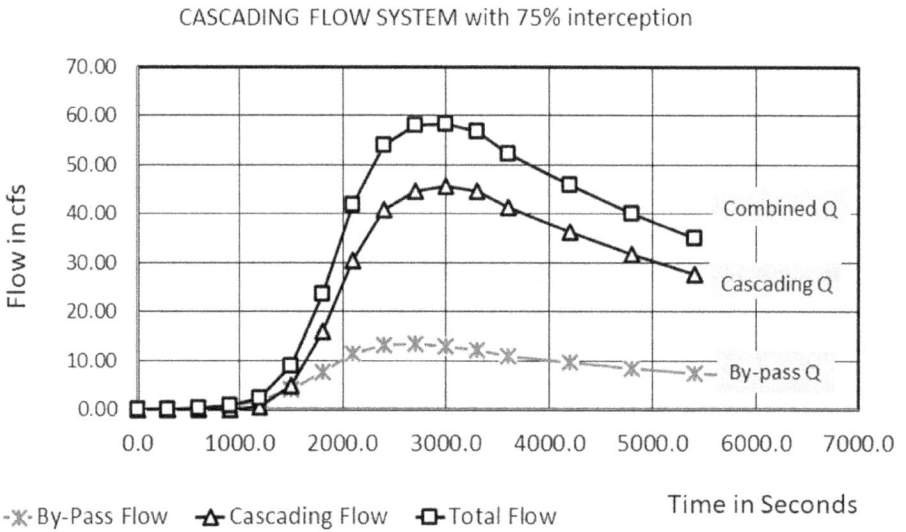

CASCADING FLOW SYSTEM with 75% interception

Combined Q

Cascading Q

By-pass Q

Flow in cfs

-✗- By-Pass Flow   ⟶ Cascading Flow   -☐- Total Flow

Time in Seconds

Figure 7.14 Hydrograph with a 75% area interception

techniques. In fact, overland flows can be treated as an input per linear reach into the channel at each time step. The response from the channel in Fig. 7.15 is described as:

$$\frac{\partial A}{\partial t} + \frac{\partial Q}{\partial x} = q \qquad (7.27)$$

$$Q = \frac{k}{N}\sqrt{S_o}\, P^{-\frac{2}{3}} A^{\frac{5}{3}} = \beta A^m \qquad (7.28)$$

$$U = \frac{Q}{A} \qquad (7.29)$$

where Q = channel flow in [L^3/T], A = flow area in [L^2], Y = flow depth in [L], U = cross-sectional average velocity in [L/T], B = channel bottom width in [L], Z = side slope in [L/L], P = wetted perimeter in [L] representing the wetted bottom length under channel flow, k = 1.486 for ft-sec units or 1.0 for meter-sec units, S_o = channel bottom slope in [L/L], x = distance in [L] along the flow direction, β = lumped parameter, m = 5/3 using Manning's equation and N = Manning's roughness for the channel lining. As recommended, $N = 0.015$ for a concrete lining, $N = 0.025$ for an earth lining, $N = 0.035$ for a grass lining and $N = 0.045$ for riprap surface.

Referring to Fig. 7.16, a channel may have inflow at the upstream entrance and linear lateral inflows along the banks. Considering the entire channel reach, i.e. $\Delta x = L_w$, the finite difference format for Eq. 7.27 is written as:

$$\frac{A(t+\Delta t) - A(t)}{\Delta t} + \frac{(\bar{Q} - \bar{Q}_i)}{L_w} = \bar{q} \qquad (7.30)$$

where Q = outflow in [L^3/T] at the downstream exit, Q_i = inflow in [L^3T] at the upstream entrance or base flow in the channel, and L_w = length of channel in [L].

The change of flow volume, DV, in the channel is:

$$DV = \frac{[q(t) + q(t+\Delta t)]}{2}\Delta t\, L_w - \Delta t \left[\frac{Q(t) + Q(t+\Delta t)}{2} - \frac{Q_i(t) + Q_i(t+\Delta t)}{2}\right] \qquad (7.31)$$

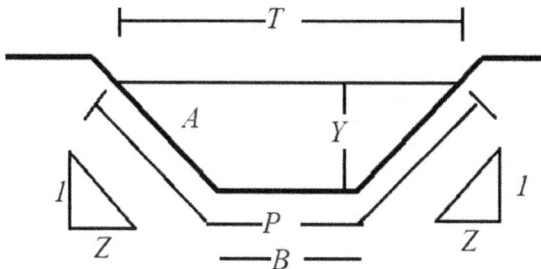

Figure 7.15 Channel cross-sectional parameters

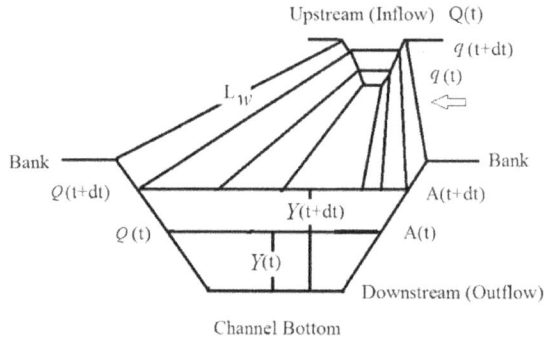

Figure 7.16 Kinematic channel routing scheme

The change of storage volume, DS, in the channel is:

$$DS = L_w \left[A(t + \Delta t) - A(t) \right] \tag{7.32}$$

$$DV = DS \tag{7.33}$$

Using Eq. 7.28, the outflow from the channel is calculated as:

$$Q(t) = \frac{k}{N} \sqrt{S_o} P(t)^{\frac{-2}{3}} A(t)^{\frac{5}{3}} \tag{7.34}$$

$$Q(t + \Delta t) = \frac{k}{N} \sqrt{S_o} P(t + \Delta t)^{\frac{-2}{3}} A(t + \Delta t)^{\frac{5}{3}} \tag{7.35}$$

The *initial condition* for Eq. 7.30 is the known base flow in the channel or a dry condition with $y(t = 0) = 0$. The *boundary condition* for Eq. 7.30 includes: (1) the given upstream inflow hydrograph and (2) the known overland inflow hydrograph along the banks. The solution for Eq. 7.33 is to balance the flow volume changes and the storage volume changes for each time step. The error tolerance for the iterative numerical procedure is determined by the criteria as in Eq. 7.12.

The reliability of numerical simulations depends on how to select the time interval, Δt. As a rule of thumb, the smaller the time interval, the more stable the simulations numerically. However, computational efficiency prefers a longer time interval for less computing time. The *Courant criterion* for numerical stability is applied to the KW channel routing scheme. As a rule of thumb, the selection of reach length and time interval should observe the following limits:

$$\frac{L_w}{2U} \le \Delta t \le \frac{2L_w}{U} \tag{7.36}$$

where U = average flow velocity in [L/T].

Table 7.10 Example for KW channel routing method

Incremental Time	$\Delta t=$	300.00 Seconds	Bottom Width	B=	5.00 feet
Channel Length	Lw=	2709.00 feet	Side Slope	Z=	1.00 ft/ft
			Longi Slope	So=	0.0100 ft/ft
			Manning's N	N=	0.0450

Tiem Seconds	Upstream Inflow Q-in cfs	Lateral Inflow q cfs/ft	GUESS Depth Y ft	Flow Area A sq ft	Wetted P-meter P ft	Outflow Q-out cfs	Flow Vol DV acre-ft	Storage Vol DS S(t + dt) – S(t) acre-ft	Balance CHECK Dv – Ds = 0 acre-ft
0.0	2.00	0.0000	0.29	1.51	5.81	2.03	0.000	0.000	0.00
300.0	2.00	0.0000	0.28	1.50	5.80	2.02	−6.953	−6.953	0.00
600.0	2.00	0.0000	0.28	1.50	5.80	2.01	−4.761	−4.761	0.00
900.0	2.00	0.0000	0.28	1.50	5.80	2.01	−3.762	−3.762	0.00
1200.0	2.00	0.0002	0.29	1.52	5.81	2.06	60.108	60.108	0.00
1500.0	2.00	0.0019	0.34	1.80	5.95	2.68	748.543	748.543	0.00
1800.0	2.00	0.0065	0.52	2.84	6.46	5.42	2820.666	2820.666	0.00
2100.0	2.00	0.0126	0.85	4.94	7.39	12.48	5698.033	5698.033	0.00
2400.0	2.00	0.0173	1.23	7.66	8.48	23.65	7362.856	7362.856	0.00
2700.0	2.00	0.0195	1.55	10.14	9.38	35.29	6725.778	6725.779	0.00
3000.0	2.00	0.0203	1.76	11.92	9.99	44.31	4819.580	4819.580	0.00
3300.0	2.00	0.0202	1.89	13.00	10.34	50.00	2909.721	2909.721	0.00
3600.0	2.00	0.0190	1.94	13.44	10.48	52.36	1185.497	1185.497	0.00
3900.0	2.00	0.0170	1.92	13.30	10.43	51.62	−368.434	−368.434	0.00
4200.0	2.00	0.0151	1.86	12.77	10.27	48.80	−1422.744	−1422.744	0.00
4500.0	2.00	0.0133	1.78	12.06	10.03	45.05	−1921.141	−1921.141	0.00
4800.0	2.00		1.52	9.90	9.30	34.12	−5852.454	−5852.454	0.00

Example 7.8: Continued from Example 7.6. The side channel is described with the parameters: B (bottom width) = 3.0 feet, Z (side slope) = 1V:1H, S_o (channel slope) = 0.01 (ft/ft), N (Manning's roughness) = 0.045. Consider Δt = 300 second, and L = 2,709 feet. This channel carries a base flow of 2.0 cfs and the overland flow described in Table 7.7. Apply the KW channel routing method to determine the outflow hydrograph.

Solution: At each time step, the computation is iterated with a guessed water depth until the volume is balanced. As shown in Table 7.10, the peak flow for this case is 52.36 cfs

In this case, a time interval of five minutes satisfies the Courant numerical stability criterion. It is noted that the channel storage is so negligible that the peak flow on the inflow hydrograph is slightly attenuated in this case.

7.6 CONCLUSIONS

Hydrologic routing methods are derived to apply a numerical procedure to verify the performance of a storage facility under design, including water reservoirs, detention basins, and storage tanks. Similarly, *hydraulic routing methods* are developed to confirm the performance of a conveyance facility such as channels, pipes, and culverts. Before a hydraulic structure is in place for service, hydraulic and hydrologic numerical methods are the only quantifiable basis for making decisions when selecting design parameters, comparing alternatives, predicting a facility's performance, and evaluating costs and damage.

All numerical methods involve empirical variables such as orifice and weir coefficients, Manning's roughness. During the design and sizing stage, the recommended empirical coefficients are adopted and applied to the numerical procedures. In many cases, the as-built hydraulic structure is equipped with monitoring devices to collect field data. With an adequate field data, the empirical coefficients in the numerical methods can be calibrated further and hence improved for future use.

HOMEWORK

Q7-1 The watershed in Fig. Q7-1 is developed into a residential subdivision. Knowing that watershed area A = 40 acres, waterway length, L_w = 1500 feet, waterway slope = 0.010, and imperviousness of 60%, convert the watershed into the equivalent rectangular KW plane.

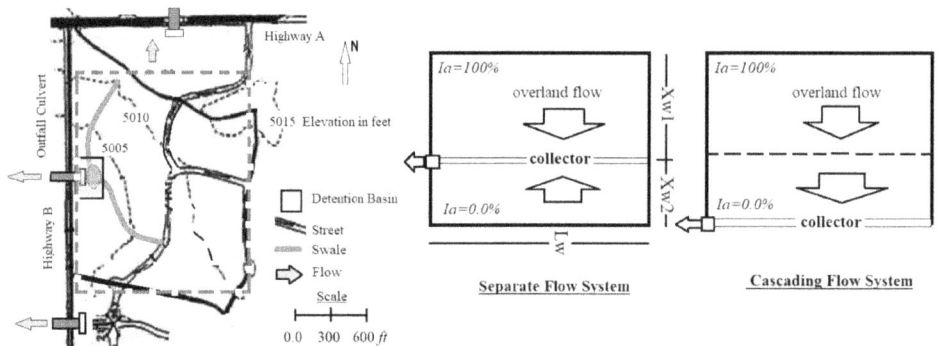

Figure Q7-1 Watershed for developing KW plane

Your tasks are:
1. Based on the impervious-area ratio, divide the KW plane into the pervious and impervious areas.
2. Under a 10-min uniform rainfall distribution, I = 10 inch/hr, and a constant infiltration rate, f = 2 inch/hr, calculate the overland hydrographs from the impervious and pervious planes.
3. Calculate the total hydrograph from a separate-flow drainage system.
4. Calculate the total hydrograph from a cascading-flow drainage system.
5. For the separate-flow system, analyze the runoff coefficient using $C = Q_p/IA$, where C = runoff coefficient, Q_p = peak flow in cfs, I = 10 inch/hr, and A = tributary area in acre.

C-*imp* = Q_p-*imp*/(10 × 40.0I_a) = ? *for the impervious area*,
C-*perv* = Q_p-*perv*/[10 ×40.0(1 – I_a)] = ? *for the pervious area, and*
C-*com* = Q_p-*total*/(10 × 40) = ? *for the entire area*.

6. For the separate-flow system, test if the area weighting method can balance C-imp, C-perv, and C-com. Explain.
7. Repeat (5) and (6) for the cascading flow system. Explain why the area weighting method fail to balance the three Cs.

Q7-2   The catchment in Fig. Q7.2 has a drainage area of 4 acres. The catchment is under a uniform rainfall intensity of 10 inch/hr for duration of 30 minutes. The central rectangular channel has a bottom width of 3 feet, bottom slope, $S_o = 0.5\%$ and Manning's roughness, $N = 0.045$. The pervious KW plane is $X_2 = 100$ ft long. The infiltration loss is defined as:

$f(t) = 1.5 + 2.5e^{-0.12t}$, where $f(t)$ = infiltration rate in inch/hr, and t = elapsed time in minutes.

The impervious KW plane is $X1 = 300$ ft long. Both KW planes are laid on a slope of 1.5% with $n = 0.025$ for overland flows. Determine the time of concentration at the outlet.

1. Determent the watershed impervious area ratio.
2. What is the length for the central channel?
3. Construct the overland hydrographs from the pervious and impervious areas.
4. Calculate the channel hydrograph at the outlet when the overland flows are collected from both the channel banks at each time step.

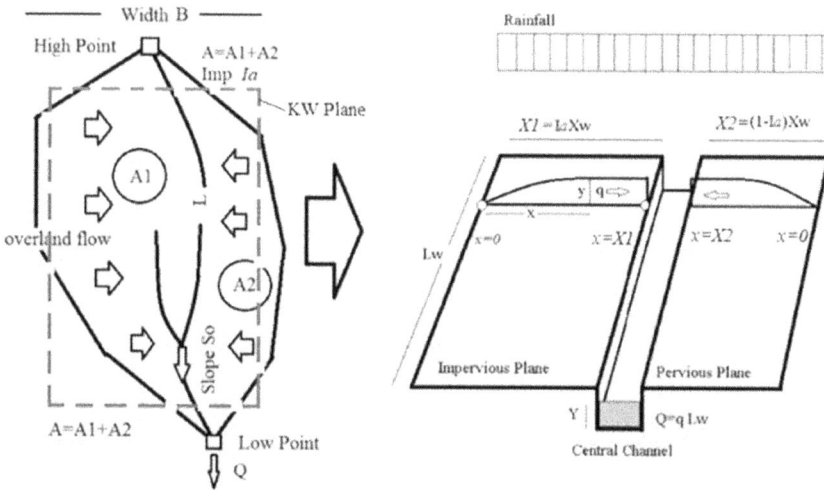

Figure Q7-2 KW layout for a central channel collection system

REFERENCES

ASCE (1994). *Design and Construction of Urban Stormwater Management System*, American Society of Civil Engineers, Manuals and Reports of Engineering Practice, no. 77, Chapter 1.

Bedient, P.B. and Huber, W.C. (1992). *Hydrology and Floodplain Analysis*, Second Edition, Addison-Wesley Publishing Company, Inc, New York.

EPA Storm Water Management Model (1983). *Inlet Design Program*, vol. 3, Federal Highway Administration, Report no. FHWA/RD-83/043, December.

Guo, J.C.Y. (2004). "Hydrology-Based Approach to Storm Water Detention Design Using New Routing Schemes", *ASCE Journal of Hydrologic Engineering*, vol. 9, no. 4, July/August.

McCuen, R.H. (1998). *Hydrologic Analysis and Design*, 2nd edition, Prentice Hall, New York.

Puls, L.G. (1928). "Construction of Flood Routing Curves", House Document 185, U.S. 70th Congress, First Session, Washington, DC.

US Army Corps of Engineers (1979). "Introduction and Application of Kinematic Wave Routing Technique Using HEC-1", Training Document 10, May.

Chapter 8

Open-channel hydraulics

8.1 CLASSIFICATION OF CHANNELS

Water flows are carried in channels or closed conduits. Channels can be *natural* or *man-made*. A natural waterway in Fig. 8.1(a) has been shaped and eroded by storm runoff flows over geologic time. Waterways vary from shallow overland flow, small brooks, and mountain streams, to rivers and tidal estuaries. The *man-made channels* in Fig. 8.1(b) are constructed for purposes such as of stormwater drainage, flood mitigation, irrigation, navigation, and power generation. A *prismatic channel* is a man-made straight channel that has a constant invert slope and uniform cross-section. Examples of prismatic channels include pipes, culverts, chutes, flumes, and canals.

A natural stream is irregular in cross-sectional width, depth, and invert slope. To estimate the flow capacity in a natural waterway, the flow cross-section is divided into several vertical rectangular slots or approximated by a trapezoidal shape.

8.2 CLASSIFICATION OF CHANNEL FLOWS

Open channel flow is three dimensional in nature. On a wide floodplain, the water flow moves *laterally, longitudinally, and vertically*. In practice, the channel flow is considered one dimensional in the longitudinal direction because the velocity component along the channel alignment dictates the flow.

At a curve section, the two-dimensional flow is manifested by the superelevation that creates significant lateral movements. To study the sediment transport, the vertical flow in the gravitational direction dominates the particles' settlement. For the purpose of channel design, the one-dimensional approach is sufficient for selecting the cross-sectional elements. As illustrated in Fig. 8.2, the channel flow is steady when the channel carries a constant discharge. The channel flow is considered *uniform* if the flow distribution remains the same at all sections, otherwise the flow is *non-uniform*.

When the acceleration is numerically insignificant in a non-uniform flow, then the flow is termed *gradually varied flow*. Otherwise, it is a *rapidly varied flow*. As illustrated in Fig. 8.3, the water flow released from a reservoir into a spillway begins with a drawdown profile at the entrance and then conforms to a uniform flow farther downstream. At the foot of the spillway, a hydraulic jump is triggered, according to the tailwater. The drawdown profile represents a gradually varied flow (GVF), and a hydraulic jump is a rapidly varied flow (RVF).

DOI: 10.1201/9781003284239-8

(a) Natural Channel   (b) Man-made Channel

Figure 8.1 Natural and man-made channels

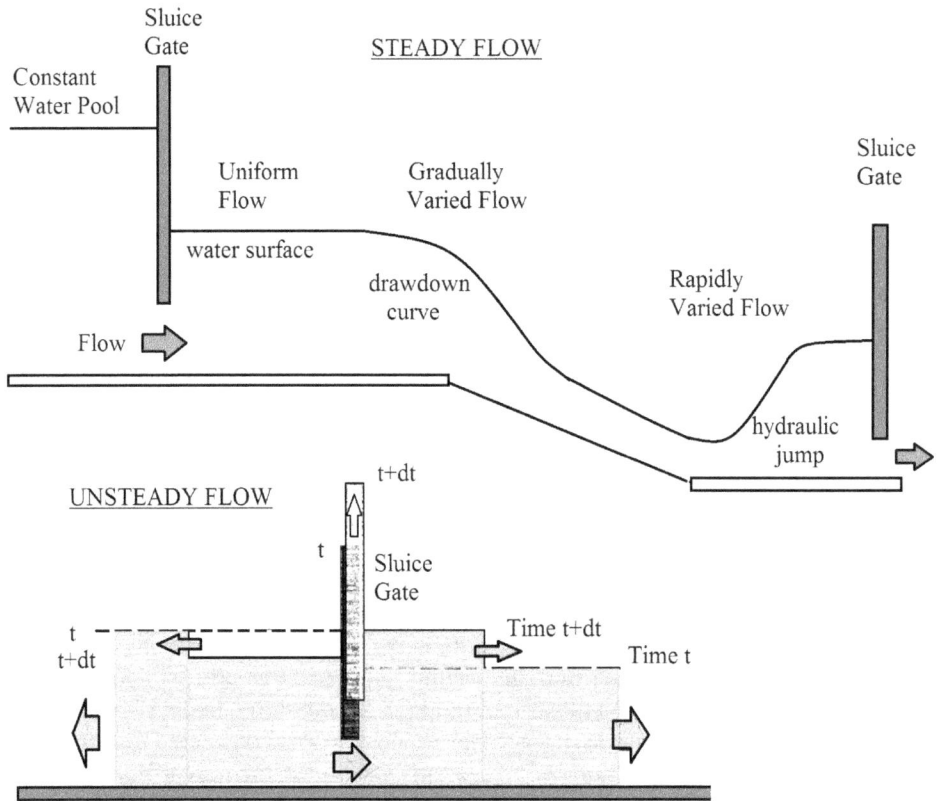

Figure 8.2 Classifications of channel flows

Using a sluice gate to control the flow release, the discharge in the channel is adjusted from time to time. When the discharge in a channel changes with respect to time, the flow is classified as *unsteady*; otherwise the flow is steady. Changes in the flow velocity are called acceleration. A *local acceleration* is the change in flow velocity at a specified cross-section over a time interval whereas a *convective acceleration* is the change in flow velocity over a distance at a specified time. The former may be caused by variations in the flow released from an upstream reservoir into the channel, and the latter may be caused by expansions or contractions in the channel width along the floodplain.

8.3 SLOPES IN CHANNEL FLOW

Channel flow is driven by the energy difference or gradient. The Bernoulli's sum at a section in a channel flow is defined as:

$$E = Y + Z + Hv = Y + Z + \frac{U^2}{2g} \tag{8.1}$$

where E = Bernoulli's sum in [L], Y = flow depth in [L], Z = channel invert elevation in [L], H_v = velocity head in [L], U = cross-sectional velocity in [L/T], and g = gravitational acceleration in [L/T^2]. As illustrated in Fig. 8.3, between two adjacent sections, the energy slope is computed as:

$$S_e = \frac{H_f}{X} = \frac{E_2 - E_1}{X} \tag{8.2}$$

Where S_e = energy slope on *energy grade line* (EGL), H_f = friction loss in [L], and X = distance in [L] between two sections. The subscript 1 means the variable associated with section 1, etc.

In a channel flow, the water surface is the *hydraulic grade line* (HGL). The slope of the HGL is calculated as:

$$S_W = \frac{(Y_2 + Z_2) - (Y_1 + Z_1)}{X} \tag{8.3}$$

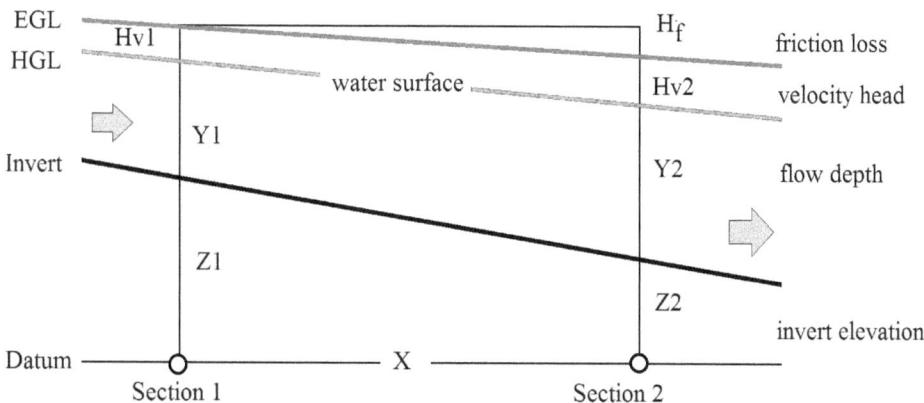

Figure 8.3 EGL and HGL in channel flow

where S_w = slope of the water surface in [L/L]. The channel invert slope is calculated as:

$$S_o = \frac{Z_2 - Z_1}{X} \tag{8.4}$$

Under the assumption that the acceleration in the channel flow is negligible, the flow is in equilibrium and called the uniform flow or normal flow. A uniform flow is characterized by:

$$S_e = S_w = S_o \tag{8.5}$$

Example 8.1: A concrete box conduit of 10-ft wide and 7 ft height, as illustrated in Fig. 8.4, carries a flow of 420 cfs. Determine the slopes associated with the flow.

Solution: At section 1, the lower elevation of the box conduit is at 7 feet and the flow depth is 5 feet. At section 2, the lower elevation of the box conduit is at 5 feet and the flow depth is 6 feet. With a distance of 100 ft between these two sections, the bottom slope, water surface slope, and flow energy slope are calculated as:

$$S_o = \frac{5.0 - 7.0}{100.0} = -0.02$$

$$S_w = \frac{(6.0 + 5.0) - (5.0 + 7.0)}{100.0} = -0.01$$

$$420 = U_1 (5 \times 10) = U_2 (6 \times 10) U_1 = 8.4 \; fps \;\; \text{and} \;\; U_2 = 7.0 \; fps$$

$$S_e = \frac{(6.0 + 5.0 + \dfrac{7.0^2}{2 \times 32.2}) - (5.0 + 7.0 + \dfrac{8.4^2}{2 \times 32.2})}{100.0} = -0.013$$

The design of a channel begins with the known ground slope. As illustrated in Fig. 8.5, the channel runs from the high point, Pt 1, to the low point, Pt 2. Between

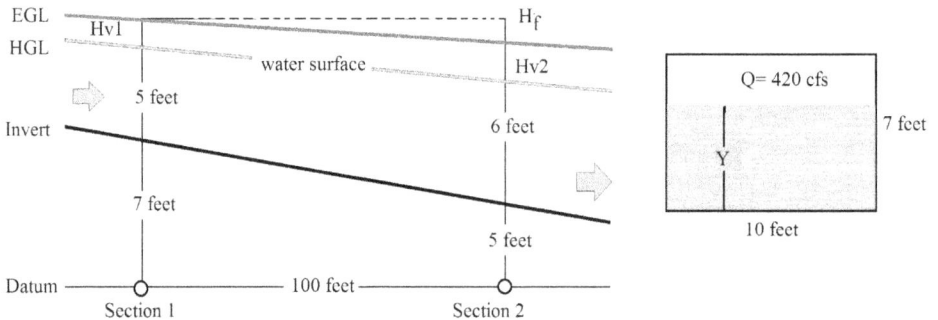

Figure 8.4 Water flow in a concrete box

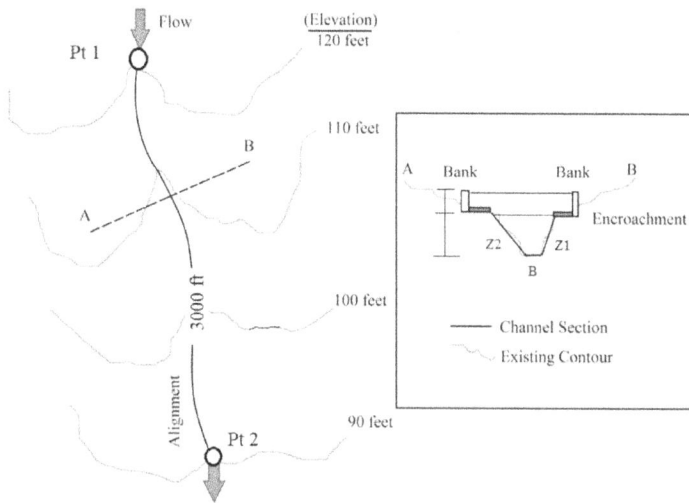

Figure 8.5 Channel alignment between points

these two points, the ground slope provides the first approximated bottom slope for the channel under design. The existing irregular cross-sections serve as guidance to select a regular cross-section for the channel under improvement. Often in urban areas, banks and levees are used to reduce the width of the floodplain. Floodplain encroachments need to observe the floodplain regulations and design criteria.

Many empirical formulas were developed to predict the cross-sectional average flow velocity in a channel. In general, they were derived under the assumption of one-dimensional uniform flow. Empirical formulas developed for channel flows are derived for depth-integrated parameters for channel flow. Therefore, a channel flow is dictated by its cross-sectional elements.

8.4 CROSS-SECTIONAL ELEMENTS

Most man-made channels are trapezoidal in shape. Both rectangular and triangular channels are special cases. The flow condition in a channel is closely related to the channel's cross-sectional geometry. The elements in a trapezoidal cross-section are illustrated in Fig. 8.6, including:

1. *flow depth*, Y, which is the vertical distance from the water surface to the lowest point on the cross,
2. *bottom width*, B,
3. *flow area*, A, which is the cross-sectional area that conveys the flow,
4. *side slope*, which is Z_1 on the left side and Z_2 on the right side,
5. *wetted perimeter*, P, which is the length of the wetted surface,
6. *top width*, T, which is the width at the water surface,

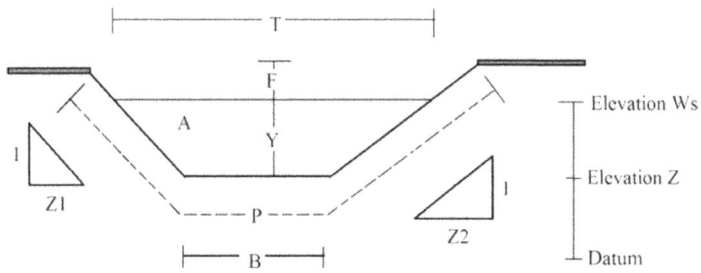

Figure 8.6 Cross-sectional elements in a trapezoidal channel

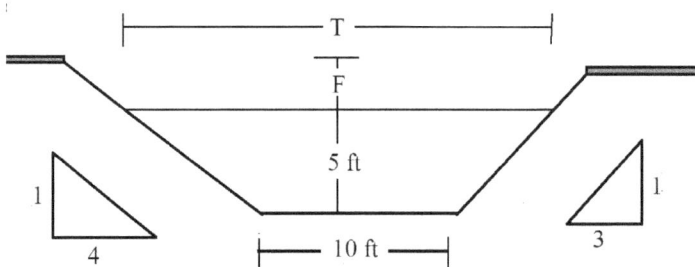

Figure 8.7 Examples of flow parameters

7. *water surface elevation (or water stage),* W_s,
8. *channel bottom elevation, Z.*
9. *freeboard, F,* which is the vertical distance above the water surface to bank top.

For a trapezoidal channel, the flow area and wetted perimeter are calculated as:

$$T = B + (Z_1 + Z_2)Y \tag{8.6}$$

$$A = \frac{1}{2}[B + T]Y = [B + \frac{1}{2}(Z_1 + Z_2)]Y \tag{8.7}$$

$$P = B + [\sqrt{1 + Z_1^2} + \sqrt{1 + Z_2^2}]Y \tag{8.8}$$

When the values of Z_1 and Z_2 are reduced to zero, we have a *rectangular channel*. Similarly, a trapezoidal channel is reduced to a *triangular channel* when the bottom width, B, vanishes.

Example 8.2: The trapezoidal channel in Fig. 8.5 is specified in Fig. 8.7. For $Y = 5$ feet, determine the flow area, A, top width, T, wetted perimeter, P, hydraulic radius, R, and hydraulic depth, D.

Solution:

$$T = B + (Z_1 + Z_2)Y = 10 + (3 + 4) \times 5 = 45 \, ft$$

$$A = \frac{B+T}{2}Y = \frac{10+45}{2} \times 5 = 137.5 \, sq \, ft$$

$$P = B + (\sqrt{1+Z_1^2} + \sqrt{1+Z_2^2})Y = 10 + (\sqrt{1+3^2} + \sqrt{1+4^2}) \times 5 = 46.43 \, ft$$

$$R = \frac{A}{P} = 2.96 \, ft$$

$$H_d = \frac{A}{T} = \frac{137.5}{45} = 3.06 \, ft$$

Example 8.2 reveals that the flow depth is neither the hydraulic depth nor the hydraulic radius. For a wide channel which is defined as $B > 10Y$, numerically, we may have $Y \approx R \approx H_d$.

8.5 EMPIRICAL FORMULA

A *normal flow* is a special case that can be developed in a long and straight prismatic channel with a constant discharge. Normal flow is also called uniform flow. In a uniform flow, its energy slope, S_e, is parallel to its bottom slope, S_o. In 1889, Robert Manning presented a formula to predict the normal flow condition in a channel. Manning's equation states:

$$U_N = \frac{K_N}{N} R_N^{\frac{2}{3}} \sqrt{S_o} \tag{8.9}$$

$$Q = U_N A_N \tag{8.10}$$

where $K_N = 1.486$ for feet-second units or 1.0 for meter-second units, N = Manning's roughness, R_N = hydraulic radius in [L], U_N = normal flow velocity in [L/T], A_N = flow area in [L²], and Q = discharge in [L³/T]. The subscript, N, represents variables associated with normal flow conditions. A channel flow is sensitive to the surface friction that is directly applied to the wetted surface. The *hydraulic radius*, R, was developed as the ratio of flow area to wetted perimeter as:

$$R_N = \frac{A_N}{P_N} \tag{8.11}$$

where P_N = wetted perimeter in [L]. The discharge carried in a channel is directly related to its water depth. An increase in the top width in the channel means an increase in the discharge. The *hydraulic depth*, H_d, is defined as the ratio of flow area to top width as:

$$H_d = \frac{A_N}{T_N} \tag{8.12}$$

where T_N = top width of water surface in [L]. As implied in the definitions of cross-sectional elements, the flow capacity in a channel is determined by the flow cross-sectional area, roughness on the wetted perimeter, and energy gradient. Substituting Eqs 8.10 and 8.11 into Eq. 8.9 yields:

$$Q = \frac{K_N}{N} A_n^{\frac{5}{3}} P_n^{\frac{-2}{3}} \sqrt{S_o} \qquad (8.13)$$

There are seven variables involved in Eq. 8.13. They are discharge, channel slope, flow depth, bottom width, right side slope, left side slope, and Manning's roughness. After selecting any six variables, the only unknown is determined by Eq. 8.13.

In practice, open channel flows are classified by two dimensionless parameters: the Reynolds number and Froude number. The *Reynolds number* is the ratio of flow inertial force to viscous force, and is defined as:

$$R_e = \frac{U_n R_n}{v} \qquad (8.14)$$

where R_e = Reynolds number and v = water kinematic viscosity in [L^2/T]. The Reynolds number serves as an indicator to differentiate *laminar* flows from *turbulent* flows. A laminar open channel flow will have a Reynolds number less than 500. In practice, it is rare to find a laminar open channel flow in nature.

The *Froude number* is the ratio of flow inertial force to weight of water. It also represents the ratio of flow velocity to wave celerity. The Froude number is defined as:

$$F_r = \frac{U_n}{\sqrt{gH_d}} = \sqrt{\frac{T_n Q^2}{gA_n^3}} \qquad (8.15)$$

Many characteristics of open-channel flows are associated with the Froude number. For instance, when $F_r < 1$, the flow is called *subcritical flow*. When $F_r > 1$, the flow is called *supercritical flow*. A *critical flow* has a Froude number equal to unity. The Froude number is often used in the similarity rules for modeling studies. Details about these three regimes of open-channel flows will be discussed later.

8.6 ROUGHNESS COEFFICIENT

Although Manning's formula is widely accepted, caution must be taken when estimating the roughness coefficient. Eq. 8.10 indicates that the channel capacity is linearly and reversely varied with respect to Manning's N. A change from $N = 0.04$ to $N = 0.02$ means the flow rate is doubled. In practice, Manning's N is mainly selected based on the lining materials. For channel design, Table 8.1 is recommended for the selection of Manning's N.

On a wide floodplain, Manning's N is varied from the left bank to the right bank or from the bottom to the water surface. For the purpose of design, a single value for the entire cross-section is acceptable for Manning's N.

Table 8.1 Recommended Manning's roughness coefficients

Channel Lining Description	Manning's Roughness, N
Cement	0.011–0.013
Concrete	0.012–0.015
Wood	0.014
Brick	0.015
Asphalt	0.016
Masonry	0.017
Earth	0.022
Vegetal lining	0.030–0.035
Rock riprap	0.035–0.045
Natural floodplain	0.045–0.050
Natural channel	0.030–0.035
Flood plain	0.050–0.080
Bushes	0.070
Trees	0.090

(a) Prismatic Channel

(b) Curve Channel

Figure 8.8 Uniform and non-uniform flows

8.7 NORMAL FLOW

Fig. 8.8(a) presents a case of a prismatic channel where a uniform flow can be developed. In contrast, the curved channel in Fig. 8.8(b) will produce non-uniform flows due to the superelevation effects along the outer bank.

A normal flow is characterized with a constant flow depth and constant velocity everywhere for the given discharge. It implies that the external forces acting on the uniform flow are in equilibrium. In practice, we can rarely find a uniform and steady flow in a natural waterway because the flow rates and cross-sections change with respect to time and to space as well. Despite this fact, the design procedure begins with normal flow.

Example 8.3: A trapezoidal channel in Example 8.2 carries a discharge of 1208 cfs. The channel is protected with rough grass linings. Determine the normal depth.

Solution: In this case, $N = 0.035$. Consider the ground slope of 0.01 in Example 8.2 for the channel design. Substituting design parameters into Eqs 8.6, 8.7, and 8.8 yields:

$$T_n = 10.0 + (3.0 + 4.0)Y_n$$

$$A_n = \frac{10.0 + T}{2}Y_n = 10.0Y_n + 3.5\ Y_n^2$$

$$P_n = 10.0 + Y_n\left[\sqrt{1 + 4.0^2} + \sqrt{1 + 3^2}\right] = 10.0 + 7.28Y_n$$

The subscript of n represents the variables associated with normal flow. Substituting the above flow variables into Eq. 8.13 yields:

$$Q = \frac{1.486}{0.035}[10Y_n + 3.5Y_n^2]^{\frac{5}{3}}(10.0 + 10Y_n)^{\frac{-2}{3}}\sqrt{0.01} = 1208\,\text{cfs}$$

The above equation is nonlinear with respect to flow depth. The solution can be found by trial and error for $Q = 1208$ cfs. In practice, it is convenient to utilize the *rating curve method* to solve Manning's equation. The rating curve in Fig. 8.9 represents the relationship between flow depths and flow rates using Eq. 8.13. A rating curve is derived for the specified cross-section, invert slope, and Manning's roughness.

Table 8.2 presents the solution for this case. The flow depth varies within the selected range. The rating curve is plotted in Fig. 8.9. The solution for this case can be determined by interpolation. So, we have $Y_n = 5.0$ ft, $T_n = 45$ ft, $A_n = 137.5$ ft^2,

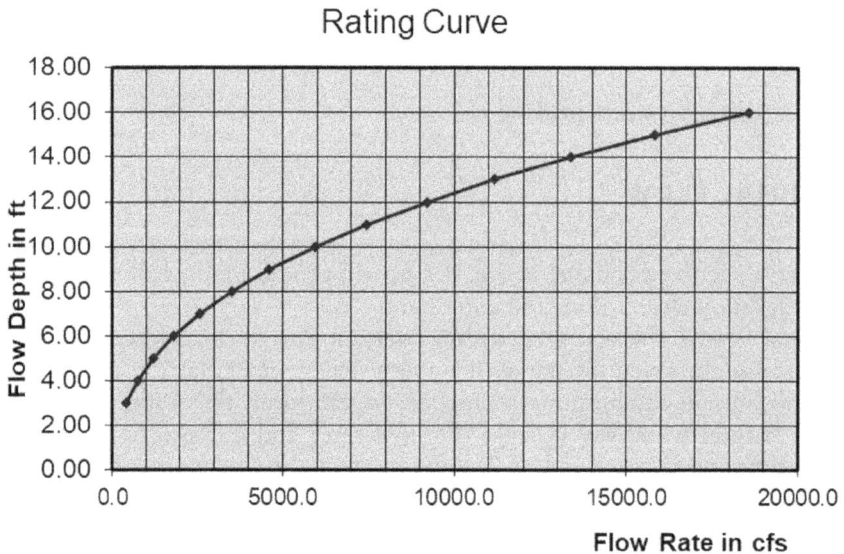

Figure 8.9 Example rating curve

Table 8.2 Example for rating curve method

	Channel Bottom Width		B=	10.00 feet				
	Left Side Slope		ZI =	3.00 ft/ft				
	Right Side Slope		Z2=	4.00 ft/ft				
	Manning's N		N=	0.035				
	Channel Bottom Slope		So=	0.0100 ft/ft				

Flow Depth Y ft	Top Width Top ft	Flow Area A sq ft	Wetted P-meter P ft	Hydraulic Radius R ft	Flow Velocity U fps	Flow Rate Q cfs	Froude Number Fr
3.00	31.00	61.50	31.86	1.93	6.60	405.7	0.83
4.00	38.00	96.00	39.14	2.45	7.74	743.5	0.86
5.00	45.00	137.50	46.43	2.96	8.79	1208.3	0.89
6.00	52.00	186.00	53.71	3.46	9.76	1815.0	0.91
7.00	59.00	241.50	61.00	3.96	10.67	2577.9	0.93

P_n = 46.43 ft, and U_n = 8.79 fps. The corresponding hydraulic depth and Froude number are:

$$F_r = \sqrt{\frac{T_n Q^2}{g A_n^3}} = 0.89$$

It is a subcritical flow because F_r = 0.89 < 1.0.

Example 8.4: An old corrugated metal pipe of 36 inches in diameter is laid on a slope of 2%. After years of service, this circular pipe needs to be replaced. The Manning's N for this old pipe is 0.035. Verify that a 27-inch *concrete* pipe can provide equivalent flowing full capacity to the old 36-inch pipe.

Solution: For a circular pipe, the flowing full area and wetted parameters are calculated as:

$$A = \frac{\pi D^2}{4} \tag{8.16}$$

$$P = \pi D \tag{8.17}$$

$$R = \frac{D}{4} \tag{8.18}$$

Substituting Eqs 8.16 and 8.17 into Eq. 8.13 yields Table 8.3 for both pipes. The proposed new pipe of 27-inch (2.25 ft) in diameter provides more flow capacity than the existing one.

Table 8.3 Comparison of full flow capacity between two pipes

Diameter feet	Pipe Line Slope ft/ft	Full-flow Area sq feet	Wetted Perimeter feet	Hydraulic Radius feet	Manning's N	Full-flow Discharge cfs
3.00	0.02	7.07	9.42	0.75	0.035	35.13
2.25	0.02	3.98	7.07	0.56	0.014	40.78

8.8 NON-UNIFORM FLOW

As the flow depth changes with respect to location, the flow in the channel is called non-uniform flow, which is dictated by the energy slope. The empirical forma for non-uniform flow is written as:

$$S_e = \frac{N^2 U^2}{K_N^2 R^{\frac{4}{3}}} \tag{8.19}$$

where S_e = energy slope in [L/L], U = flow velocity in [L/T], and R = hydraulic radius in [L]. The energy slope in Eq. 8.19 is the unknown or it implies that the flow depth and discharge are known. An energy slope is measured between two adjacent sections. The Bernoulli's sum at a cross-section is defined as:

$$H = E_s + \frac{U^2}{2g} \tag{8.20}$$

$$W_S = Y + Z \tag{8.21}$$

$$E_S = Y + \frac{U^2}{2g} \tag{8.22}$$

where H = Bernoulli's sum in [L], Z = invert elevation in [L], Y = water depth in [L] from water surface to the lowest point in the cross-section, U = cross-sectional flow velocity in [L/T], E_S = specific energy in [L], and W_S = hydraulic head or water surface elevation in [L]. The Bernoulli's sum at a cross-section consists of a *hydraulic head* and a *dynamic head*. In practice, the energy principle is often used to determine the length of a backwater profile. Referring to Fig. 8.10, the water depth, Y_2, at a downstream bridge section is 3 feet deeper than the normal depth, ($Y_2 = Y_N + 3$ ft). We want to know how far the upstream flow depth would be asymptotically merged to the normal depth ($Y_1 = Y_N$).

Using Eqs 8.1, 8.2, and 8.4, the energy grade line is defined with the energy difference over the distance as:

$$\Delta E = \Delta \left(Y + \frac{U^2}{2g} \right) + \Delta Z = \Delta E_s + \Delta Z \tag{8.23}$$

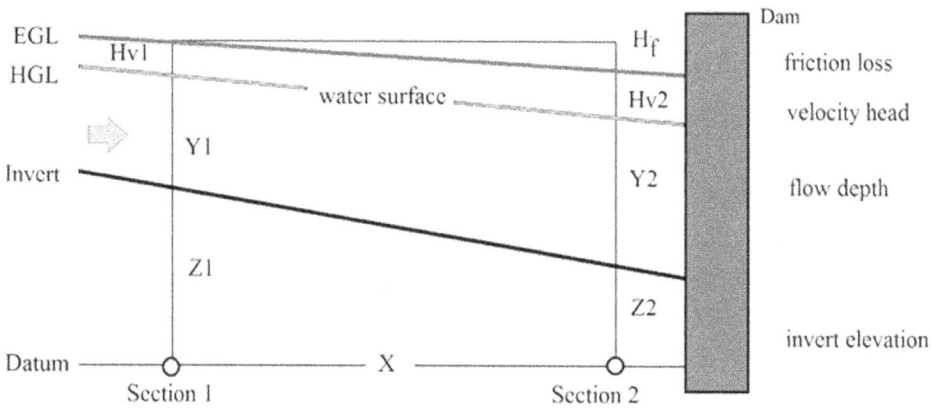

Figure 8.10 Computation of a backwater water surface profile

$$E_s = Y + \frac{U^2}{2g} \tag{8.24}$$

$$S_a = \frac{S_1 + S_2}{2} \tag{8.25}$$

$$X = \frac{\Delta E_s}{|S_o - S_a|} \tag{8.26}$$

where X = distance in [L] between the upstream and downstream sections, S_a = average energy slope between two sections, S_1 = energy slope at the upstream section, and S_2 = energy slope at the downstream section. Both S_1 and S_2 are respectively determined with the known depths using Eq. 8.19. All items with a delta (Δ) represent the difference between the upstream and downstream values.

Example 8.5: Continued from Example 8.3. A trapezoidal channel carries a discharge of 1208 cfs and drains into the reservoir in Fig. 8.10. Knowing that the water depth at the dam is 3 feet above the normal depth, determine the distance to the normal depth upstream.

Solution: Based on Example 8.3, we have $Y_1 = 5.0$ ft defined by the normal flow condition at Section 2. Setting $Y_2 = 8.0$ ft, Table 8.4 gives a summary of the upstream and downstream flow conditions.

$Y_1 = 5$ ft, $A_1 = 137.5$ sq ft, $P = 46.4$ ft Use Manning's equation to confirm: $U_1 = 8.79$ fps.

$Y_2 = 8$ ft, $A_2 = 304.0$ sq ft, Use the continuity principle to calculate: $U_2 = Q/A_2 = 3.97$ fps

Table 8.4 Backwater profile computations

Flow Depth Y ft	Flow Area A sq ft	Wetted P-meter P ft	Hy- Radius R ft	Flow Velocity U fps	Sp Energy Es ft	Energy Slope Se ft/ft
5.00	137.5	46.4	2.96	8.79	6.20	0.01002
8.00	304.0	68.3	4.45	3.97	8.24	0.0000971

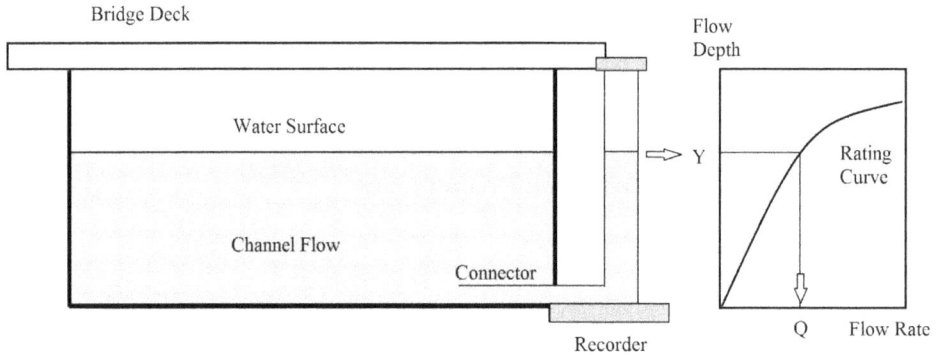

Figure 8.11 Rating curve at a stream gauge

$$\Delta E_s = 8.24 - 6.20 = 2.04\,ft$$

$$S_2 = \frac{0.035^2 \times 3.97^2}{1.486^2 \times (4.45)^{\frac{4}{3}}} = 0.0000971$$

$$S_a = \frac{S_1 + S_2}{2} = \frac{0.01 + 0.0000971}{2} = 0.005$$

In this case, the tailwater effect diminishes over a distance of 408 ft upstream.

8.9 CONCLUSIONS

Channel design begins with normal flow. Normal flow is defined by Manning's formula under the assumption that $S_e = S_o$. Solutions for a normal flow are achieved using the rating curve method, which is a one-to-one increasing function between flow rates and flow depths. The normal flow approach provides the first estimate of channel dimension and capacity. As always, refinements are required when the channel alignment is curved or/and the channel width is changed.

In the field, a stage-flow stream gauge as illustrated in Fig. 8.11 and observed in Fig. 8.12, is installed to measure and to record the variations of flows in the channel. Often a stream gauge only reports the instantaneous water depth. The corresponding flow rate is then read from the rating curve.

(a) Stage-Flow Recorder   (b) Stage Station on Bank Wall

Figure 8.12 **Stage-flow measurement**

A stream gauge, shown in Fig. 8.12b, shall be installed at a stable cross-section. If the cross-section is subject to severe erosion or backwater effects, the rating curve will not accurately represent the true flow conditions. Calibrations are required to maintain the integrity of a stream gauge. In addition to a stream gauge, we can also install weirs and orifices to record water depths. A time series of water depths can be converted into its flow hydrographs using the rating curve pre-developed for the measurement instrument.

HOMEWORK

Q8-1   A concrete trapezoidal channel carries $Q = 1000$ cfs with $S_o = 0.015$. Knowing that the bottom width is 10 ft and the side slopes are $Z_1 = Z_2 = 1.0$, determine the normal flow condition. (Answer: $Y_n = 3.32$ ft)

Q8-2   Knowing that the critical flow is defined with $F_r = 1.0$, determine the critical flow depth in Q8-1. (Answer: $Y_c = 5.58$ ft)

Q8-3   The Manning's roughness in Q8-1 is increased from 0.012 for concrete linings to 0.035 for grass linings. Find the normal and critical flow depths. Explain why the critical depth remains the same.

Q8-4   The trapezoidal channel in Fig. Q8-4 is built with $B = 5$ feet, $S_o = 1.0\%$, $N = 0.045$ and $Z_1 = Z_2 = 2.0$. A flow of 486.5 cfs is caried in the channel. At the downstream bridge, the water depth is raised to 6.0 feet. (1) Verify the normal flow depth $Y_n = 5$ ft (2) Determine the backwater distance. (Hint: 145 feet)

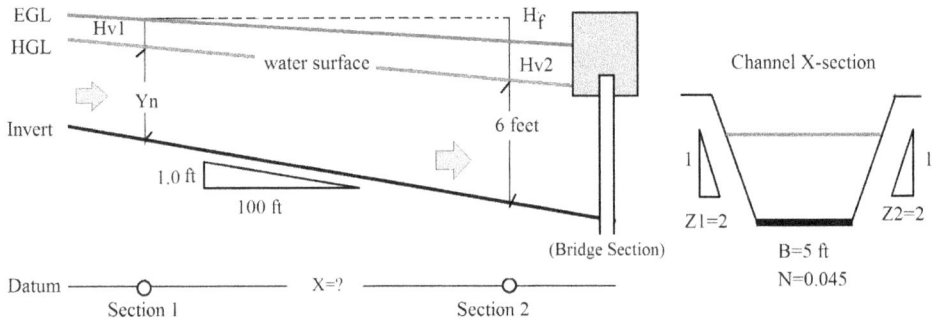

Figure Q8-4 Backwater profile in channel flow

Table Q8-4 Fill in your answers for backwater profile computations

Flow Depth Y ft	Flow Area A sq ft	Wetted P-meter P ft	Hy- Radius R ft	Flow Velocity U fps	Sp Energy Es ft	Energy Slope Se ft/ft	Computed Flow Q cfs	Froude Numer Fr
5.00	75.0		2.74		5.65	0.01000		0.66
6.00		31.8			6.35			

Energy Dissipatioin Δ Es =	ft
Average Energy Slope Sa=	ft/ft
Back Water Distance =	ft/ft

REFERENCES

ADOWR (2000). *Design Guideline for Diversion Channels*, Flood Warning and Dam Safety Section, Arizona Department of Water Resources, Arizona.

ASCE and WEF (1992). "Design and Construction of Urban Stormwater Management Systems", *American Society of Civil Engineers and Water Environment Federation, Reports of Engineering Practice*, no. 77 and WEF Manual of Practice FD-20, ASCE, New York.

Barnes, H.H. (1967). *Roughness Characteristics of Natural Channels*, US Department of the Interior, Geologic Survey, Washington, DC.

Chow, Ven T. (1959). *Open Channel Hydraulics*, McGraw-Hill, New York.

City and County of Sacramento, California (1992). *Hydrologic Standards*, Brown and Caldwell, California.

Clark County (1995). *Hydrologic Criteria and Drainage Design Manual*, Clark County Regional Flood Control District, Las Vegas, NV.

French, R.H. (1985). *Open Channel Hydraulics*, McGraw Hill, New York.

Guo, J.C.Y. (2009). *Grade Control for Urban Channel Design*, Elsevier Science, 10.1016/j.jher.2009.01.001, February/March, pp. 1–4.

Little, W.C. and Murphey, J.B. (1982). "Model Study of Low Drop Grade Control Structures", *Journal of Hydraulic Division*, 108 (HY10).

Robert Manning (1891). "On the flow of water in open channel and pipes", *Transactions, Institution of Civil Engineers of Ireland*, vol. 20, pp. 161–207.

US ACOE (2008). *Hydraulic Design of Flood Control Channels*, Department of the Army, US Army Corps of Engineers, Washington DC.

USWDCM (2010). *Urban Storm Water Design Criteria Manual*, vols 1 and 2, Denver Urban Drainage and Flood Control District.

Chapter 9

Street conveyance hydraulics

9.1 STREET HYDRAULIC CONVEYANCE CAPACITY

The *street hydraulic capacity* (SHC) consists of two aspects: the stormwater convey-ance capacity in Fig. 9.1a and the stormwater storage capacity in Fig. 9.1b. The *street hydraulic conveyance capacity* (SHCC) is related to the stormwater flowing on the street with a positive slope.

As indicated in Hydraulic Engineering Circulars No. 12 and No 22, entitled *Drainage of Highway Pavements*, the SHCC for a given street depends on the street cross-sectional geometry and hydraulic roughness on the pavement surfaces (HEC 12 in 1984; HEC 22 in 2010). In general, the water spread in Fig. 9.2 for the design storm event should not encroach into the emergency traffic lane and the gutter flow depth should not exceed the curb height (Gallaway et al. 1979).

9.1.1 Straight cross-section

As illustrated in Fig. 9.3, the street cross-section is triangular in shape. Stormwater flowing through such a triangular gutter section can be described by the revised Manning's equation as (HEC 22 in 2010):

$$Q = \frac{K}{n} S_x^{1.67} T^{2.67} \sqrt{S_o} \tag{9.1}$$

where Q = SHCC in cfs or cms, K = 0.56 for foot-second units or 0.376 for meter-second units, n = Manning's roughness of the street surface, S_x = street transverse slope, S_o = street longitudinal slope in [L/L], and T = water spread width in [L] on the street. In general, a value of 0.016 is recommended for street the Manning's roughness for the street. The standard street transverse slope is 2%, but a slope of 1% is acceptable.

The gutter flow depth, Y, at the curb is calculated as:

$$Y = TS_x \tag{9.2}$$

The cross-sectional area, A, for the street flow is calculated as:

$$A = 0.5YT \tag{9.3}$$

DOI: 10.1201/9781003284239-9

(a) Water Flow on Street   (b) Water Pool At Street Corner

Figure 9.1 Potential water hazards to vehicles during storm event

(a) Gutter-Full Flow   (b) Water Spread into Traffic Lane

Figure 9.2 Water carried in a street gutter

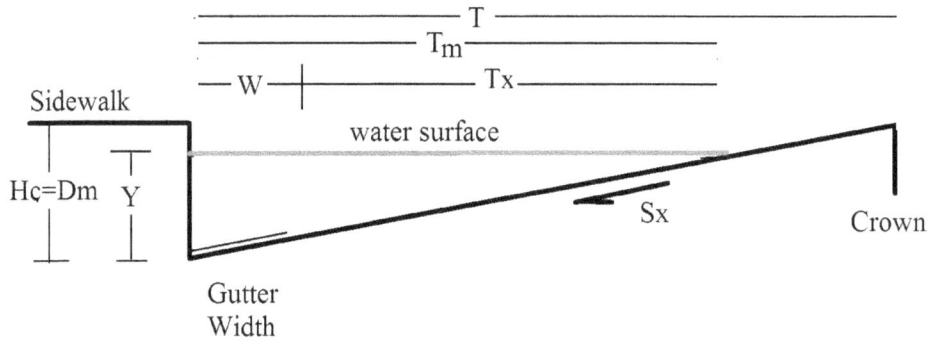

Figure 9.3 Triangular cross-section of gutter flow

The cross-sectional average flow velocity, V, is:

$$V = \frac{Q}{A}$$ (9.4)

9.1.2 Composite street section

In practice, a standard depression of 2 inches in Fig. 9.4 is often introduced at the street curb in order to increase the gutter hydraulic conveyance capacity. With an additional depression, the transverse slope across the gutter is:

$$S_w = S_x + \frac{D_s}{W}$$ (9.5)

where S_w = gutter cross slope, W = gutter width, such as two feet, and D_s = gutter depression, such as 2 inches. The water depth at the curb is the sum of the flow depth and gutter depression as:

$$D = Y + D_s$$ (9.6)

where D = water depth in the gutter.

With the gutter depression shown in Fig. 9.4, the water flow is carried in a composite street-gutter cross-section that is divided into *gutter flow and side flow*. The gutter flow is carried within the gutter width, W, and the side flow is carried by the water spread, T_x in [L], encroaching into the traffic lanes. For design, the water depth at a gutter must not exceed the curb height, which is the maximum gutter depth, D_m in [L]. For convenience, the water spread width, T_s in [L], across the gutter width is calculated as:

$$T_s = \frac{D}{S_w}$$ (9.7)

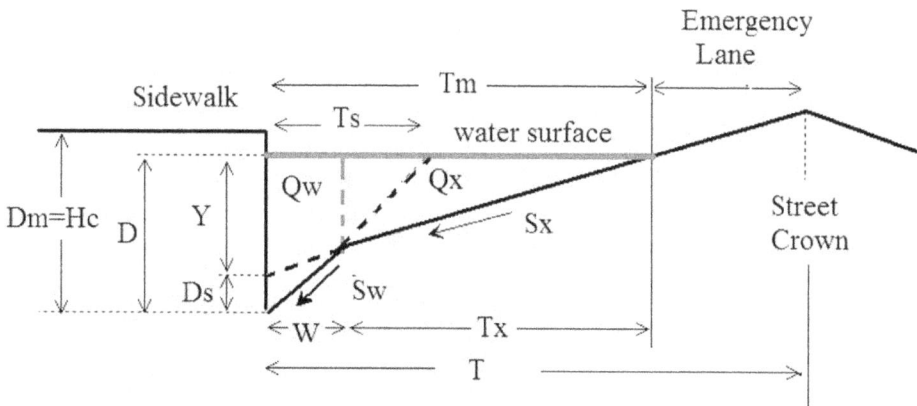

Figure 9.4 Composite street cross-section

The total water spread, T in [L], is the sum as:

$$T = W + T_x \qquad (9.8)$$

where T_x = water spread in [L] for the side flow. Applying Eq. 9.1 to both the gutter and side flow, we have:

$$Q_x = \frac{K}{n} S_x^{1.67} T_x^{2.67} \sqrt{S_o} \qquad (9.9)$$

$$Q_w = \frac{K}{n} S_w^{1.67} \left[T_s^{2.67} - (T_s - W)^{2.67} \right] \sqrt{S_o} \qquad (9.10)$$

where Q_x = side flow in [L³/T], Q_w = gutter flow in [L³/T], W = gutter width in [L], and T_s = water spread in [L] defined in Fig. 9.4. The total flow, Q in [L³/T], on the street is:

$$Q = Q_x + Q_w \qquad (9.11)$$

The flow cross-sectional area in [L²/T] for a composite street is:

$$A = 0.5YT + 0.5WD_s \qquad (9.12)$$

Aided by Eqs 9.11 and 9.12, the flow velocity on a composite street is calculated using Eq. 9.4.

Example 9.1: Street BC in Fig. 9.5 has T_m = 15 feet, n = 0.016, W = 2 feet, D_s = 2 inches, S_o = 1.50 %, and S_x = 2%. The maximum gutter depth, D_m, for this section is equal to the curb height, H_c, of 6 inches. During a minor event, the street crown must be free from water, so the maximum water spread is 15 feet in this case. Determine the *street-full capacity* for this street.

Solution: Referring to Fig. 9.5, the specified water spread, is T = 15 feet, in this case. The street flow is calculated as:

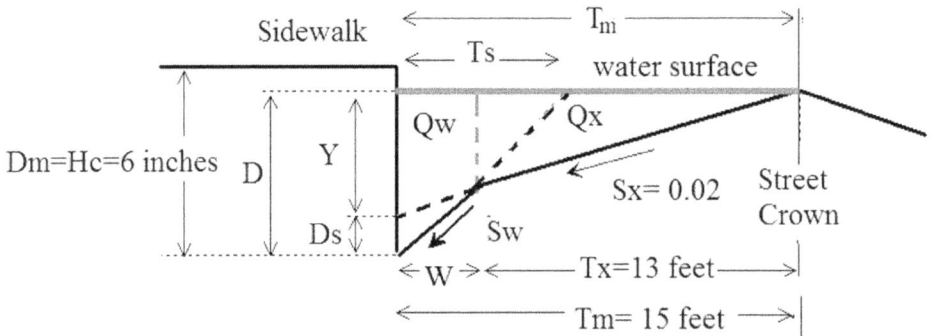

Figure 9.5 Street full flow capacity

$S_w = S_x + D_s /(12W) = 0.02 + (2/12)/2 = 0.103$ ft/ft,

$$T = \frac{(D_m - D_s)/12}{S_x} = \frac{(Y-2)/12}{0.02} = 15ft, So\, Y = 0.3 ft \, and \, D = 0.47 ft < D_m (Ok)$$

$T_x = T - W = 15.0 - 2.0 = 13.0$ ft,

$T_s = 6.0 /(12 \times 0.103) = 4.52$ feet

$$Q_w = \frac{0.56}{0.016}(0.103)^{1.67}[4.52^{2.67} - (4.52 - 2.0)^{2.67}]\sqrt{0.015} = 5.88\,cfs$$

$$Q_x = \frac{0.56}{0.016}0.02^{1.67}13.0^{2.67}\sqrt{0.015} = 4.28\,cfs$$

$Q = Q_x + Q_w = 10.16$ cfs

Example 9.2: Referring to Fig. 9.6, Street BC has a composite section: T_m = 15 feet, n = 0.016, W = 2 feet, D_s = 2 inches, S_o = 1.50 %, and S_x = 2%. The maximum gutter depth, D_m, for this section is equal to the curb height, H_c, of 6 inches. During a minor event, the street crown can be under water in this case. Determine the gutter full capacity of Street BC.

Solution: Let $D = D_m = H_c = 6.0$ inches. In this case, the gutter full condition results in a water spread > T_m (15 ft). It means that the street crown is under water and a *virtual flow*, Q_v, as illustrated in Fig. 9.6, needs to be subtracted from the side flow.

$S_w = S_x + D_s /(12W) = 0.02 + (2/12)/2 = 0.103$ ft/ft,

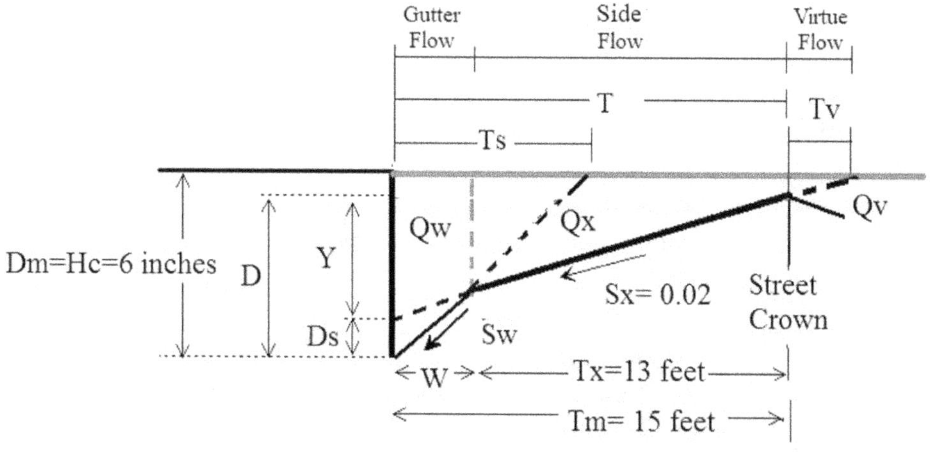

Figure 9.6 Gutter full flow capacity

The gutter full water spread width is calculated as:

$$T = \frac{(D_m - D_s)/12}{S_x} = \frac{(6-2)/12}{0.02} = 16.67\,ft, > 15\,feet$$

$T_c = 16.67 - 15.0 = 1.67$ *feet.* So, a virtual flow exists.

$T_x = T - W = 14.67$ *ft,*

$T_s = 6.0\,/(12 \times 0.103) = 4.85$ *feet*

$$Q_W = \frac{0.56}{0.016}(0.103)^{1.67}[4.85^{2.67} - (4.85 - 2.0)^{2.67}]\sqrt{0.015} = 4.95\,cfs\,(gutter\,flow)$$

$$Q_X = \frac{0.56}{0.016}0.02^{1.67}14.67^{2.67}\sqrt{0.015} = 8.11\,cfs\,(side\,flow)$$

$$Q_V = \frac{0.56}{0.016}0.02^{1.67}1.67^{2.67}\sqrt{0.015} = 0.03\,cfs\,(virtual\,flow)$$

$Q = Q_X + Q_W - Q_V = 13.03$ *cfs (subtracting the virtual flow from the side flow.)*

9.2 TRAFFIC SAFETY WITH STREET RUNOFF

The hydroplaning effect is associated with the spinning acceleration of a vehicle on a thin layer of water (Huebner, Reed and Henry 1986). The hyperbolic relationship between the flow velocity, V, and the water depth, D, sets a safety limit for the VD product. With $D_s = 0$, the VD product can be derived from Eq. 9.1 as:

$$VD = \frac{2K}{n}(TS_x)^{1.67}\sqrt{S_0} \tag{9.13}$$

$$Q = VA \tag{9.14}$$

Using Eq. 9.3, Eq. 9.14 is converted to:

$$Q = V\left(\frac{1}{2}DT\right) = \frac{1}{2}(VD)T \tag{9.15}$$

Eq. 9.15 describes the relationship between the SHCC and VD product. Using Eq. 9.15, the unit-width flow, q in [L^2/T], across the water spread is:

$$q = \frac{Q}{T} = \frac{VD}{2} \tag{9.16}$$

The momentum impulse of the gutter flow consists of both static and dynamic force components. In contrast, the static force in a gutter flow is negligible because of the shallow depth. Aided by Eqs 9.1 and 9.15, the dynamic force, M, associated with a gutter flow is

$$M = \rho Q V = \frac{\rho}{2S_x}(VD)^2 \tag{9.17}$$

where M = dynamic force, and ρ = density of water. Eqs 9.16 and 9.17 indicate that both the SHCC and dynamic force in a gutter flow are proportional to its VD product. A reduction of the VD product can directly impose a limit on both the unit-width capacity and momentum impulse in a gutter flow. Consider that the VD product of a gutter flow should not exceed a limit, L, defined by safety as:

$$VD \leq L \tag{9.18}$$

where L = permissible VD product. For instance, L = 6.0 cfs/ft recommended by Clark County in Las Vegas, Nevada for a 10-year event. According to Eq. 9.15, Clark County imposes a limit to the gutter flow of 3.0 cfs per foot, and the momentum force carried by the gutter flow section becomes no more than 1,728.00 pounds when S_x = 0.02 ft/ft, and ρ = 1.92 slug/cubic ft. Substituting Eq. 9.13 into Eq. 9.18 yields the permissible water spread, T_L, for the specified L as (Guo 2000):

$$T_L \leq \frac{1}{S_x}\left(\frac{nL}{2K\sqrt{S_0}}\right)^{0.6} \tag{9.19}$$

Using Eq. 9.15 and Eq. 9.18, the allowable street conveyance capacity (ASHCC) is equal to a flow reduction as:

$$Q_L = \frac{1}{2}LT_L \tag{9.21}$$

where Q_L = allowable street hydraulic capacity in $[L^3/T]$.

9.3 DISCHARGE REDUCTION METHOD

In parallel to the approach of a permissible VD product, a discharge reduction method is also developed and is recommended for street drainage design. For instance, the City and County of Denver converts the limiting VD product into a set of reduction factors. The allowable SHCC is equal to the gutter-full capacity multiplied by a reduction factor that is defined as (Guo 1997):

$$R = \frac{Q_L}{Q_{GF}} \tag{9.22}$$

where Q_{GF} = gutter full SHCC and R = reduction factor. Substituting Eqs 9.19 and 9.21 into Eq. 9.22 yields:

$$R = \frac{\frac{1}{2}LT_L}{V\left(\frac{1}{2}D_m T\right)} = \frac{1}{(TS_x)^{2.67}}\left[\frac{nL}{2K\sqrt{S_0}}\right]^{1.60} \quad \text{for } 0 < R < 1 \tag{9.23}$$

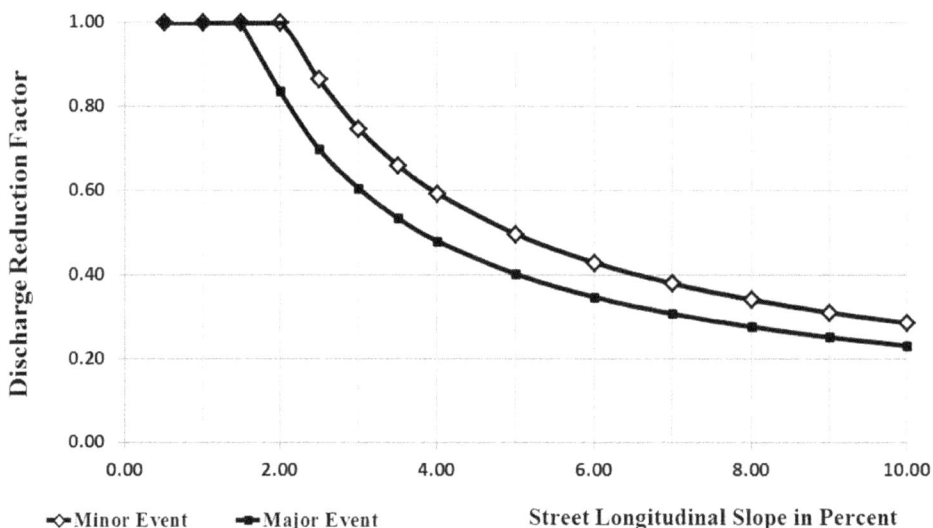

Figure 9.7 Reduction factors for street flow capacity

As shown in Eq. 9.23, the flow reduction factor is inversely proportional to the street longitudinal slope for a specified L and T. In the metro Denver area, Eq. 9.23 was used to produce a regional safety chart for flow reduction on steep streets. Considering $T = 12.5$ feet and $L = 1.0$ cfs/ft for a minor event and $T = 20.5$ feet and $L = 2.0$ cfs/ft for a major event, Fig. 9.7 was produced for street longitudinal slopes ranging from 0.5 to 10% (Guo 1997). For a steep street, Eq. 9.23 results in $R < 1.0$, indicating that the design flow on the street shall be reduced. For streets with a mild slope, Eq. 9.23 may result in $R > 1.0$, indicating that the street gutter can carry more than the design discharge without exceeding the permissible VD product.

9.4 ALLOWABLE STREET HYDRAULIC CONVEYANCE CAPACITY

For a given street cross-section, the street full flow capacity, Q_{SF}, is firstly determined by the specified water spread. Secondly, the gutter full capacity, Q_{GF}, shall be calculated based on the curb height, and the flow reduction factor is determined from Fig. 9.7. The allowable SHCC is determined as (UDFCD Manual 2010):

$$Q_a = \min(Q_{SF}, RQ_{GF}) \tag{9.24}$$

In practice, it is critical to know the allowable SHCC at the street site because wherever the water flow on the street exceeds the allowable amount defined in Eq. 9.24, an inlet or a culvert must be installed to remove the storm runoff from the street.

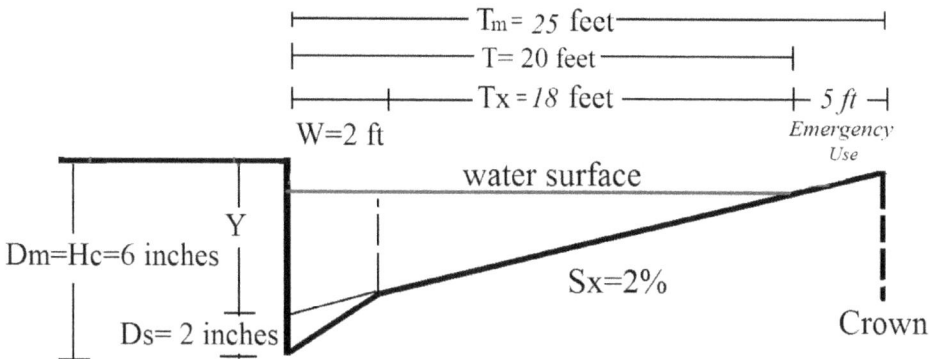

Figure 9.8 Street cross-section

Example 9.3: Street BC was evaluated for both its street full hydraulic conveyance capacity in Example 9.1 with T_m = 15 feet and its gutter full hydraulic conveyance capacity in *Example 9.2* with D_m = 6 inches. Determine the allowable flow capacity for this street.

Solution:

Q_{SF} = 10.16 cfs from *Example 9.1*

Q_{GF} = 13.03 cfs from *Example 9.2*

R = 1.0 based on S_o = 0.015 from Fig. 9.7

The conclusion is that the allowable storm runoff flow on Street BC is:

Q_a = *Min (10.16, 1.0 × 13.03)* = *10.16 cfs*

Example 9.4: A half street section is given in Fig. 9.8. The design parameters are: W = 2 feet, T = 25 feet, H_c = D_m = 6 inches, D_s = 2 inches, n = 0.016, S_x = 2%, and S_o = 2.5%. As a collector street, the middle lane of 5-feet in each traffic direction is reserved for emergency use. (1) Determine the allowable flow capacity. (b) Determine the flow condition for Q_s = 12 cfs on the street.

Solution: The half width of the street is 25ft. The available water spread width is 20 ft after subtracting 5 feet for emergency traffic use. For S_o = 3%, the flow reduction factor is R = 0.75 in Fig. 9.7 for a minor event. In this example, we shall quantify the allowable SHCC and then determine the design flow condition for the street.

(1) Determining the allowable hydraulic conveyance capacity.

A.	Street Geometry			*ft*
	A Half Width of the Street	T=	25.00	
	Width of Emergency Traffic Lanes	T-emergency=	5.00	ft
	Curb Height (6" for standard)	Hc=	6.00	inches
	Gutter Width (2 ft for standard)	W=	2.00	ft
	Gutter Depression (2" for standard)	Ds =	2.00	inches
	Street Transverse Slope	Sx=	0.0200	ft/ft
	Street Longitudinal Slope	So=	0.0300	ft/ft
	Manning's Roughness	n=	0.0160	
	Select 1 or 2: (1) Minor event or	Event =	1	
	(2) Major Event			
B.	Determination of Reduction Factor			
	Gutter Full Water Depth (Dm=Hc)	Dm=Hc=	6.00	inches
	Max Water Spread on the Street=T-T emergency	Tm=	20.00	feet
	Reduction factor	R=	0.75	
C.	Street Gutter-Full Conveyance Capacity and Reduction			
	Gutter-full water depth (D=Dm=Hc)	D=Dm=	6.00	inches
	Gutter-Full Water Spread Width	T=	16.67	ft
	Water Depth above Gutter Depression: $Y = T^*Sx =$	Y=	0.33	ft
	Spread for Side Flow: Tx =T − W=	Tx=	14.67	ft
	Gutter Cross Slope: Sw = Sx + Ds/W =	Sw	0.103	ft/ft
	Spread for Gutter Flow: Ts = D/Sw =	Ts=	4.84	ft
	Gutter Flow: $Qw = 0.56/n^*Sw^{\wedge}1.67^* (Ts^{\wedge}2.67-(Ts-W)^{\wedge}2.67)^*So^{\wedge}0.5=$	Qw=	7.00	cfs
	Side Flow: $Qx = 0.56/n^* Sx^{\wedge}1.67^* Tx^{\wedge}2.67^*So^{\wedge}0.5=$	Qx=	11.47	cfs
	Virtual Flow: $Qv = 0.56/n^* Sx^{\wedge}1.67^* (T-Tm)^{\wedge}2.67^*So^{\wedge}0.5=$	Qv=	0.00	cfs
	Gutter Full Flow: Q-full =Qw+Qx−Qv	Q-full=	18.47	cfs
	Reduced Gutter-Full Flow: Q-full*R =	R*Q-full	13.85	cfs
D.	Street-Full Conveynace Capacity			
	Street Allowable Water Spread	T=	20.00	ft
	Water Depth above Gutter Depression: $Y = T^*Sx =$	Y=	0.40	ft
	Gutter-Full Water Depth at Curb: D=Y+Ds =	Dm=	0.57	ft
	Spread for Side Flow: Tx =T − W=	Tx=	18.00	ft
	Gutter Cross Slope: Sw = Sx + Ds/W =	Sw	0.103	ft/ft
	Spread for Gutter Flow: Ts = D/Sw =	Ts=	5.48	ft

Gutter Flow: Qw = 0.56/n*Sw ^1.67*(Ts^2.67–(Ts–W)^2.67) *So^0 5=   Qw=   9.04   cfs

Side Flow: Qx = 0.56/n* Sx^1.67* Tx^2.67*So^0.5=   Qx=   19.81   cfs

Street-Full Conveynace Capacity   Qs=   28.85   cfs

E.   Street Allowable Hydraulic Conveynace Capacity   Q-allow=   13.85   cfs

(2) Determining the flow condition when the design discharge is 12 cfs.

Street Geometry

Curb Height	Hm=	6.00	inches
Gutter Width	W=	2.00	ft
Gutter Depression	Ds =	2.000	inches
Street Transverse Slope	Sx=	0.020	ft/ft
Street Longitudinal Slope	So=	0.0300	ft/ft
Manning's Roughness	n=	0.016	
Maximum Gutter Water Depth	Dm=	6.00	inch
Maximum Water Spread	Tm=	20.00	feet
Design Flow on the Street	Qo=	12.00	cfs
Water Spread Width (Guess it)	T=	13.88	ft
Gutter Cross Slope: Sw = Sx + Ds/W =	Sw	0.1033	ft/ft
Water depth Y = T*Sx =	Y=	0.28	ft
Gutter Water Depth at Curb: D=Y+Ds =	D=	0.44	ft
Spread for Side Flow: Tx =T – W=	Tx=	11.88	ft
Spread for Gutter Flow: Ts = D/Sw =	Ts=	4.30	ft
Qws = 0.56/n* Sw^1.67* Ts^2.67*So^0.5=	Qws=	6.73	cfs
Qww = 0.56/n*Sw^1.67*(Ts–W)^2.67*So^o. 5=	Qww=	1.27	cfs
Gutter Flow: Qw=Qws–Qww =	Qw=	5.46	cfs
Side Flow: Qx = 0.56/n* Sx^l. 67* Tx^2.67*So^0.5=	Qx=	6.54	cfs
Calculated Total Flow: Qs = Qx + Qw =	Qs=	12.00	cfs
Flow Difference to seek the solution	dQ=	0.00	
Check on the VD product			
Flow Area As =0.5*(YT+DsW)=	As=	2.09	sq ft
Flow Velocity Vs = Q/As =	Vs=	5.73	fps

9.5 CONCLUSIONS

Stormwater movement on the street is open channel flow in nature. For convenience, Manning's formula was re-arranged for s triangular cross-section. It is noted that the internal friction between the gutter and side flows in a street section was ignored. Therefore, a slightly higher Manning's roughness coefficient is recommended. On a continuous grade, the hydraulic conveyance capacity dictates the water spread into the traffic lanes and the water depth in the street gutter, whichever is smaller. On a steep slope, a flow reduction factor is recommended to set the upper limit for the street flow capacity. In practice, the first step is to identify the allowable SHCC, and then to analyze the design flow on the street. The design flow velocity and depth are required information for sizing a street inlet.

HOMEWORK

Q9-1   Referring to Fig. Q9-1, the street conveyance capacity is confined with $T_m = 25$ feet and $D_m = 6$ inches. Under a minor storm event, the water spread on the street is 20 feet. Conduct the following sensitivity study.

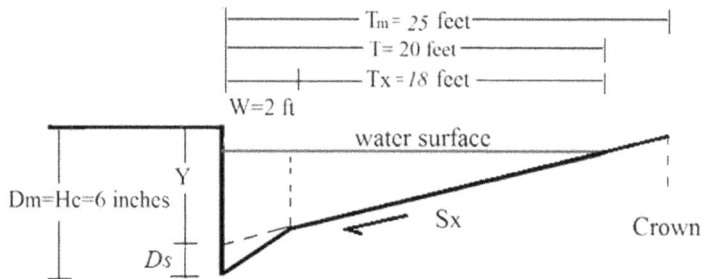

Figure Q9-1 Street cross-section

(1) Sensitivity test on Q to D_s and S_o

Street Slope	Discharge (cfs)	Discharge (cfs)
S_o	$D_s = 0$ inches and $S_x = 0.02$	$D_s = 2$ inches and $S_x = 0.02$
0.01		
0.03		

(2) Sensitivity Test on Q to S_x and S_o

Street Slope	Discharge (cfs)	Discharge (cfs)
S_o	$D_s = 2$ inches and $S_x = 0.01$	$D_s = 2$ inches and $S_x = 0.02$
0.01		
0.03		
0.05		

REFERENCES

Gallaway, B.M., Ivey, D.L., Hayes, G.G. Ledbetter, W.G., Olson, R.M., Woods, D.L., and Schiller, R.E. (1979). "Pavement and Geometric Design Criteria for Minimizing Hydroplaning", Federal Highways Administration, Research Report no. FHWA-RD-79-31.

Guo, J.C.Y. (1997). *Street Hydraulics and Inlet Sizing*, Water Resources Publications, LLC, Littleton, CO.

Guo, J.C.Y. (2000). "Street Storm Water Conveyance Capacity", *ASCE Journal of Irrigation and Drainage Engineering*, vol. 126, no. 2, March/April.

HEC12 (1984). *Drainage of Urban Highway Drainage*, US Department of Transportation, Federal Highways Administration.

HEC22 (2010). *Urban Drainage Manual*, US Department of Transportation, Federal Highways Administration.

Huebner, R.S., Reed, J.R., and Henry, J.J. (1986). "Criteria for Predicting Hydroplaning Potential", *ASCE Journal of Transportation Engineering*, vol. 12, no. 5, September.

UDFCD Manual (2010). *Storm Water Drainage Design Criteria Manual*, vols 1 and 2, Urban Drainage and Flood Control District Denver, CO.

Chapter 10

Inlet hydraulics

10.1 TYPES OF INLET

Inlets are often placed on streets, roadways, and highways to collect the surface runoff flows into the sewer or river system. For convenience, various types of inlets are developed to cope with street topography, urban debris, and traffic safety.

10.1.1 Grate inlet

A *grate inlet* unit is composed of an *inlet box* with its opening covered by a *grate*. Grate inlets are often installed horizontally along the flow line in a street gutter. Stormwater falls into the inlet box through the grate and then drains into storm sewers downstream. Grates are formed with steel bars as shown in Fig. 10.1(a) or vanes, as shown in Fig. 10.1(b). A grate is described by its length, L, and width, W, type of bars or vanes, and area opening ratio after subtracting the bar and vane areas from the total surface area.

10.1.2 Curb-opening inlet

A unit of a *curb-opening inlet* is composed of a *vertical opening* on the street curb, a *depression pan* within the street gutter, and an underground *inlet box* to transfer stormwater to storm sewers downstream. As shown in Fig. 10.2a, a curb-opening unit is described by the height of the opening, H, the length of the opening, L, and the wing width, W_p, of the depression pan. Although a depression pan can enhance the hydraulic efficiency, it does interfere with traffic such as bikes and snow plows. Therefore, a *depression pan* is optional with a curb opening inlet. However, a curb-opening inlet has a high hydraulic efficiency and is the least susceptible to debris clogging.

10.1.3 Combination inlet

As shown in Fig. 10.2b, a combination inlet is composed of a curb opening inlet and a grate inlet along the flow line within the street gutter. A combination inlet should be installed in places where a high stormwater interception is required or there is a high risk of debris clogging.

DOI: 10.1201/9781003284239-10

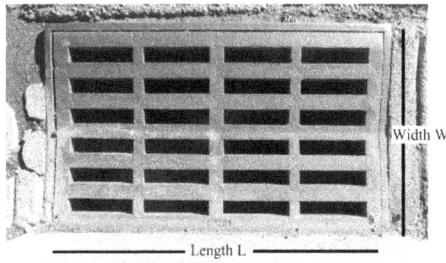

(a) Bar Grate

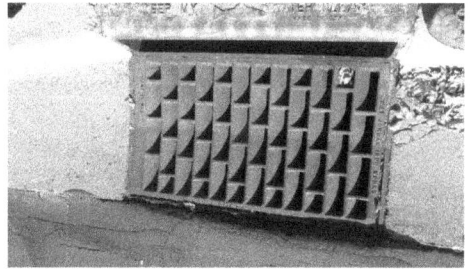

(b) Vane Grate

Figure 10.1 Grate inlet

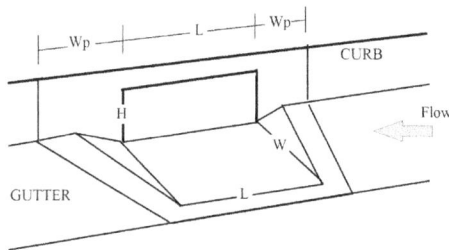

(a) Curb Opening

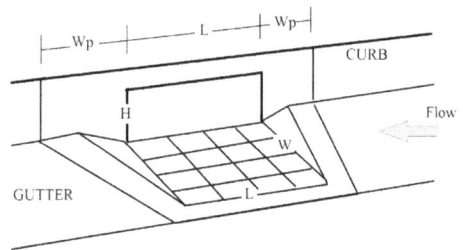

(b) Combo – Opening and Grate

Figure 10.2 Curb opening inlet

10.1.4 Slotted inlet

A slotted inlet in Fig. 10.3 is a strip of horizontal grating placed on top of an underground trench. A slotted inlet can be either placed in parallel to the curb or perpendicular to the curb, depending on stormwater spread, traffic interference, possible snow plowing, and other maintenance considerations.

10.2 INLET HYDRAULICS

From a hydraulic point of view, inlets are classified into two categories: (1) on a continuous grade, and (2) in a sump. Sumps are formed by local topography or street curbs and crowns. For instance, at a street interception, the street crowns may confine the flow to the street corner. On a continuous grade, the inlet is designed to capture 70 to 80% of the flow on the street because the significant diminishing collection capacity when more inlets are added on a sloping street. At a sump, the inlet is designed to capture either 100% of the peak flow under the design headwater depth or the available flow; whichever is less. **Fig. 10.4** illustrates how to identify the hydraulic conditions at various inlet locations. Fig. 10.5 are examples observed in the field.

(a) Slotted Inlet   (b) Trench Inlet

Figure 10.3 Slotted inlet

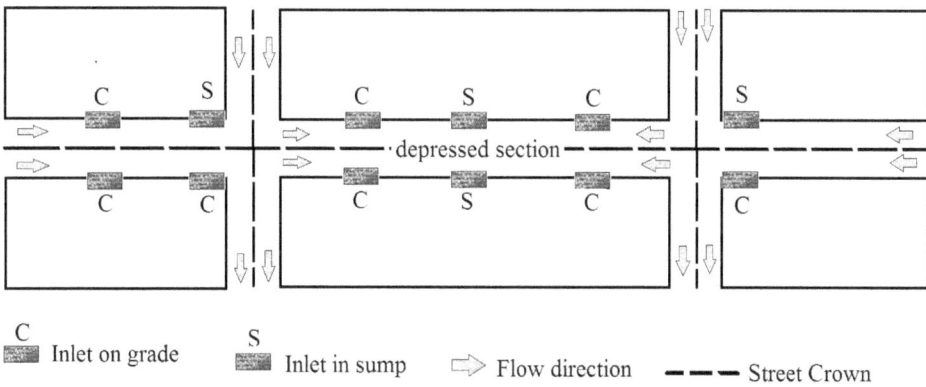

C
▨ Inlet on grade   ▨ Inlet in sump   ⇨ Flow direction   — — — Street Crown

Figure 10.4 Inlets on a grade versus inlets in a sump

(a) Inlet in Sump   (b) Inlet on Grade

Figure 10.5 Inlets on a grade and in a sump

10.3 DETERMINATION OF DESIGN DISCHARGE

On a continuous grade, an inlet is designed to capture 70 to 80% of the design flow on the street. At the design point, the design flow consists of local flow and the carryover flow from the upstream inlet(s) as:

$$Q_S = Q_B + Q_{CO} \qquad (10.1)$$

where Q_S = design flow on the street in [L³/T], Q_B = local flow in [L³/T], and Q_{CO} = carryover flow in [L³/T].

Example 10.1: The catchment in Fig. 10.6 is the one studied in Example 3.2 in Chapter 3. The flow was assumed to take the flow path from Points A to B and then from B to C. The 5-year peak discharge at Point C was predicted to be 15.3 cfs. Street BC was also evaluated in Example 4.3 to have an allowable flow capacity of 10.16 cfs. Knowing that the carryover flow at Point B is 1.3 cfs, determine the design flow on Street BC.

Solution: In this case, the peak flow produced from the catchment is greater than the allowable flow capacity on the street. The strategy is to split the flow of 15.3 cfs into two halves. The first half of 7.7 cfs will be carried through Street BC, and the second half will take Street DC. At Point C, two inlets will be installed to collect two flows from Streets BC and DC.

The design flow, Q_S, for Street BC is the sum of the two flows: Q_{CO} = 1.3 cfs bypassed from the upstream inlet of Point B and the local flow, Q_B = 7.7 cfs as:

Q_S = 7.7 + 1.3 = 9.0 < 10.16 cfs, which is the allowable flow capacity.

With a longitudinal slope of 1.5%, the design flow on Street BC is analyzed in Table 10.1 as:

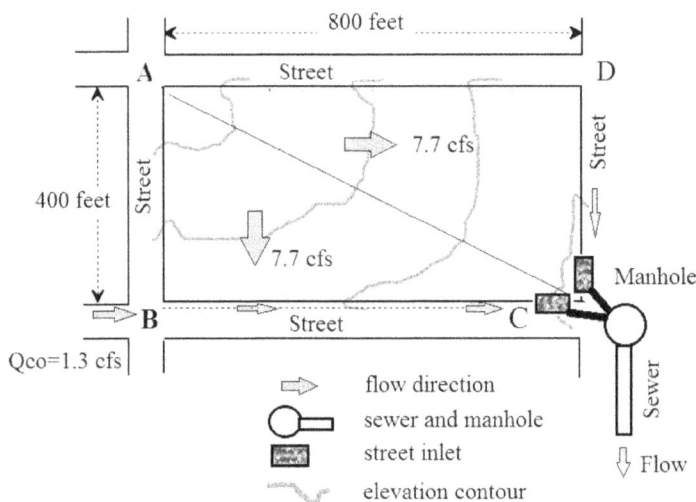

Figure 10.6 Street flow on street BC

Table 10.1 Design flow on street BC

Street Geometry

Curb Height	Hm=	6.00	inches
Gutter Width	W=	2.00	ft
Gutter Depression	Ds =	2.000	inches
Street Transverse Slope	Sx=	0.020	ft/ft
Street Longitudinal Slope	So=	0.0150	ft/ft
Manning's Roughness	n=	0.016	
Maximum Gutter Water Depth	Dm=	6.00	inch
Maximum Water Spread	Tm=	15.00	feet
Design Flow on the Street	Qo=	9.00	cfs

Water Spread Width (Guess it)

	T=	14.24	ft
Gutter Cross Slope: Sw = Sx + Ds/W =	Sw	0.1033	ft/ft
Water depth Y = T*Sx =	Y=	0.28	ft
Gutter Water Depth at Curb: D=Y+Ds =	D=	0.45	ft
Spread for Side Flow: Tx =T − W=	Tx=	12.24	ft
Spread for Gutter Flow: Ts = D/Sw =	Ts=	4.37	ft
Qws = 0.56/n* Sw^1.67*Ts^2.67*So^0.5=	Qws=	4.96	cfs
Qww = 0.56/n*Sw^1.67*(Ts−W)^2.67*So^0.5=	Qww=	0.97	cfs
Gutter Flow: Qw=Qws−Qww =	Qw=	4.00	cfs
Side Flow: Qx = 0.56/n* Sx^1.67* Tx^2.67*So^0.5=	Qx=	5.00	cfs
Calculated Total Flow: Qs = Qx + Qw =	Qs=	9.00	cfs
Flow Difference to seek the solution	dQ=	0.00	

Check on the VD product

Flow Area As =0.5*(YT+DsW)=	As=	2.19	sq ft
Flow Velocity Vs = Q/As =	Vs=	4.10	fps
The VD product = VsD =	VsD=	1.85	
Check If T<=Tm?		Yes	
Check if D<=Dm?		Yes	

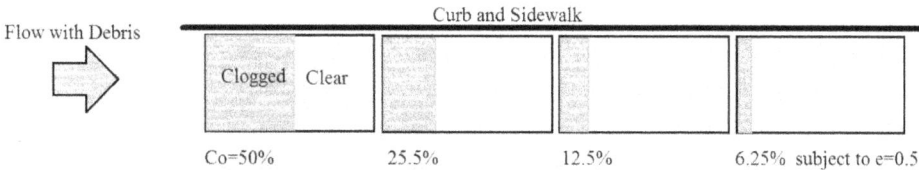

Figure 10.7 Decay of inlet clogging percentage

10.4 CLOGGING FACTOR

It is recommended that a clogging factor of 50% be applied to a single grate, and 10% be applied to a single curb-opening unit. As illustrated in Fig. 10.7, for a series of inlets, it is reasonable to assume that the clogging percentage decays as the number of inlets increases.

As a result, the clogging factor for multiple inlets in series is equal to the total clogging percentage divided by the number of inlet units as (Guo 2000c, Guo 2006):

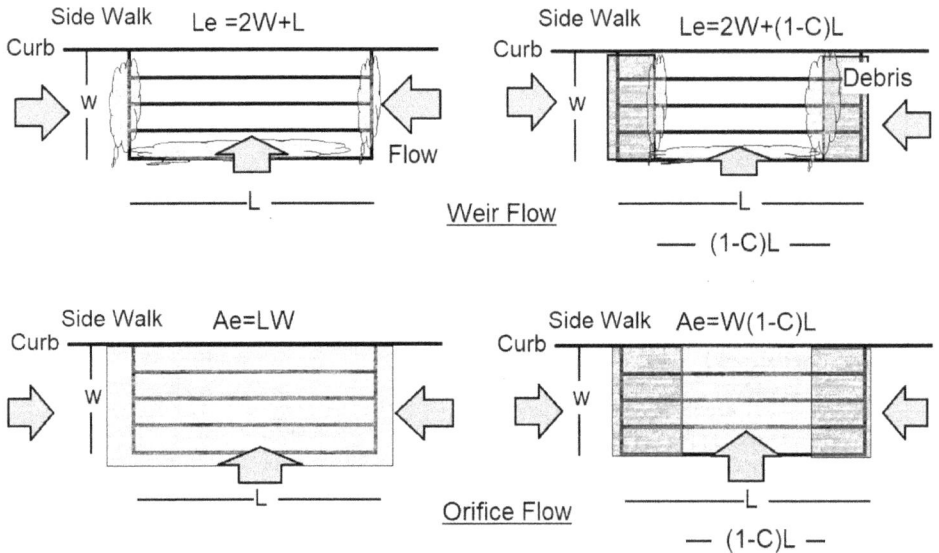

Figure 10.8 Illustration of a clogged area and clogged width

$$C_g = \frac{1}{N}\left(C_o + eC_o + e^2C_o + e^3C_o + ... + e^{N-1}C_o\right) = \frac{C_o}{N}\sum_{i=1}^{i=N}e^{i-1} \tag{10.2}$$

where C_g = multiple-unit clogging factor, $0 \le C_g \le 1$, C_o = single-unit clogging factor, e = decay ratio less than unity, and N = number of inlets. Field data suggests $e = 0.25$ to 0.5. When N becomes large, Eq. 10.2 converges to:

$$C_g = \frac{C_o}{N(1-e)} \tag{10.3}$$

The interception of an inlet on a grade in Fig. 10.8 is proportional to the inlet length, and in a sump as in Fig. 10.8 is proportional to the inlet opening area. Therefore, a clogging factor should be applied to the length of an inlet on a continuous grade as:

$$L_e = \left(1-C_g\right)L \tag{10.4}$$

where L_e = effective (unclogged) length in [L]. Similarly, a clogging factor should be applied to the opening area of an inlet in a sump as:

$$A_e = \left(1-C_g\right)A \tag{10.5}$$

where A = opening area in [L^2], and A_e = unclogged opening area in [L^2].

10.5 GRATE INLET ON A CONTINUOUS GRADE

As illustrated in Fig. 10.9, the storm water moving on a street can be divided into the *gutter flow*, which is carried by the street gutter within the gutter width, and the *side flow*, which is the water spread into the traffic lanes.

The interception of the gutter flow by a grate is determined by the length of the grate, average cross-sectional water velocity, and water splash velocity due to the interference of the grate. A regression analysis using the laboratory data reported by HEC-12 results in the following empirical formula for determining the splash-over velocity, V_o, as a function of the grate length and type. The splash velocity in Fig. 10.10 is associated with a supercritical flow which tends to jump over a grate placed on a steep slope. The empirical formula to estimate the splash velocity is:

$$V_o = \alpha + \beta L_e - \gamma L_e^2 + \eta L_e^3 \tag{10.6}$$

where V_o = water splash velocity in feet/second on the grate, L_e = effective length in feet, $\alpha, \beta, \gamma,$ and η = constants, depending on the type of grate as shown in Table 10.2.

The interception capacity of a grate is separately determined for the gutter and side flows. The interception percentage of a gutter flow, Q_w, is estimated as:

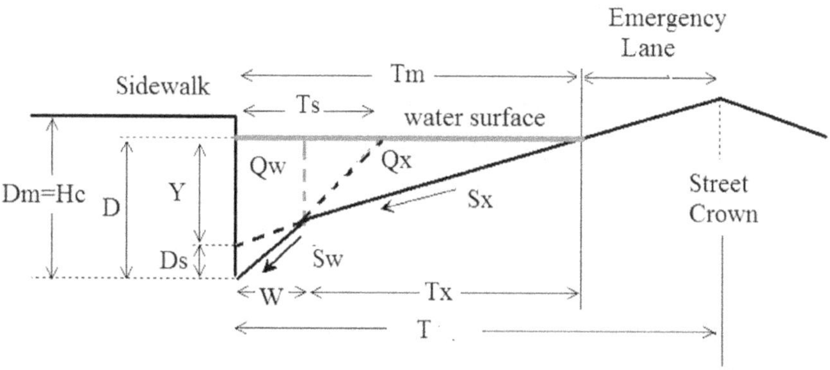

Figure 10.9 Parameters for street flow

(a) Shallow Water Splash Velocity   (b) Splash Velocity Over Grate

Figure 10.10 Splash velocity over a grate observed in the field

Table 10.2 Coefficients to determine splash velocity over inlet grates

Type of Grate	A	β	γ	η
Bar P-1-7/8	2.22	4.03	0.65	0.06
Bar P-1-1/8	1.76	3.12	0.45	0.03
Vane grate	0.30	4.85	1.31	0.15
45-degree bar	0.99	2.64	0.36	0.03
Bar P-1-7/8-4	0.74	2.44	0.27	0.02
30-degree bar	0.51	2.34	0.20	0.01
Reticuline	0.28	2.28	0.18	0.01

$$R_w = 1 - 0.09(V_s - V_o) \text{ if } V_s \leq V_o, \text{otherwise } R_w = 1 \qquad (10.7)$$

where R_w = interception ratio of gutter flow, and V_s = average cross-sectional water velocity in feet/second which can be obtained from the street runoff flow analysis. For most cases, the condition, $V_s < V_o$, prevails, or $R_w = 1.0$. The interception percentage of the side flow, Q_x, is expressed by

$$R_x = \frac{1}{\left(1 + \dfrac{K_R V^{1.8}}{S_x L_e^{2.3}}\right)} \qquad (10.8)$$

where R_x = interception ratio of side flow, $K_R = 0.15$ for feet-second units or 0.083 for meter-second units. As a result, the total interception capacity, Q_a, for a grate inlet is equal to

$$Q_a = R_w Q_w + R_x Q_x \qquad (10.9)$$

Example 10.2: Try bar grate on a grade for the location in Fig. 10.11. A bar grate unit has a width of 1.58 feet and a length of 2.96 feet. Assuming a clogging factor C = 0.5 for a single unit and a decay factor of 0.25 is applied to the clogging factor for multiple grates, determine the number of bar grates required to intercept approximately 80% of the design flow in Example 10.1.

Solution: As shown in Fig.10.11, the location for this inlet under design is in a sump. In order to place a grate on a grade at this location, we need a concrete pan across the street intersection. The carryover flow will flow through the concrete pan to the next inlet. The interception capacity of a grate in grade depends on the incoming gutter flow. Referring to Table 10.1, we have Q_w = 4 *cfs*, Q_x = 5 *cfs*, and V_s = *4.10 feet/sec*. Consider four bar grates. The total grate length is:

$$L_g = nL = 4 \times 2.96 = 11.84 \text{ ft}$$

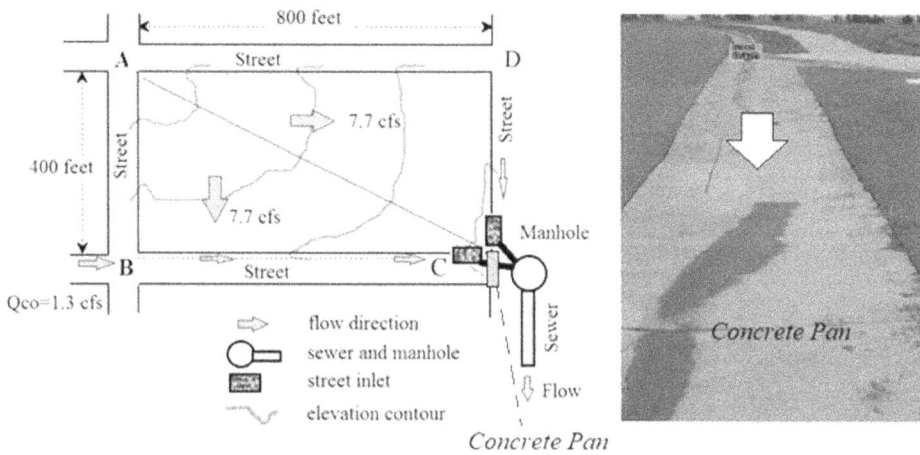

Figure 10.11 Grate inlet with a concrete pan

The clogging factor is 0.5 for a single unit. With $e = 0.25$, the clogging factor for four units is computed by Eq. 10.3 as:

$$C_g = 0.5/[4 \times (1 - 0.25)] = 0.17$$

The effective grate length free from clogging is:

$$L_e = (1 - 0.17) \times 11.84 = 9.87 \ ft$$

Assuming that the empirical coefficients developed for Bar P-1-7/8 are suitable for the bar grate in this case, we have

$$V_o = 2.22 + 4.03L_e - 0.65L_e^2 + 0.06L_e^3 = 36.3 \ feet/second > V_s \ So, R_f = 1.0$$

$$R_x = \frac{1}{1 + \dfrac{0.15 \times 4.10}{0.02 \times 9.87^{2.3}}} = 0.67$$

The interception is calculated as:

$$Q_c = 1.0 \times 4.0 + 0.67 \times 5.0 = 7.35 \ cfs$$

$$Q_{co} = 9.0 - 7.35 = 1.65 \ cfs$$

As summarized in Table 10.3, using four grates on a grade, the interception ratio for this example is: $7.35/9.0 = 81.7\%$, with carryover flow is of 1.65 cfs.

Table 10.3 Design of grate on grade

Design Discharge on the Street	Qo=	9.00	cfs
Unit Length for each Grate	Unit L=	2.96	ft
Number of Grate	N=	4.00	
Clogging Factor for Single Grate	Co=	0.500	
Clogging Factor for Grate Inlet (Decay Factor=0.25)	Clog=	0.167	
Total Length for the Grate Inlet	L=	11.84	ft
Effective Length of Grate Inlet = (1−Clog)*L=	Le=	9.87	ft
Interception Rate of Gutter Flow=1.0	Rf=	1.00	
Interception Rate of Side Flow=1/(1+0.15Vs^1.8/(SxLe^2.3))=	Rx=	0.670	<1.0
Interception Capacity=Rf*Qw+RxQx=	Qa=	7.35	cfs
Carry-over Flow = Qs−Qa =	Q−co=	1.65	cfs
Capture percentage = Qa/Qs =	C% =	81.7%	

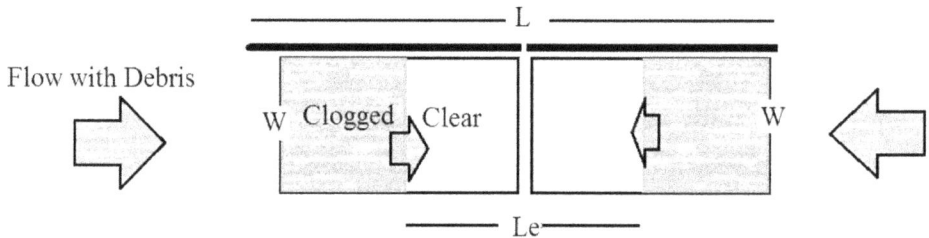

Figure 10.12 Effective wetted parameter for clogged grates

10.6 GRATE INLET IN A SUMP

At a sump, the inlet is installed with an additional inlet depression in order to accommodate the concrete thickness on top of inlet box. A grate inlet operates like a weir under a shallow depth. Its capacity is estimated as:

$$Q_w = \frac{2}{3}C_d\sqrt{2g}P_eY_s^{1.5} = C_wP_eY_s^{1.5} \tag{10.10}$$

where Q_o = interception capacity in [L³/T], C_d = discharge coefficient, such as 0.6, C_w = weir coefficient, such as 3.0 for feet-second units or 1.77 for meter-second, Y_s = water depth in [L], and P_e = effective weir length in [L]. As illustrated in Fig. 10.12, water flows may overtop the clogged areas from the sides. As a result, the effective wetted parameter for a weir flow is estimated as:

$$P_e = m\left[(1-C_g)L+2W\right] \tag{10.11}$$

where m = length opening ratio, such as 0.7, after subtracting the length of the steel bars.

When a submerged grate as in Fig. 10.13 operates like an orifice, its capacity is estimated as:

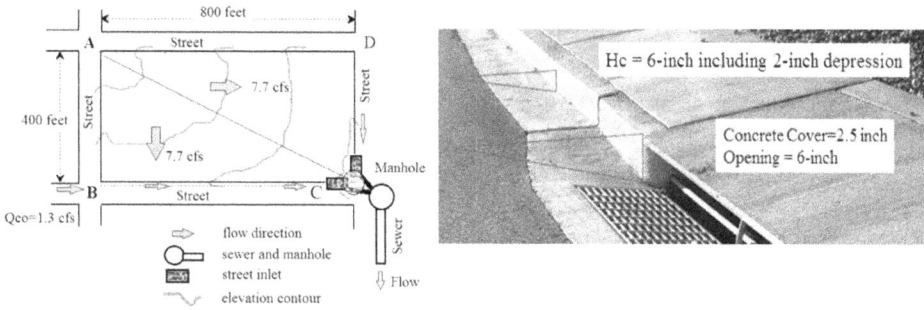

Figure 10.13 Grates in a sump

$$Q_o = C_d A_e \sqrt{2gY_s} \tag{10.12}$$

$$A_e = n(1 - C_g)WL \tag{10.13}$$

where C_d = discharge coefficient for orifice flow such as 0.60, g = gravitational acceleration in [L/T²], W = grate width in [L], L = grate length in [L], and n = area opening ratio, such as 0.5 to 0.6, on the grate after subtracting the area of the steel bars. The transition between weir flow and orifice flow is not clearly understood. In practice, for a specified water depth on a grate, the interception capacity is evaluated by both Eqs 10.10 and 10.12, and the smaller one dictates.

$$Q_a = \min(Q_w, Q_o) \tag{10.14}$$

Example 10.3: Try bar grate in sump for the location in Fig. 10.13. A bar grate in sump has a length of 2.96 ft and width of 1.58 ft. The length opening ratio $m = 0.7$ and the area opening ratio $n = 0.45$. The clogging factor is $C_o = 0.50$ for a single unit and decays with $e = 0.50$. To intercept the design flow of 9 cfs, determine the number of grates for Street BC under a water depth of 0.5 feet.

Solution: An inlet placed in a sump needs to have an additional inlet depression in order to accommodate the thickness of the concrete cover on top of the inlet box. As illustrated in Fig. 10.14, the thickness of concrete cover is 2.5 inches in this case. As a result, the available ponding depth, Y_a, at this location is:

$$Y_a = 2.5 + H_c = 8.5 \ inches = 0.71 \ ft$$

Next, we need to select the proper number of grates to have a headwater, Y_s, close to but less than Y_a. The flow condition for this case is presented in Table 10.4 using two (2) grates, $Y_s = max(0.69, 0.38) = 0.69 < Y_a$. Fig.10.14 shows the observed weir flow and orifice flow at an inlet.

Table 10.4 Design of grate in sump

Design Discharge on the Street	$Qo=$	9.00	cfs
Thickness of Concrete Cover	$d-cover=$	2.50	inches
Curb Height (6 inches for standard)	$Hm=$	6.00	inches
Unit Length for each Grate	Unit L=	2.96	ft
Unit Width of Grate Inlet	Unit W=	1.58	ft
Number of Grate	$N=$	2.00	
Clogging Factor for Single Grate	$Co=$	0.500	
Clogging Factor for Grate Inlet (Decay Factor=0.5)	$Clog=$	0.500	
Length for Grate Inlet	$L=$	5.92	ft
Unclogged Length of Grate Inlet	$Le=$	2.96	ft
Availabe Water Depth ($Ya=Hm+d$)	$Ya=$	0.71	ft
As a Weir			
Weir Coefficient	$Cw=$	3.00	
Length Opening Ratio of the Grate	$n-length=$	0.70	
Design Water Depth as a Weir Flow (Guess)	$Ys=Yweir=$	0.69	ft
Flow Capture as a Weir=$Cw[nLe+2W]Ys^1.5=$	$Qweir=$	9.00	cfs
Carry-over Flow (must be zero)	$Q-co=$	0.00	cfs
Is Ys<Ya?	Check	YES	
As an Orifice			
Orifice Coefficient	$Cd=$	0.65	
Areal Opening Ratio of the Grate	$m-area=$	0.60	
Design Water Depth as an Orifice Flow (Guess)	$Ys=Yorifice=$	0.38	ft
Flow Interception as an Orifice =$Cd\ m\ LeW(64.4*Ys)^0.5=$	$Qorifice=$	9.00	cfs
Carry-over Flow (must be zero)	$Q-co=$	0.00	cfs
Is Ys <Ya?	Check	YES	
Interception Capacity for Design=max(Yweir, Yorifice)	$Qa=$	9.00	cfs
Design Water Ponding Depth	$Ys=$	0.69	ft

(a) Orifice Flow on Grate   (b) Weir Flow around Grate

Figure 10.14 Orifice and weir flows at a grate inlet

10.7 CURB OPENING ON A GRADE

To install a curb opening inlet on a continuous grade, the required curb opening length, L_t, for a complete interception of the design storm runoff, Q_s, on the street is calculated by:

$$L_t = K_t Q^{0.42} S_0^{0.30} \left(\frac{1}{nS_e} \right)^{0.6} \tag{10.15}$$

$$S_e = S_x + S_w \frac{Q_w}{Q_s} \tag{10.16}$$

where L_t = required length for a 100% runoff interception, K_t = 0.6 for feet-second units or 0.82 for meter-second units, S_0 = street longitudinal slope in [L/L], n = Manning's roughness of 0.016, and S_e = equivalent transverse street slope in [L/L]. The curb-opening inlet should have a length less than, but close to, L_t. The interception capacity of a curb-opening inlet is calculated as:

$$Q_a = Q \left[1 - \left(1 - \frac{L_e}{L_t} \right)^{1.80} \right] \tag{10.17}$$

where Q_a = inlet capacity in [L³/T], and L_e = effective length in feet of the curb-opening inlet.

Example 10.4: A curb-opening inlet unit has a length of five feet. Considering a clogging factor $C_o = 0.10$ for a single unit and $e = 0.25$ for debris decay, determine the interception rate to apply to the three units in Fig. 10.15 for the design flow in Example 10.1. The solution is summarized in Table 10.5.

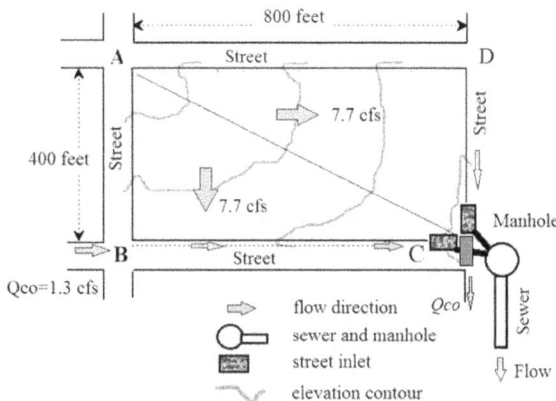

Figure 10.15 Curb-opening inlet on a grade

Table 10.5 Design of curb opening on grade

Design Discharge on the Street	Qs=	9.00	cfs
Unit Length for Curb Opening Unit	Unit L=	5.00	ft
Number of Curb Opening Unit	N=	3.00	
Clogging Factor for Single Curb Opening Unit	Co=	0.10	
Clogging Factor for Curb Opening Inlet (Decay Factor=0.25)	Clog=	0.044	
Length of the Curb Opening Inlet (must be <Lt)	L=	15.00	
Unclogged Length Le = (1-Clog)L =	Le=	14.33	ft
Gutter Flow to Design Flow Ratio	Eo=	0.444	ft
Equivalent Slope Se = Sx + Sw. Eo =	Se=	0.1033	ft/ft
Length Lt =0.60*Qo^0.42*So^0.30*[1 /(nSe)1^0.6=	Lt=	19.99	ft>L
Interception Capacity for given Length L= Qs(1-(1-L/Lo)^1.8)=	Qa=	8.07	cfs
Carryover flow = Qs-Qa =	Q-co=	0.93	cfs
Capture percentage for this inlet = Qa/Qs =	C% =	89.70%	

10.8 CURB-OPENING INLET IN A SUMP

Referring to Fig. 10.2, a curb-opening inlet in a sump operates like a weir. Its interception capacity is estimated as:

$$Q_w = \frac{2}{3}C_d\sqrt{2g}P_eY_s^{1.5} = C_wP_eY_s^{1.5} \qquad (10.26)$$

$$P_e = (1-C)(L+kW_p)+2W \qquad (10.27)$$

where P_e = effective weir length in [L] around the depressed pan in front of the curb-opening inlet, W_p = width of depressed pan in [L], and k = 1.8 to 2.0 for two sides of the pan in Fig. 10.2. When the water depth gets deeper, a curb-opening inlet operates like an orifice and can be modeled as:

$$Q_o = C_dA_e\sqrt{2g(Y_s - 0.5H)} \qquad (10.28)$$

$$A_e = (1-C)HL \qquad (10.29)$$

where H = height of curb opening in [L]. The transition of an inlet from weir flow to orifice flow is not clearly defined. In practice, both Eqs 10.26 and 10.28 should be applied to the water depth and the smaller of these should be used for the inlet interception.

Example 10.5: Try a curb-opening inlet in a sump. Considering a clogging factor of 12% and a debris decay factor of 0.25, determine the number of the curb-opening units in Fig. 10.16 for the flow in Example 10.1.

The solution is summarized in Table 10.6.

For this case, it takes one unit under a ponding depth of 0.43 foot the design flow of 9.0 cfs. Fig10.17 presents the observed curb-opening inlets in field.

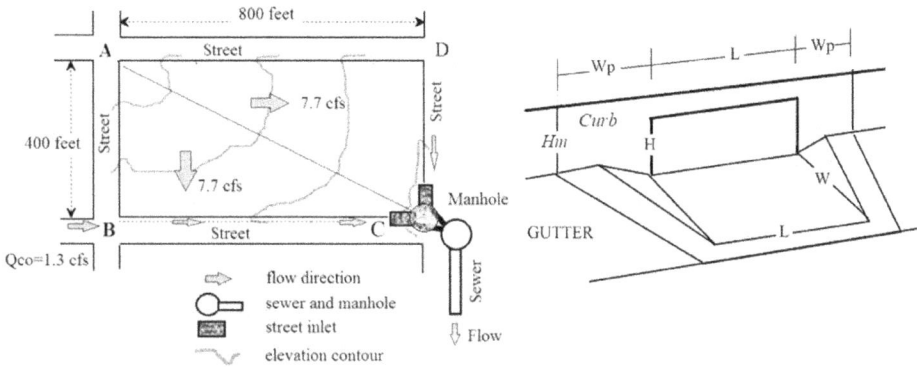

Figure 10.16 Example of a curb opening inlet in a sump

Table 10.6 Design of curb opening in sump

Design Discharge on the Street	Qs=	9.00	cfs
Thickness of Concrete Cover	d–cover=	2.50	inches
Height of the Curb Opening	Hm=	6.00	inches
Height of Curb Opening Unit Inlet	Unit H=	6.00	inches
Length of Curb Opening Inlet Unit	Unit L=	3.00	ft
Wing Width of Depression Pan	Wp=	1.00	ft
Number of Curb-Opening Inlet Unit	N=	1.00	
Clogging Factor for Single Curb-Opening	Co=	0.100	
Clogging Factor for Curb-Opening Inlet (Decay Factor=0.5)	Clog=	0.100	
Availabe Water Depth (Ya=Hm+d)	Ya=	0.71	ft
As a Weir			
Weir Coefficient	Cw=	3.00	
Design Water Depth as a Weir Flow (Guess)	Ys=Yweir=	0.43	ft
Length for Curb-Opening Inlet	L=	3.00	ft
Capacity as a Weir =Cw[(1–Clog)(L+2Wp)+2W)]Ys^1.5=	Qweir=	9.00	cfs
Carry-over Flow (must be zero)	Q–co=	0.00	cfs
Is Ys<Ya?	Check	YES	
As an Orifice			
Orifice Coefficient	Cd=	0.65	ft
Design Water Depth as an Orifice Flow (Guess)	Ys=Yorifice=	0.26	ft
Length for Curb-Opening Inlet	L=	3.00	ft
Capacity as an Orifice=Co(1–Clog)HL(64.4(Ys–0.5H))^0.5=	Qorifice=	9.00	cfs
Carry-over Flow (must be zero)	Q–co=	0.00	cfs
Is Ys<Ya?	Check	YES	
Interception Capacity for Design=max(Yweir, Yorifice)	Qa=	9.00	cfs
Design Water Ponding Depth	Ys=	0.43	ft

(a) Shallow water at Sump   (b) Deep Water at Sump

Figure 10.17 Flow at a sump curb opening inlet

10.9 SLOTTED INLET

Hydraulically, a slotted inlet is similar to a curb-opening inlet. As a result, design formulas developed for curb-opening inlets are also applicable to slotted drain inlets.

10.10 COMBINATION INLET

The combination inlet in Fig. 10.18 consists of a curb opening and a grate inlet. During a storm event, if one is clogged, as shown in Fig. 10.18, the other can still function. Empirical formulas for sizing a grate or a curb opening were developed under the conditions that the inlet operates independently. The interference between the grate and the curb opening in a combination inlet is not yet fully understood yet. The assumption of independent operations implies that the curb-opening inlet is placed immediately downstream of the grate inlet. In other words, the curb-opening inlet receives the carryover flow from the grate inlet. If there is 100% interception by the grate, the curb opening inlet will intercept no flow at all.

In theory, the capacity of a combination inlet is the sum of the intercepted discharges by the grate and the curb opening. To be conservative, the capacity of a combination inlet is the greater of the grate and the curb-opening capacities (Guo, 1997)(Guo, McKenzie and Mommandi 2008).

10.11 CARRYOVER FLOW

When the street flow on a grade is not completely collected by a street inlet, the residual flow is carried to the downstream inlet. The carryover flow is defined as:

$$Q_c = Q_s - Q_a \tag{10.30}$$

where Q_c = carryover flow. At a sump, the flow is supposed to be completely intercepted, except that a crossing concrete pan in Fig. 10.19 is available to bypass the excess water flow.

Clogged Grate   Clogged Curb-opening

Figure 10.18 Clogged combination inlet

(a) Carryover Flow   (b) Crossing Pan

Figure 10.19 Crossing concrete

The carryover flow must be treated as a part of the design flow for the immediately downstream inlet. As mentioned above, the carryover flow can be directly added to the local flow.

10.12 CONCLUSIONS

On a grade, the flow interception at an inlet is sensitive to the street longitudinal slope. The diminishing return of adding more inlets discourages a high flow interception rate. Therefore it is important in practice that an inlet on a grade be designed to capture 70 to 80% of the flow on the street. Obviously, a sump inlet has to be designed to have a complete interception within the gutter width. An undersized inlet at a sump will inundate the adjacent sidewalks. Flow interception at an inlet is dominated by either weir or orifice flow, depending on the flow depth. On top of the hydraulic complexity, inlet flows are further complicated by debris clogging. The design procedures discussed in this chapter are semi-theoretical. As expected, more research efforts are required to improve the understanding of the mixing flow between orifice and weir flows.

HOMEWORK

Q10-1 a. Street inlets are installed at the street corner as shown in Fig. Q10.1. The design information is given as follows) The 5-yr design rainfall intensity, I (inch/hr), is given as:

$$I(\text{inch/hr}) = \frac{28.5 P_1}{(10+T_d)^{0.789}} \text{ where } P_1 = 1.35 \text{ inch and } T_d = \text{duration in minutes.}$$

b. The overland flow time, T_o, should be calculated by the airport formula for a flow length ≤300 feet. The swale and gutter flow velocities, V_f, are estimated by the SCS upland method with a conveyance coefficient of 20.

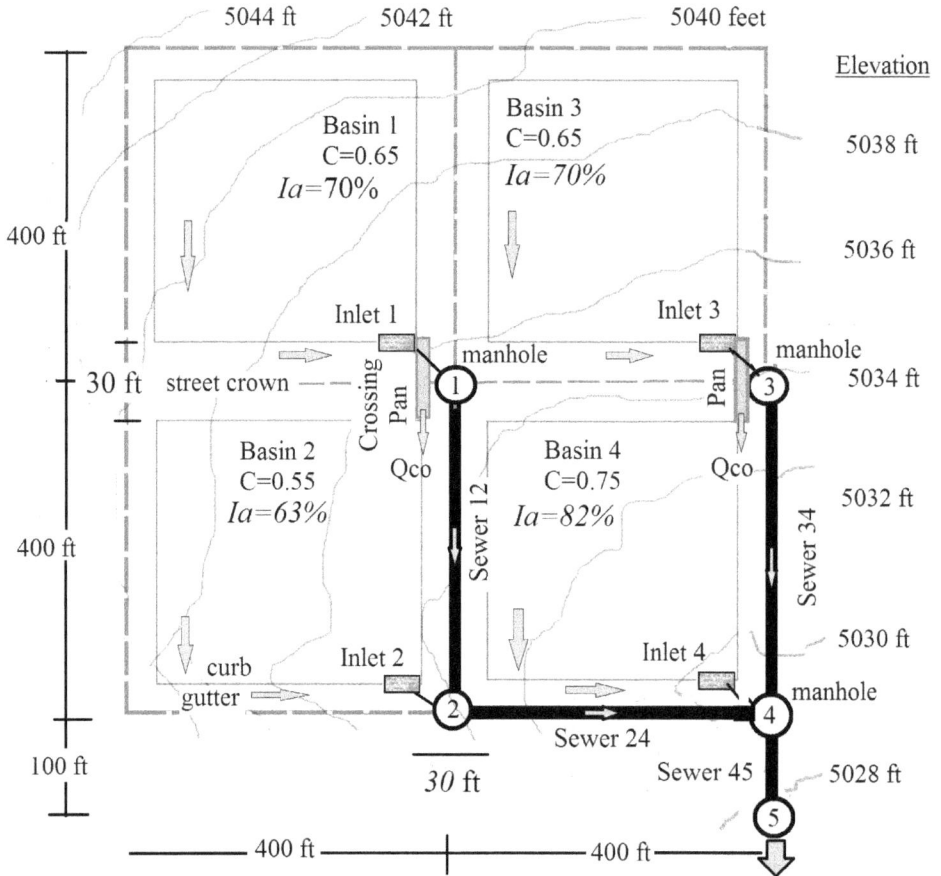

Figure Q10.1 Design of street inlets

Your tasks are:

1. Inlet 1 is placed on a grade because it has a downstream crossing pan. Size Inlet 1 using grate inlets.
2. Inlet 3 is placed on a grade because it has a downstream pan. Size Inlet 3 using curb opening inlets.
3. Inlet 2 is in sump. Size Inlet 2 using combination inlets.
4. Inlet 4 is in sump. Size Inlet 4 using combination inlets.

Solution for Inlet 1

Step 1: Design flow

		Overland Flow			Swale				Flow				Gutter Flow Existing
5-yr Coefficient C5	Slope % So	Length ft Lo	Time min To	Slope % S2	Length ft L2	SCS K fps K2	Time min T2	Slope % S3	Length ft L3	SCS K fps K3	Time min T3	Tc min Tc-comp
0.65	1.00	300.00	14.07	1.00	100.00	20.00	0.83	1.00	400.00	20.00	3.33	18.24

Drainage Area acre A	Imp Ratio Ia	Design Runoff Coef G-design	Total Flow Length ft L	Average Slope % S	Initial Time min T*	Convey Factor fps K*	Flow Velocity fps V*	Flow Time min Tf	Future Tc min Tc-Reg	Design Storm Duration min Td	Rainfall Intensity 5-yr in/ hr I	Peak Flow cfs Qp
3.67	0.70	0.65	800.00	1.00	7.50	28.80	2.88	4.63	12.13	12.13	3.34	7.98

Step 2: Design flow on the street

A. **Street Geometry**

Curb Height	H=	6.00	*inches*
Gutter Width	W=	2.00	*ft*
Gutter Depression	Ds =	2.00	*inches*
Street Transverse Slope	Sx=	0.0200	*ft/ft*
Street Longitudinal Slope	So=	0.0100	*ft/ft*
Manning's Roughness	n=	0.0160	
Design Discharge on the Street	Qs=	7.98	*cfs*

B. **Street Conveynace Capacity**

Guess Water Spread Width for Design Qs	T=	14.75	*ft*	**(Guess)**
Water depth Y = T*Sx =	Y=	0.30	*ft*	
Water Depth at Curb: D=Y+Ds =	D=	0.46	*ft*	
Spread for Side Flow: Tx =T – W=	Tx=	12.75	*ft*	
Gutter Cross Slope: Sw = Sx + Ds/W =	Sw	0.103	*ft/ft*	
Spread for Gutter Flow: Ts = D/Sw =	Ts=	4.47	*ft*	
Gutter Flow	Qw=	3.42	*cfs*	
Side Flow	Qx=	4.56	*cfs*	
Total Flow from the Guessed T	Qs=	7.98	*cfs*	
Check on Design Flow ΔQ=0	ΔQ=	0.00	*cfs*	**(Check)**
Flow Area As = 0.5YT+0.5WDs =	As=	2.34	*sqft*	
Flow Velocity Vs = Q/As =	Vs=	3.41	*fps*	
The VD product VsD =	VD=	1.57	*ft^2/s*	

Step 3: Size Inlet 1 using grates

Design Discharge on the Street	Qs=	7.98	*cfs*
Length of Unit Grate	L=	2.96	*ft*
Clogging Factor for Single Grate	Co=	0.50	
Number of Unit Grate	N=	4.00	
Clogging Factor for Grate Inlet	Cg=	0.17	
Effective Length of Grate Inlet = (1–Clog)*L=	Le =	9.87	*ft*
Interception Rate of Gutter Flow	Rf=	1.00	
Design Flow Velocity	Vs=	3.41	*cfs*
Interception Rate of Side Flow=1/(1+0.15Vs^1.8/(SxLe^2.3))=	Rx=	0.74	
Interception Capacity=Rf*Qw+RxQx=	Qa =	6.79	*cfs*
Carry-over Flow = Qs–Qa =	Q–co=	1.19	*cfs*
Capture percentage = Qa/Qs=	C%=	85.1%	

Repeat the above procedure for Inlets 2, 3, and 4.

REFERENCES

Guo, J.C.Y. (1997). *Street Hydraulics and Inlet Sizing*, Water Resources Publications, LLC, Littleton, CO.

Guo, J.C.Y. (2000a). "Street Storm Water Conveyance Capacity", *ASCE Journal of Irrigation and Drainage Engineering*, vol. 126, no. 2, March/April.

Guo, J.C.Y. (2000b). "Street Storm Water Storage Capacity", *Water Environmental Research Journal*, vol. 72, no. 5, September/October.

Guo, J.C.Y. (2000c). "Design of Grate Inlets with Clogging Factor", in *Advances in Environmental Research*, vol. 4, Elsevier Science, Dublin.

Guo, J.C.Y. (2006). "Decay-Based Clogging Factor for Curb Inlet Design", *ASCE Journal of Hydraulic Engineering*, vol. 132, no. 11, November.

Guo, J.C.Y. and McKenzie, K., and Mommandi, A. (2008). "Sump Inlet Hydraulics", *ASCE Journal of Hydraulic Engineering*, vol. 135, no. 1, November.

HEC12 (1984). *Drainage of Urban Highway Drainage*, Hydraulic Engineering Circular no. 12, US Department of Transportation, Federal Highways Administration.

HEC22 (2010). *Urban Drainage Manual*, Hydraulic Engineering Circular no. 22, US Department of Transportation, Federal Highways Administration.

UDFCD Manual (2010). *Urban Storm Water Design Criteria Manual*, vols 1 and 2, Urban Drainage and Flood Control District, Denver, CO.

Chapter 11

Roadway storage basin

11.1 STORMWATER DETENTION VOLUME

Along a straight street, the hydraulic conveyance capacity is determined by Manning's formula (Guo 2000a). At a street intersection as shown in Fig. 11.1, the two street crowns form a depressed pool at the street corner. At a sump, the water depth is firstly determined by the weir or orifice formula for the design flow, but the water spread at a street intersection is solely dictated by the stored water volume and the local geometry at the sump. In this chapter, a *volume-based method* is derived to predict the water spread at a sump based on the volume balance between (1) the *stormwater detention volume*, and (2) the *storage capacity* around the sump inlet (Guo 2000b).

In an urban area, streets and roadways often cross waterways, natural streams, and irrigation canals. Stormwater detention is a common policy to avoid negative impacts on the highway water environment. As illustrated in Fig. 11.2, an *in-stream detention basin* is placed within the right-of-way along the roadside ditches or/and within the widened floodplain. An *off-stream detention basin* is used to divert excess stormwater to adjacent open space outside of the floodplain, such as a lowland park or sports field.

Highway drainage designs are often related to small tributary areas (≤100 to 150 acres). The assumption of the Rational method using a uniform rainfall distribution is acceptable for runoff volume predictions. As a result, the required storage volume for detention basin design can be estimated directly using the Rational method. Since the Rational method is mainly applied to flow prediction, it is therefore necessary to modify the Rational method for volume predictions.

11.1.1 In-stream detention volume

A trapezoidal hydrograph is a reasonable approximation when using the Rational method. Such a hydrograph has not only the same peak flow as predicted by the Rational method, but also the same runoff volume as the excess rainfall depth. As shown in Fig. 11.3, the rainfall hyetograph is uniform for the selected rainfall duration, and the corresponding runoff hydrograph has three segments: (1) a linear rising limb over the time of concentration, T_c, of the tributary watershed, (2) the peaking portion as a plateau from the time of concentration to the end of the rainfall event, T_d, and (3) the linear recession over the time of concentration. For the specified rainfall duration in Fig. 11.3, the excess rainfall volume, V_i, is calculated as:

$$V_i = CAI_d T_d \tag{11.1}$$

DOI: 10.1201/9781003284239-11

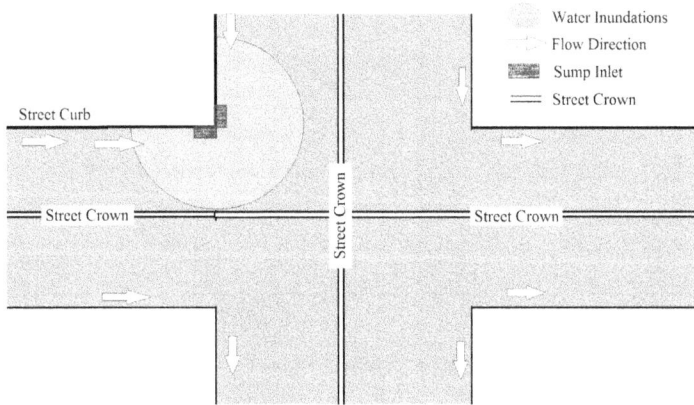

Figure 11.1 Water inundations at a sump at a street intersection

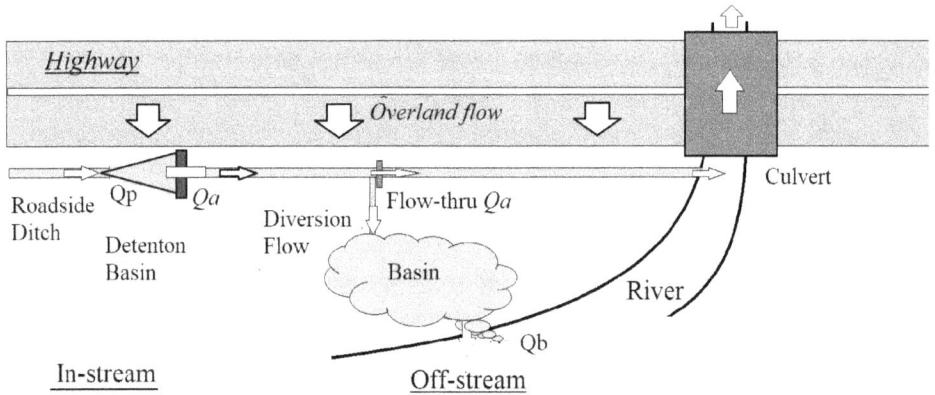

Figure 11.2 In-stream and off-stream detention basins

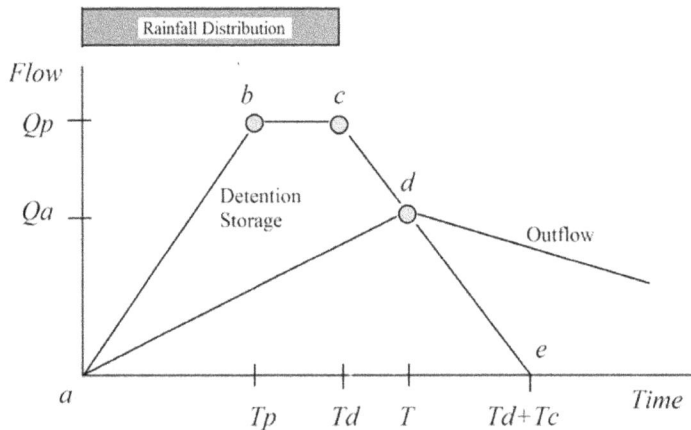

Figure 11.3 Detention volume by volume method

where V_i = excess (or effective) rainfall volume or runoff volume in [L³], C = runoff coefficient, A = tributary area in [L²], I_d = rainfall intensity in [L/T], and T_d = rainfall duration in [T]. Care must be taken when using the variables with different units in Eq. 11.1. Make sure all variables in Eq. 11.1 are converted into feet-second or meter-second for further calculations.

According to the local design criteria, the allowable release, Q_a, is set to be the pre-development peak flow. The detention basin should be sized to provide an adequate storage volume to reduce the post-development flow to the allowable. As shown in Fig. 11.3, the storage volume is the area $abcda$, which is the difference between the excess rainfall volume and the outflow volume, $adea$, as:

$$V_d = V_i - V_o \tag{11.2}$$

where V_d = detention volume in [L³] and V_o = outflow volume in [L³]. The outflow volume is calculated as a triangular volume as:

$$V_o = \frac{1}{2}Q_a\left(T_d + T_c\right) \tag{11.3}$$

For convenience, the outflow volume is calculated by an average outflow over the rainfall duration as (FAA in 1970, US Army and Air Force in 1977):

$$V_o = QT_d \tag{11.4}$$

The allowable peak flow is varied with respect to the rising hydrograph. To modify Eq. 11.4 with a variable outflow, a linear increasing variable is suggested (Guo 1999) as:

$$Q = mQ_a \tag{11.5}$$

Where m = adjustment factor. Aided by Eqs 11.3, 11.4, and 11.5, the value of m is found to be:

$$m = \frac{1}{2}\left(1+\frac{T_c}{T_d}\right) \quad \text{for } T_d \geq T_c \tag{11.6}$$

As implied in Eq. 11.6, the value of m decays from 1 to 0.5 as the rainfall duration increases. The detention volume in Eq. 11.3 varies with respect to the rainfall duration. The key variable for the solution is to find the design rainfall duration that maximizes the volume difference between the inflow and outflow volumes. The storage volume for the detention basin is determined as:

$$V_d = \max\left(V_i - V_o\right) = \max\left(CAI_dT_d - mQ_aT_d\right) \quad \text{for } T_d \geq T_c \tag{11.7}$$

Eq. 11.7 should be applied to a range of rainfall durations. As suggested in Eq. 11.6, start from $T_d = T_c$, and then use a small increment for storm duration to compute the inflow and outflow volumes until the storage volume is maximized (Guo 1999).

Example 11.1: As illustrated in Fig. 11.4, the highway catchment has $A = 8.25$ ac, $C = 0.75$, and $T_c = 15$ min. The allowable release rate is $Q_a = 2.89$ cfs, defined by the downstream ditch's capacity. The design rainfall intensity–duration–frequency (IDF) curve is given as:

$$I(\text{inch/hr}) = \frac{38.5}{(10 + T_d)^{0.789}}$$
(11.8)

where T_d = rainfall duration in minutes. Determine the detention storage volume.

Solution: (1) Try $T_d = 90$ min

$$I_d = \frac{38.5}{(10 + 90)^{0.789}} = 1.02 \text{ inch/hr}$$

$$V_i = CI_d A T_d = 0.75 \times \frac{1.02}{12}(43560 \times 7.5) \times \frac{90}{60} = 34252.5 \text{ cubic ft}$$

(2) The outflow runoff volume is calculated as:

$$V_o = mQ_a T_d = \frac{1}{2}\left(1 + \frac{15}{90}\right) \times 2.89 \times (90 \times 60) = 9095.6 \text{ cubic ft}$$

(3) The stormwater storage volume, V_d, for the 90-minute rain storm is the volume difference as:

$$V_d = V_i - V_o = 34252.5 - 9095.6 = 25156.8 \text{ cubic ft}$$

(a) Roadway Detention Basin

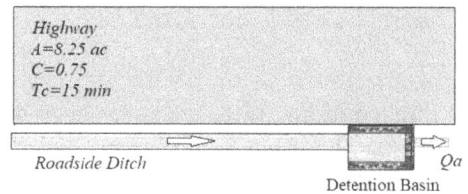

Highway
A=8.25 ac
C=0.75
Tc=15 min

Roadside Ditch

Detention Basin

Qa

(b) Case for Design

Figure 11.4 Highway detention basin

Table 11.1 Detention volume by volumetric method

			10-yr Event		
Duration minutes	Rainfall Intensity inch/hr	Inflow Volume cubic ft	Adjustment Factor m	Outflow Volume cubic ft	Storage Volume cubic ft
90.00	1.02	34252.46	0.58	9095.63	25156.83
100.00	0.94	35301.27	0.58	9961.88	25339.40
110.00	0.88	36255.00	0.57	10828.13	25426.87
120.00	0.83	37130.36	0.56	11694.38	25435.98
130.00	0.78	37940.02	0.56	12560.63	25379.39
140.00	0.74	38693.79	0.55	13426.88	25266.91
150.00	0.70	39399.42	0.55	14293.13	25106.29
				Detention Vol=	25435.98

Repeating this process, as shown in Table 11.1, the maximal storage volume for this highway detention basin is 25,436 cubic ft for an event with duration of 120 minutes.

This case is a typical example for highway stormwater detention basin designs. It implies that the critical storm event for detention volume sizing is much longer than the time of concentration used for peak-flow predictions.

11.1.2 Off-stream detention volume

When the highway drainage system is overloaded, the excess stormwater may be diverted into an off-stream detention basin. As illustrated in Fig. 11.5, the down-stream inlet can collect up to Q_1, which is the straight-through capacity in the roadside ditch. In practice, a diversion system is installed along the roadside ditch. This system includes an overtopping weir, which is placed on top of the channel bank, and a downstream culvert, which is designed to pass up to Q_1. The head-water at the entrance of the culvert serves as the backwater profile to stabilize the overtopping weir flow. The detention basin should be designed to release its peak flow, Q_2. The sum of Q_1 and Q_2 must not exceed the total allowable release, Q_a, based on the pre-development conditions. Because the base flow of Q_1 is chosen as the straight-through capacity, the flow diversion begins at the time determined by the pre-selected flow rate, Q_1. The outflow volume, the area of _abefg_ in Fig. 11.5, is the sum of the triangle _befb_ and the trapezoid _abfga_, which can be calculated as (Guo 2012):

$$V_o = \frac{Q_2}{2}(T_d + T_c - 2T_1) + \frac{Q_1}{2}\left[(T_d + T_c - 2T_1) + (T_d + T_c)\right] \qquad (11.9)$$

where V_o = outflow volume in [L³], and T_1 = time to begin flow diversion in [T].

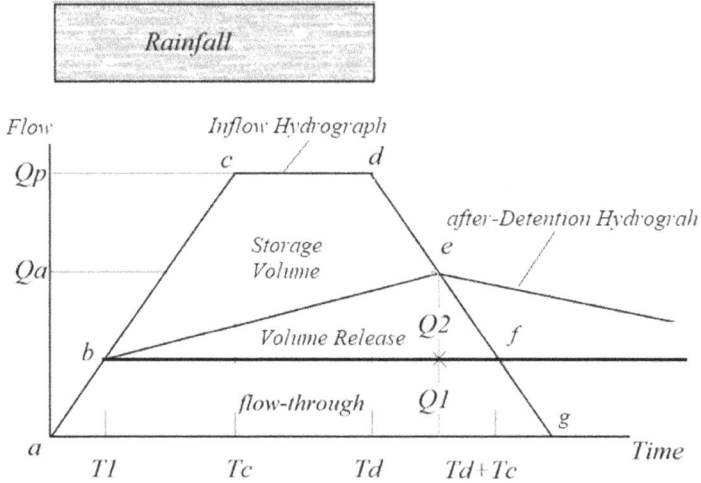

Figure 11.5 Off-stream detention volume

On the linear rising limb, the peak inflow occurs at the time of concentration, T_c, and the diversion flow is triggered at:

$$T_1 = \frac{Q_1}{Q_p} T_c \qquad (11.10)$$

where Q_1 = straight-thru capacity in [L^3/T] and Q_P = peak inflow in [L^3/T] at T_c. The peak inflow is calculated as:

$$Q_p = CIA \qquad (11.11)$$

where C = runoff coefficient and A = watershed tributary area in [L^2]. For mathematical convenience, the outflow volume, V_o in [L^3], is expressed by the average release over the rainfall duration, T_d, as:

$$V_o = mQ_a T_d \qquad (11.12)$$

where m = volume adjustment factor. Equating Eq. 11.9 to Eq. 11.12 yields:

$$m = \frac{1}{2}\left[1 + \frac{(T_c - 2T_1)}{T_d}\right] + \frac{Q_1}{Q_2}\left[1 + \frac{(T_c - T_1)}{T_d}\right] \qquad \text{for } T_d \geq T_c \qquad (11.13)$$

For an in-stream detention basin, $Q_1 = 0$ and $T_1 = 0$. As a result, Eq. 11.13 is reduced to Eq. 11.6. When using Eqs 11.10 to 11.13 to calculate water volumes, make sure that all variables are converted to the units of feet-second or meter-second.

Example 11.2: The roadside ditch in Fig. 11.6 collects stormwater generated from a tributary area of 62 acres located in Denver, Colorado. The runoff coefficient for the tributary area is $C = 0.68$. The time of concentration of the tributary area is $T_c = 20$ minutes. The total allowable stormwater release is: $Q_a = 62$ cfs, which is divided into the straight-though capacity, $Q_1 = 15$ cfs, and the allowable peak release: $Q_2 = 47$ cfs from the off-stream basin. Determine the 100-yr detention volume for this off-stream detention basin.

Solution:
For this case, the allowable release from the basin is $Q_2 = 47$ cfs ($62 - 15 = 47$). Try $T_d = 50$ minutes using Denver's rainfall IDF formula. The calculations are summarized as follows:

(1) *Inflow volume*

$$I\left(\frac{inch}{hr}\right) = \frac{74.1}{(10+T_d)^{0.789}} = \frac{74.1}{(10+50)^{0.789}} = 2.97 \ \ inch/hr$$

$$Q_p = CIA = 0.68 \times 2.97 \times 62 = 125.2 \ cfs$$

$$V_i = CIAT_d = 0.68 \times 2.97 \times 62 \times 50 = 8.68 \ \ acre\text{-}ft$$

(2) *Outflow volume*

$$T_l = \frac{Q_l}{Q_p} T_c = \frac{15}{125.2} \times 20 = 2.40 \, minutes$$

$$m = \frac{1}{2}\left[1 + \frac{20 - 2 \times 2.4}{50}\right] + \frac{15}{47}\left(1 + \frac{20 - 2.4}{50}\right) = 1.1$$

$$V_o = mQ_aT_d = 1.1 \times 47 \times 50 \times 60 / 43560 = 3.51 \ \ acre\text{-}ft$$

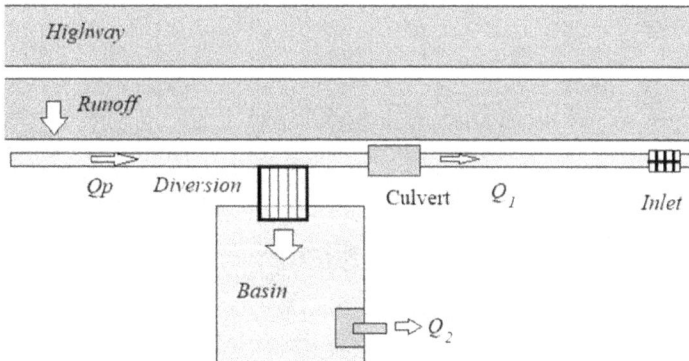

Figure 11.6 Off-stream detention in a highway system

Table 11.2 Example for off-stream detention volume

Duration minutes	Rainfall Intensity inch/hr	Inflow Volume acre-ft	Peak Runoff cfs	Diversion Time T_I minutes	Factor M	Outflow Volume acre-ft	Storage Volume acre-ft
40.00	3.42	8.02	144.32	2.08	1.16	3.00	5.01
50.00	2.97	8.68	125.05	2.40	1.08	3.51	5.18
60.00	**2.63**	**9.23**	**110.78**	**2.71**	**1.03**	**4.01**	**5.22**
70.00	2.37	9.70	99.75	3.01	1.00	4.52	5.18
80.00	2.16	10.10	90.93	3.30	0.97	5.02	5.08

Note: 1 inch = 25.4 mm, 1 ft = 0.305 m, 1 acre = 0.4 hectare.

(3) *Stormwater detention volume*, V_d, *for the 50-minute rainstorm:*

$$V_d = 8.68 - 3.15 = 5.18\,acre\text{-}ft$$

Repeating this process for the range of rainfall durations from 40 to 80 minutes, Table 11.2 summarizes the variation of detention storage volumes. The maximum storage volume is identified to be 5.22 acre-ft with a storm duration of 60.0 minutes, in comparison to a time of concentration of 20 minutes in this case.

11.2 STORAGE VOLUME AT A STREET SUMP INLET

In an urban area, the distance between two adjacent street inlets is approximately 300 to 400 feet. The catchment area for a sump inlet is approximately one to three acres. To predict the peak runoff from a small urban catchment, the Rational method states:

$$Q_d = KCI_d A \tag{11.14}$$

The rainfall intensity in Eq. 11.14 can be described as:

$$I_d = \frac{\alpha}{(T_d + \beta)^\eta} \tag{11.15}$$

where Q_d = peak runoff rate in cfs (cms), K = unit conversion factor, equal to 1 for English units and 1/360 for SI units, C = runoff coefficient, A = watershed area in acres (hectares), I_d = rainfall intensity in inch/hr or mm/hr, T_d = rainfall duration in minutes, and α, β, and η = constants of the intensity–duration–frequency (IDF) formula.

11.2.1 Storage volume at sump inlet

The basic concept of the volume-based method is to find the maximum volume difference between the inflow and outflow volumes under a series of storm events with different duration (Department of the Army and the Air Force in 1977, Guo in 1999). For specified rainfall duration, the inflow runoff volume is determined as:

$$V_i = 60KCAI_dT_d \qquad (11.16)$$

where V_i = inflow volume in $[L^3]$. The factor of 60 is to convert the duration, T_d, from minutes into seconds. The outflow volume can be estimated by the sump inlet capacity as:

$$V_o = 60QT_d \qquad (11.17)$$

where V_o = outflow volume in $[L^3]$ and Q = sump inlet capacity in $[L^3/T]$ designed to pass a 2- or 5-year peak discharge. Aided by Eqs 11.14 to 11.17, the stormwater detention volume, V_d, is obtained as:

$$V_d = 60\left[KCA\frac{\alpha}{(T_d + \beta)^\eta}T_d - QT_d \right] \qquad (11.18)$$

To maximize the storage volume, the first derivative of Eq. 11.18 with respect to rainfall duration is set to be zero. It yields:

$$\frac{dV_d}{dT_d} = \left[\frac{-\eta}{(T_d + \beta)^{\eta+1}} + \frac{1}{(T_d + \beta)^\eta} - \frac{Q}{K\alpha CA} \right]_{T_d = T_m} = 0 \qquad (11.19)$$

The solution for Eq. 11.19 is:

$$T_m = \frac{1}{\eta}\left[(T_m + \beta) - \frac{Q}{K\alpha CA}(T_m + \beta)^{\eta+1}\right] \qquad (11.20)$$

where T_m = design rainfall duration in minutes. Eq. 11.20 can be solved for T_m by trial-and-error. Substituting $T_d = T_m$ into Eq. 11.20, the design stormwater detention storage volume can be maximized. Of course, such a maximal solution can also be achieved by a finite difference approach. For instance, use a small increment of rainfall duration to test if Eq. 11.18 is maximized.

11.2.2 Sump street storage capacity

As illustrated in Fig. 11.7, the storage volume around a sump inlet consists of two portions: volumes below and above the curb height. The depression storage volume at a street corner is similar to an inverted conic volume with the street transverse slope as the conic side slope. The truncated conic volume between two adjacent depths is calculated as:

$$V_h = \frac{1}{3}h_1 A_1 + \frac{(h - h_1)}{3}\left(A + A_1 + \sqrt{AA_1}\right) \tag{11.21}$$

where h = water depth in [L], A = water surface area in [L^2], V_h = conic volume in [L^3] at depth h. The subscript 1 represents the variables at depth, h_1. When $h_1 = 0$, i.e. on the ground, its area, $A_1 = 0$. Eq. 11.21 is reduced to:

$$V_h = \frac{1}{3}hA \tag{11.22}$$

As illustrated in Fig. 11.7, the storage volume around a curb is sum of two portions: (a) the *volume below the curb height*, H_c, and (b) the *volume above the curb height*.

Volume for $h < H_c$

As illustrated in Fig. 11.8, when the depth, h, is less than the curb height, the water surface area can be approximated as a fraction of a circle, depending on the configuration of the street section. For instance: $k = 1/4$ at a 90-degree depressed street section, $k = 1/2$ in Fig. 11.9a for a straight depressed street section, $k = 3/4$ in Fig. 11.9b for a 270-degree intersection area of two sloping streets, and $k = 1$ under the condition of no curb. With a specified value for k, the water surface area at a depth, $h < H_c$ is:

$$A_h = k\pi R_h^2 \quad \text{for } h < H_c \tag{11.23}$$

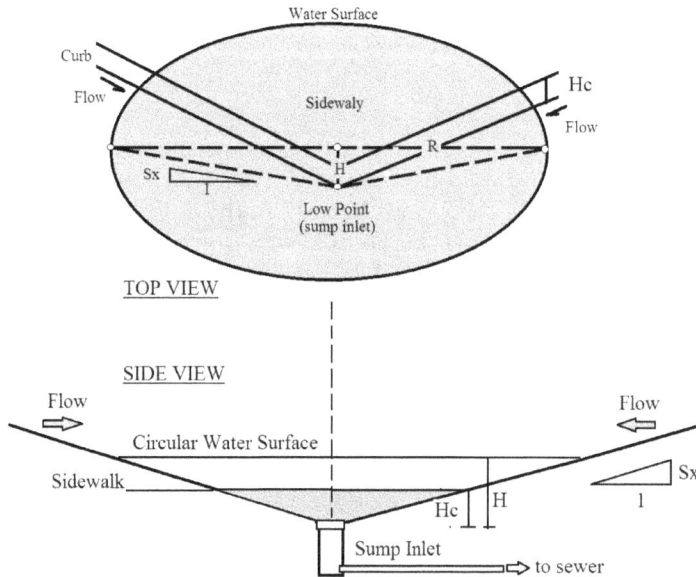

Figure 11.7 Illustration of street depression storage volume

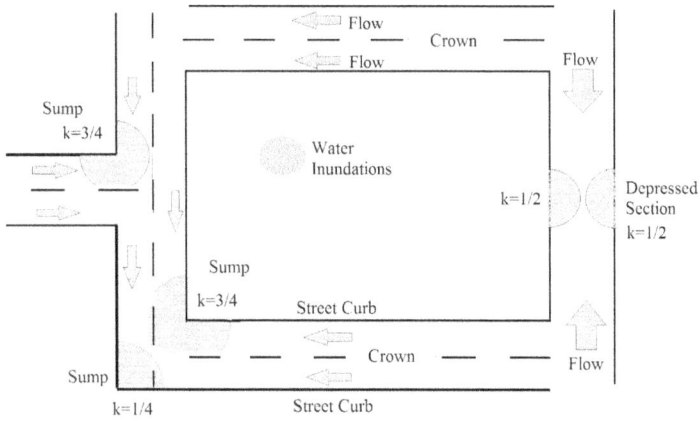

Figure 11.8 Storage parameters at sumps and depressions

(a) Ponding water with K=1/2   (b) K=3/4 for the pool water

Figure 11.9 Storage parameters at sumps and depressions

where k = a fraction of a circle and R_h = radius in [L] of the circular water area at a sump.

The radius of the water surface is approximated by the water depth, h, and the street transverse slope, S_x, as:

$$R_h = \frac{h}{S_x} \tag{11.24}$$

Substituting Eqs 11.23 and 11.24 into Eqs 11.22 yields:

$$V_{h \le Hc} = \frac{k \, \pi h^3}{3 \, S_x^2} \quad \text{for } h \le H_c \tag{11.25}$$

Volume for h > H_c

Above the curb height, the additional volume between h and H_c is estimated by Eq. 11.21 as:

$$V_{h>Hc} = \frac{\pi(h - H_c)}{3S_x^2}(H_c^2 + h^2 + H_c h) \quad \text{for } h > H_c \tag{11.26}$$

Aided by Eq. 11.25, the total storage volume for a depth, h, above the curb height is:

$$V = \frac{k}{3}\frac{\pi H_c^3}{S_x^2} + \frac{\pi(h - H_c)}{3S_x^2}(H_c^2 + h^2 + H_c h) \tag{11.27}$$

where $V_{h<Hc}$ = water volume in [L³] below H_c, $V_{h>Hc}$ = water volume in [L³] above H_c, and V_h = water volume at depth h in [L].

After knowing the detention volume by Eqs 11.18 and 11.20, the corresponding water depth is predicted by Eqs 11.26 or 11.27, depending on the water depth relative to the curb height. The corresponding spread, R_h, can be predicted by Eq. 11.24. Considering the standard transverse slope of 0.02, Fig. 11.10 presents a plot of Eqs 11.26 and 11.27 with H_c = 6 inches (Guo 2000b).

Example 11.3: As shown in Fig. 11.11, the depressed section on South Speer Avenue in the City of Denver, Colorado, has a width of 100 feet and length of 1000 feet. At the low point, a sump trench collects storm runoff from both directions. The capacity of this trench inlet is 6.5 cfs. For safety, the maximum water depth at the sump is no more than 12 inches. Evaluate if the water ponding at the low point

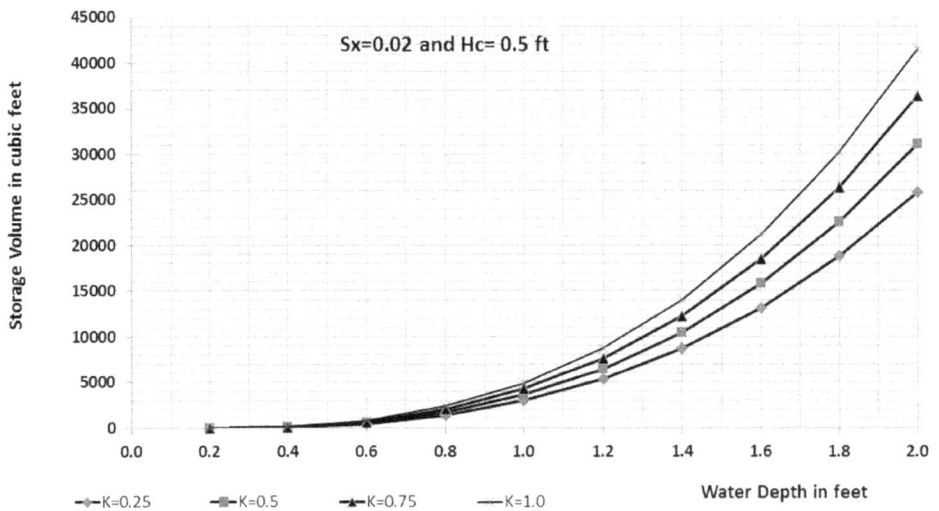

Figure 11.10 Sump-inlet storage capacity with S_x = 0.02 and H_c = 0.5 ft.

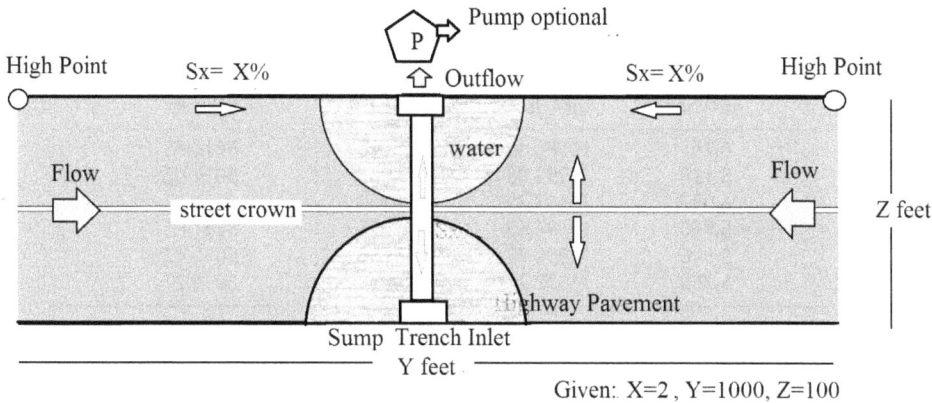

Given: X=2 , Y=1000, Z=100

Figure 11.11 Depressed street section along South Speer Avenue, Denver, Colorado

under a 100-yr event satisfies the design criteria. The design rainfall intensity formula is given as:

$$I = \frac{74.1}{(10+T_d)^{0.789}}, \text{ where } T_d = \text{duration in minutes}$$

Solution: Q = 6.5 cfs, K = 1.0 for English units, C = 0.90, and A = 2.3 acres. Eq. 11.20 becomes:

$$T_m = \frac{1}{0.789}[(T_m+10)] - \frac{6.5}{74.1 \times 1.0 \times 0.90 \times 2.3}(T_m+10)^{0.789+1}]$$

By trial and error, T_m = 14.6 minutes. Aided by Eq. 11.18, the stormwater detention volume is calculated as:

$$V_d = 1.0 \times 0.90 \times 2.3 \times \frac{74.1}{(14.6+1.0)^{0.789}} \times 14.6 \times 60 - 6.5 \times 14.6 \times 60 = 5042 \text{ ft}^3$$

Table 11.3 presents another approach using an incremental duration to maximize the detention volume. At T_d = 14.5 min, the maximal volume is determined to be 5041.85 cubic feet, which agrees with the analytical solution.

The storm water volume of 5042 ft³ will be accumulated at the sump inlet. Next, we need to estimate the available storage volume at the site. Using Eq. 11.14, the water depth at the pool is:

$$5042 = 2\left[\frac{0.50\,\pi \times 0.5^3}{3} \frac{1}{0.02^2} + \frac{\pi(h-0.5)}{3 \times 0.02^2}(0.5^2 + h^2 + 0.5 \times h)\right]$$

So, h = 0.9 ft < 1.0 ft.

Table 11.3 Incremental duration to maximize detention volume

Duration Min	Intensity inch/hr	Volume cubic ft	Volume cubic ft	Volume cubic ft
13.00	6.243	10080.45	5070	5010.45
13.50	6.138	10292.03	5265	5027.03
14.00	6.037	10497.39	5460	5037.39
14.50	5.940	10696.85	5655	5041.85
15.00	5.846	10890.72	5850	5040.72
15.50	5.755	11079.27	6045	5034.27

The water depth for this case is 0.9 ft, and the corresponding radius for the water pool is:

$$R_h = \frac{0.9}{0.02} = 45 < 50\text{ft (OK!)}$$

For comparison, Eq. 11.14 was also analyzed for various side slopes. It was predicted to have a ponding depth of 1.07 ft for $S_x = 0.03$ or 0.6 ft for $S_x = 0.01$.

11.3 CONCLUSIONS

The major concern of traffic safety during a storm event is directly related to the water spread and depth on the street. In comparison, more traffic accidents occur at street intersections because stormwater is entrained to form a pool at the sump inlet. The conventional approach using the conveyance-based method fails to predict the water spread at a sump. The method presented in this chapter provides an effective means to evaluate the water ponding and maximum accumulated depth at a street intersection. In particular, a depressed street section must be evaluated for the possibility of water ponding under a major event. This method indicates that the flatter the surface slope, the shallower the accumulated water depth.

HOMEWORK

Q11-1 A curb-opening inlet as shown in Fig. Q11-1 is located at a street corner. The tributary area to this inlet is 2.58 acres, with a runoff coefficient of 0.55. This inlet is protected from urban debris with blocks around the curb opening. The flow into the inlet is 2.5 cfs before the water flow overtops the blocks. Determine the height of the blocks for a 5-yr Denver storm event.

Solution: For this case, the surface area of the pooled water is $K = 3/4$ of a circle. The rainfall IDF is prescribed with $a = 38.5$, $b = 10$, and $m = 0.789$ for a 5-yr event. The critical rainfall duration is solved for $T_d = 9.07$ minutes and the detention volume is 817.54 ft³. Considering a transverse slope of 0.02, the block should be designed with a height of 0.75 feet, which results in a water spread of 37.5 feet.

(a) Street Inlet with Blocks

K=1 for ft or K=1:360 for meter			K=	0.75	
Tributary Area			A=	2.58	acre
Runoff Coefficient			C=	0.550	
Outflow from Sump Inlet			Q=	2.50	cfs
Rainfall Information			a=	38.50	
			b=	10.00	
$I_d = \dfrac{a}{(T_d+b)^m}$			m=	0.789	
Guess Duration Tm			Tm=	9.07	minutes
Check if Tm is acceptable			Balance=	0.000	
Design Rainfall Intensity Td=Tm			I=	3.76	inch/hr
Stormwater Detention Volume			V-design=	817.54	cubic ft
Storage Volume in Sump					
Street Transverse Slope			Sx=	0.02	ft/ft
Curb Height			Hc=	1	feet
Incremental Depth			Dy=	0.25	foot
Depth		STORAGE	VOLUME		
h (ft)		Vh	cubic ft		
K=	0.25	0.50	0.75		
0.00			0.00		
0.25			30.68		
0.50			245.44		
0.75			815.17		

(b) Ponded Water Depths and Volumes

Figure Q11-1 Design of a sediment bay at a street corner

Q11-2 A highway runs through an alluvial fan in Fig. Q11-2 that has an area of 4 acres with a runoff coefficient of 0.6. The rainfall IDF formula at this site is:

$$I = \frac{45.5}{(6.5+T_d)^{0.65}},$$ where I = intensity in inch/hr and T_d = duration in minutes.

A culvert is installed at the sag to pass storm runoff across the highway. The entrance pool in front of the culvert is shaped with a slope of 5% in all directions toward a highway. The crossing culvert is designed to release an average flow of 7.50 cfs before the entrance pool becomes full. Determine the maximum water depth at the entrance pool before the highway is overtopped.

Solution: For this case, $k = 0.5$ (or a half circle) is the cross-sectional geometry at the entrance pool. Let $H_c = 1.0$ feet. Try $T_m = 16.7$ min to determine the maximal water depth of 0.8 feet.

Alluvial Fan Area

Sx=5%

Sx=5%   Sx=5%

Highway

PLAN VIEW   Culvert

Flow Q

Top of Road

Q

Hc

VERTICAL VIEW   Culvert

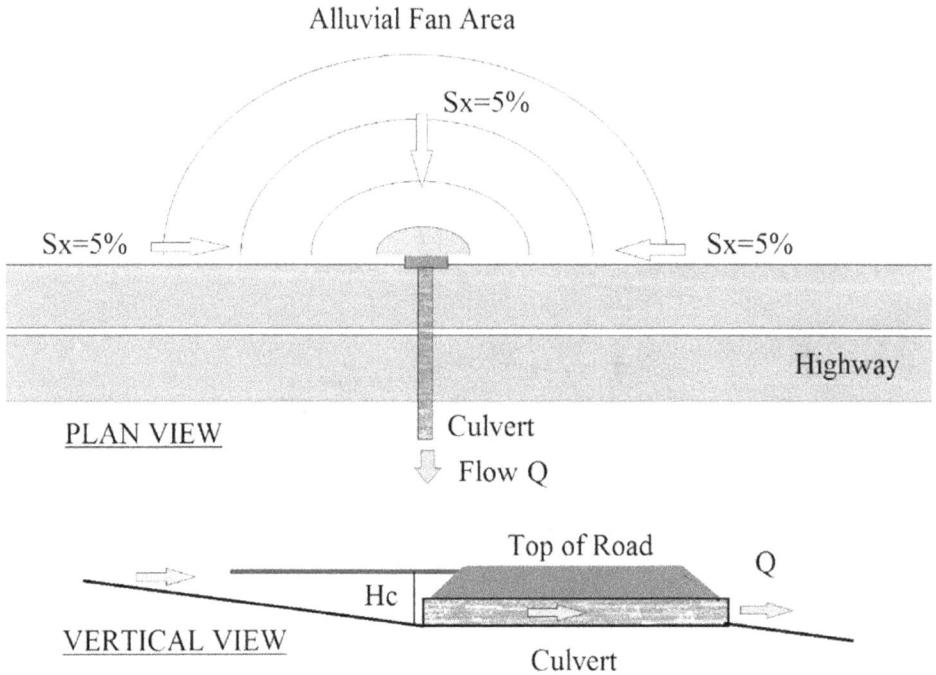

Figure QI1-2 Crossing culvert under a highway

STREET DETENTIONA AND STORAGE AT SUMP

Detention Volume for Design

K=1 for ft or K=1/360 for meter	K=	1.00	
Tributray Area	A=	4.00	acre
Runoff Coefficient	C=	0.600	
Outflow from Sump Inlet	Q=	7.50	cfs
Rainfall Information	a=	45.50	
$I_d = \dfrac{\alpha}{(T_d+b)^m}$	b=	6.50	
	m=	0.650	
Guess Duration Tm	Tm=	16.79	minutes
Check if Tm is acceptable	Balance=	0.000	
Design Rainfall Intensity Td=Tm	I=	5.88	inch/hr
Stormwater Detention Volume	V-design=	6660.07	cubic ft
Storage Volume in Sump			
Street Transverse Slope	Sx=	0.05	ft/ft
Curb Height	Hc=	1	foot
Incremental Depth	Dy=	0.2	foot

Depth h (ft)		STORAGE Vh	VOLUME cubic ft	
k =	0.25	0.50	0.75	1.00
0.00	0.00	0.00	0.00	0.00
0.20	20.94	41.89	62.83	83.78
0.40	167.55	335.10	502.66	670.21
0.60	1518.44	2083.93	2649.42	3214.90
0.80	5317.96	6643.36	7968.75	9294.14

REFERENCES

FAA (1970). *Airport Drainage*, Report 150/5320-5B, US Department of Transportation, Federal Aviation Administration, Washington, DC.

FAA (1977). *Surface Drainage Facilities for Airfields and Heliports*, Technical Manual no. 5-820-1, US Department of the Army and Air Force, Washington, DC.

Guo, J.C.Y. (1999). "Detention Storage Volume for Small Urban Catchment", *ASCE Journal of Water Resources Planning and Management*, vol. 125, no. 6, November/December.

Guo, J.C.Y. (2000a). "Street Storm Water Conveyance Capacity", *ASCE Journal of Irrigation and Drainage Engineering*, vol. 126, no. 2, March/April.

Guo, J.C.Y. (2000b). "Street Storm Water Storage Capacity", *Water Environmental Research Journal*, vol. 72, no. 5, September/October.

Guo, J.C.Y. (2012). "Off-stream detention design for stormwater management", *ASCE Journal of Irrigation and Drainage Engineering*, vol. 138, no. 4, April 1.

Chapter 12

Culvert hydraulics

12.1 CULVERT LAYOUT AND DESIGN CONSIDERATIONS

A culvert is often laid under a highway in order to maintain the continuity of water flows. The hydraulic capacity of a culvert varies with respect to the water depth at the entrance. As shown in Fig. 12.1, a culvert structure includes three elements: the *inlet work*, *closed conduits*, and *outlet work*.

Culvert hydraulics are sensitive to the length of the conduit. In general, the length of a culvert system is limited to 400–450 ft (130–150 m). When a culvert is so short (see Fig. 12.2b) that the entrance is daylighted, the trash rack may be ignored; otherwise, the culvert entrance must be protected with a trash rack placed with a proper inclined angle (see Fig. 12.2a). The surface area of the rack must be at least four times the open area of the conduit. The trash rack prevents trash and objects from being washed into the conduit.

As a culvert is laid along a natural stream that crosses a highway, its alignment should be located as close to the natural waterway as possible. At the entrance, the *headwater depth*, H_w, is referred to the vertical distance between the entrance inlet to the water surface. The *available depth*, H_a, is the vertical distance from the entrance invert to the top of the roadway. Although the hydraulic capacity of a culvert is directly related to the headwater depth, a high headwater depth may damage the pipe due to uplift as the seepage flow develops beneath the pipe line. To be conservative, Table 12.1 presents the recommended allowable headwater depth to circular pipe diameter (or height of box culvert).

In operation, trash and debris carried in stormwater impose the risk of pipe clogging at the entrance. Table 12.2 presents the recommended minimum culvert sizes for various applications. The length of culvert pipe used for highway crossing should be in multiples of two feet, and the length of a concrete box culvert should be to the nearest foot.

A concrete *headwall*, shown in Fig. 12.3a, is recommended for culverts with 50 square feet of crossing area (>96-inch in diameter) and larger. The headwall should be placed perpendicular to the culvert centerline. All concrete box culverts in Fig. 12.3a and pipes with full concrete headwalls require concrete *wing walls*. Wing walls should maintain a streamlined angle between 45 and 75 degrees to guide the water into the entrance of the culvert. Concrete aprons, shown in Fig. 12.3b, are recommended to be placed at the culvert exit for scour protection.

DOI: 10.1201/9781003284239-12

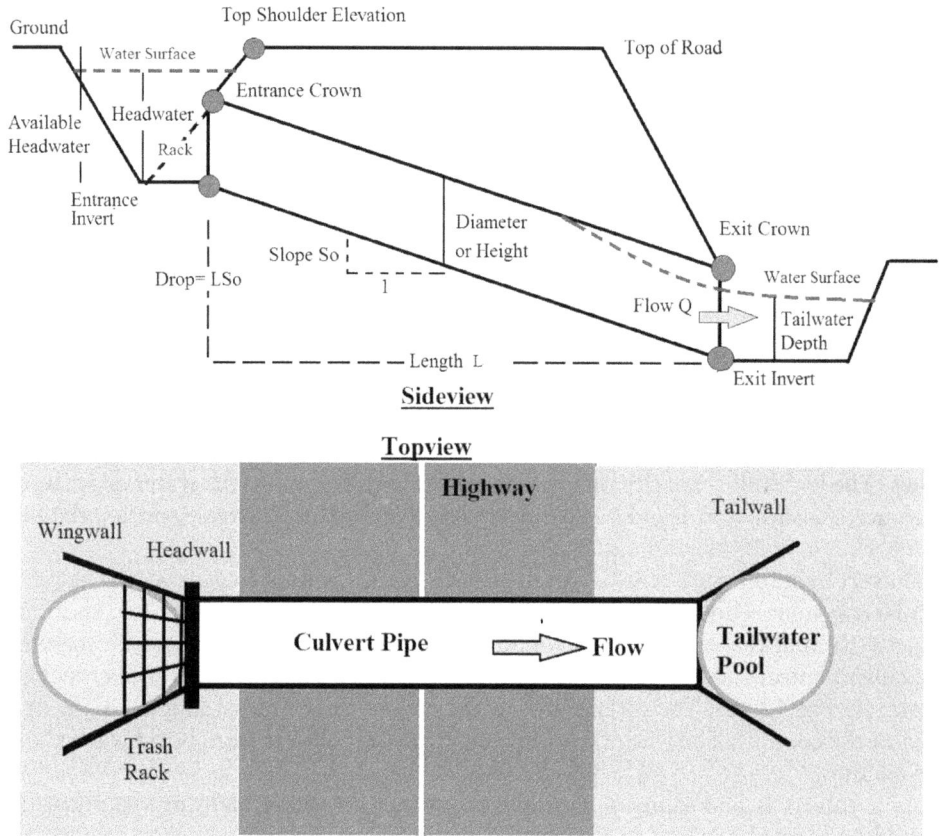

Figure 12.1 Elements of a culvert

(a) Culvert with Rack   (b) Day-lighted Crossing Culvert

Figure 12.2 Examples of a culverts

Table 12.1 Allowable headwater depth to culvert diameter ratios

Culvert Diameter or Height (feet)	Maximum Ratio Headwater/Diameter H_w/d
less than 3	1.5
from 3 to 5	1.3
from 5 to 7	1.2
larger than 7	1.0

Table 12.2 Recommended minimum culvert sizes

Type of Culvert	Minimum Diameter
Cross culvert	24 inches
Side drain	18 inches
Median drain	18 inches
Storm sewer trunk line	18 inches
Storm sewer connections	15 inches
Irrigation crossing	18 inches

(a) Head Wall and Wing Walls   (b) Outlet Apron

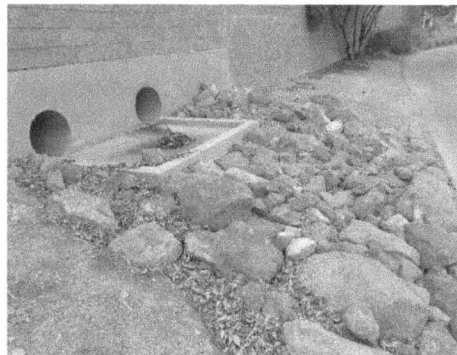

Figure 12.3 Culvert elements

12.2 CULVERT SIZING

Although a culvert is sized to pass the design flow, the service of the culvert is subject to a wide range of flows. Therefore, the concept of multiple frequencies (events) is important to culvert design. The barrel (conduit) is often sized based on the assumption of normal flow for the design flow. The normal flow in a partially full circular pipe is described as:

$$A = \frac{d^2}{4}(\theta - \sin\theta\cos\theta) \tag{12.1}$$

$$P = d\theta \tag{12.2}$$

$$R = \frac{d(\theta - \sin\theta\cos\theta)}{4\theta} \tag{12.3}$$

$$Y = \frac{d}{2}(1 - \cos\theta) \tag{12.4}$$

$$T = d\sin\theta \tag{12.5}$$

$$Q = \frac{k}{N}P^{-\frac{2}{3}}A^{\frac{5}{3}}\sqrt{S_o} \tag{12.6}$$

$$V = \frac{Q}{A} \tag{12.7}$$

$$F_r = \sqrt{\frac{Q^2 T}{gA^3}} \tag{12.8}$$

where A = flow area in $[L^2]$, d = diameter of the pipe in $[L]$, P = wetted perimeter in $[L]$, R = hydraulic radius in $[L]$, θ = half of the central angle shown in Fig. 12.4, k = 1.486 for ft-second units or 1.0 for meter-second units and Y = flow depth in $[L]$. It is noted that when the pipe is full, the top width of the flow area is reduced to zero, or the Froude number for full flow vanishes.

For a box conduit in Fig. 12.4 with a height of H in $[L]$ and width of W in $[L]$, its hydraulic radius is computed by:

$$A = WY \tag{12.9}$$

$$P = W + 2Y \tag{12.10}$$

$$R = \frac{WY}{W + 2Y} \tag{12.11}$$

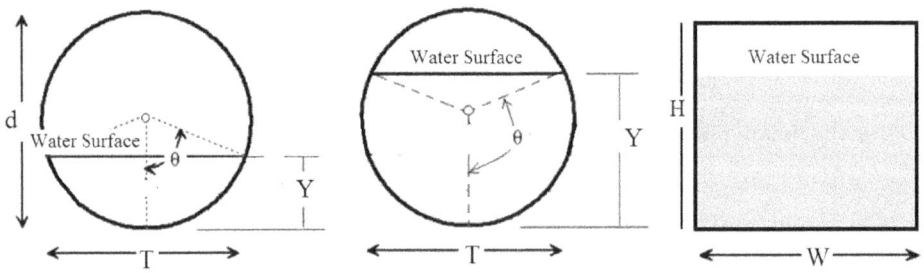

Figure 12.4 Hydraulic parameters in a closed conduit

When both entrance and exit are submerged, the flow in the culvert is pressurized. The flowing full hydraulic radius of a circular pipe becomes:

$$R = \frac{A}{P} = \frac{d}{4} \tag{12.12}$$

With a half of central angle of 180 degrees, Eq. 12.6 is reduced to:

$$d = \left(\frac{NQ}{K\sqrt{S_o}} \right)^{\frac{3}{8}} \tag{12.13}$$

where N = Manning's roughness, Q = design flow in $[L^3/T]$, S_o = conduit invert slope, d = hydraulically required circular diameter in $[L]$, and K = 0.462 for feet-second units or 0.311 for meter-second units. Eq. 12.13 provides a hydraulically required pipe diameter in feet or meters. In practice, the next larger commercially available pipe should be selected, and the normal flow condition for the design flow in the selected pipe is solved by trial-and-error to find the proper central angle.

Example 12.1: Determine the normal flow condition for Q = 23.5 cfs in a 24-inch concrete pipe that is laid on a slope of 1.5% with N = 0.014.

Solution: Try θ = 120 degrees = 3.1416 × 120/180 = 2.09 radians

$$Y = \frac{2}{2}(1 - \cos 2.09) = 1.50 \text{ ft}$$

$$A = \frac{2^2}{4}(2.09 - \sin 2.09 \times \cos 2.09) = 2.53 \text{ sq ft}$$

$$P = 2 \times 2.09 = 4.18 \text{ ft}$$

$$R = \frac{2(2.09 - \sin 2.09 \times \cos 2.09)}{4 \times 2.09} = 0.60 \text{ ft}$$

$$T = 2 \sin 2.09 = 1.73 \text{ ft}$$

$$Q = \frac{1.486}{0.014} \times 4.18^{\frac{-2}{3}} \times 2.53^{\frac{5}{3}} \sqrt{0.015} = 23.46 \text{ cfs}$$

$$V = \frac{23.46}{2.53} = 9.28 \text{ fps}$$

$$F_r = \sqrt{\frac{23.46^2 \times 1.73}{32.2 \times 2.53^3}} = 1.35$$

For this case, a half of the central angle is confirmed to be 120 degrees. In practice, this trial-and-error procedure may take 3 to 4 times to find the solution for the given flow.

Example 12.2: A concrete culvert is placed on a slope of 1.5% to carry a design flow of 200 cfs. Size a commercial pipe for this case.

Solution:

$$d = \left(\frac{NQ}{K\sqrt{S_0}}\right)^{\frac{3}{8}} = \left(\frac{0.014 \times 200}{0.462\sqrt{0.015}}\right)^{3/8} = 4.32 \text{ ft} = 51.8 \text{ inches}$$

Use $d = 54$ inches for the design.

Example 12.3: Recommend a commercial concrete pipe placed on a slope of 1.5% for a design flow of 40 cfs.

Design Flow		Q= 40.00	cfs							
Manning's roughness		N= 0.0140								
Pipe slope		So= 0.0150								
Required Diameter		D-required 28.35	inches 2.36	ft						
Diameter used		D-used = 30.00	inches 2.50	ft						

A half Central Angle degree	Angle radian	Flow Depth Y ft	Flow Area A ft^2	Wetted Perimeter P ft	Hydraulic Radius R ft	Top Width T ft	Depth Centroid Y_c ft	Flow Velocity V fps	Flow Rate Q fps	Froude Number F_r
0.00	0.00	0.00	0.00	0.00	0.00	0.00	0.00	0.00	0.00	NP
115.23	2.01	1.78	3.74	5.03	0.74	2.26	0.79	10.68	40.00	1.46
180.00	3.14	2.50	4.91	7.85	0.62	0.00	1.25	9.50	46.65	NP

Example 12.4: Continue Example 12.3 to determine the critical flow condition for $Q = 40$ cfs in a circular pipe of $d = 30$ inches.

Solution: By definition, the critical flow depth is defined with its flow Froude number = 1 in Eq. 12.8. By trial and error, for this case, $\theta = 2.35$ radians (134.72 degrees). Details are listed below:

Flow Variables	Cricial Flow	
Half Central Angle for Critical Flow	$\theta-c$=	2.35 *rad*
Flow depth	Y_c=	2.13 *ft*
Flow area	A_c=	4.45 *sq ft*
Wetted perimeter	P_c=	5.88 *ft*
Check on Flow Froude Number	If Fr=1	1.00
Flow Variable	V_c=	8.98 *fps*

12.3 CULVERT HYDRAULICS

A culvert flow is controlled by either the orifice flow hydraulics at the entrance or the pipe flow through the culvert pipe, whichever requires a higher headwater depth for the given design flow. Therefore, the culvert flow must be analyzed for the two following conditions as:

1. an *inlet-control culvert*, and
2. an *outlet-control culvert*.

The headwater required by the orifice flow formula at the entrance is called the *inlet-control headwater depth*. The headwater required by the energy balance between the entrance and the known tailwater depth is called the *outlet-control headwater depth*. When the inlet-control headwater depth is greater than the outlet-control headwater depth, the culvert operates under inlet control; otherwise, it is under outlet-control. A culvert may behave like an inlet control to pass one discharge and then switch to outlet control to pass another discharge. Therefore, for a given discharge, the culvert must be examined by both the orifice formula and the energy grade line in order to know which one dictates the operation.

12.3.1 Culvert hydraulics under inlet control

When a culvert is laid on a steep slope, the culvert capacity is controlled by the headwater depth at the entrance and the critical flow at the entrance. Such a culvert is called an inlet-control culvert. As illustrated in Fig. 12.5, the balance of flow energy between the approaching flow and the culvert entrance is written as:

$$H_1 = H_c + \frac{U_c^2}{2g} \text{ (if elevations are used in the computation)} \tag{12.14}$$

$$Y_1 = Y_c + \frac{U_c^2}{2g} + LS_0 \text{ (if water depth are used)} \tag{12.15}$$

$$U_c = \sqrt{2g(H_1 - H_c)} = C_o\sqrt{2g\left(Y_1 - LS_0 - \frac{d}{2}\right)} \tag{12.16}$$

$$Q = C_o A \sqrt{2g\left(Y_1 - LS_0 - \frac{d}{2}\right)} = C_o A \sqrt{2g\left(H_w - \frac{d}{2}\right)} \tag{12.17}$$

where H = water surface elevation in [L], Y = water depth in [L] relative to the culvert exit invert, U = flow velocity in [L/T], L = length of culvert in [L], S_0 = slope of culver pipe in [L/L], C_o = discharge coefficient 0.55 to 0.65, and A = cross-sectional area of culvert pipe in [L²]. The subscript 1 means the variables associated with Section

1. Similarly, the subscript c represents the variables associated with the critical flow at the culvert entrance.

For a given design flow, Q, Eq. 12.17 defines the required inlet-control headwater depth, $H_{w\text{-}IN}$.

$$H_{W-IN} = H_w \text{ defined by Eq. 12.17} \tag{12.18}$$

As shown in Eq. 12.17, the capacity through an inlet-control culvert is independent of the culvert length, barrel slope, and roughness.

12.3.2 Culvert hydraulics under outlet control

For a flat culvert, the culvert hydraulics are dominated by the energy balance between the entrance, Section 1 in Fig. 12.6, and the exit section. Obviously, the friction loss in this case depends on the tailwater depth and the culvert pipe's roughness, the flow line slope, and the culvert length.

Applying the energy principle between Sections 1 and 3 in Fig. 12.6 yields:

$$Y_1 = H_w + LS_o = Y_3 + \frac{U_3^2}{2g} + h_e + h_{23} \tag{12.19}$$

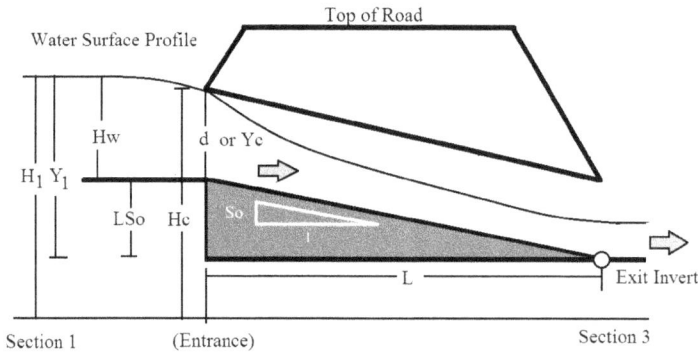

Figure 12.5 Culvert hydraulics under Inlet control

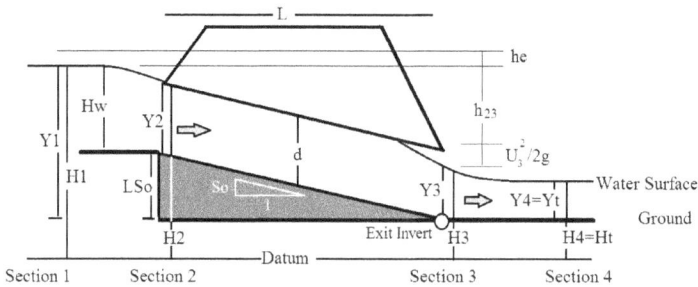

Figure 12.6 Culvert with a submerged inlet and outlet

Under the full flow condition $U_2 = U_3 = U_f$ and $R = d/4$.

$$h_e = K_e \frac{U_2^2}{2g} \approx K_e \frac{U_f^2}{2g} \tag{12.20}$$

$$h_{23} = S_f L = \left[\frac{N^2 U_f^2}{k^2 R^{\frac{4}{3}}} \right] L = \left[\frac{2g}{k^2} \frac{N^2}{R^{\frac{4}{3}}} L \right] \frac{U_f^2}{2g} = K_f \frac{U_f^2}{2g} \tag{12.21}$$

$$Y_3 = \max\left[Y_t, \frac{Y_c + d}{2} \right] \tag{12.22}$$

$$Q = A\left[\frac{2g(Y_1 - Y_3)}{1 + K_e + K_f} \right]^{0.5} = A\sqrt{\frac{1}{1 + K_e + K_f}} \sqrt{2g(H_w + LS_o - Y_3)} \tag{12.23}$$

where h_{23} = friction loss in [L] between Sections 2 and 3, U_f = full flow velocity in [L/T], K_e = entrance loss coefficient, Y_t = tail water depth in [L], Y_c = critical flow depth in [L], K_f = roughness coefficient, S_f = friction slope in [L/L], and k = 1.486 for feet-second units or 1.0 for meter-second units. The subscript 1 represents the variables associated with Section 1. For a given design flow, Q, Eq. 12.23 defines the required outlet-control headwater depth, $H_{w\text{-OUT}}$.

$$H_{w-OUT} = H_w \quad \text{defined by Eq. 12.23} \tag{12.24}$$

12.3.3 Determination of culvert capacity

As discussed before, for a given design flow, we shall analyze the culver hydraulics for both inlet and outlet control conditions. The design headwater depth is determined by:

$$H_w = \max(H_{w-IN}, H_{w-OUT}) \quad \text{for a given flow } Q \tag{12.25}$$

In contrast, for a given headwater depth at the entrance, the culvert capacity is determined by the smaller value between Eqs 12.17 and 12.23.

$$Q = \min(Q_{IN}, Q_{OUT}) \quad \text{for a given headwater } H_w \tag{12.26}$$

where Q_{IN} = discharge determined by Eq. 12.27 and Q_{OUT} = discharge determined by Eq. 12.23 for a given H_w.

Example 12.5: As shown in Fig. 12.7, a discharge of 40 cfs is carried by a circular culvert laid on a slope of 1.5%. The roughness coefficient of the sewer is 0.014. The water depth at the exit of the culvert is 2.0 feet. Suggest a pipe size and headwater for this case.

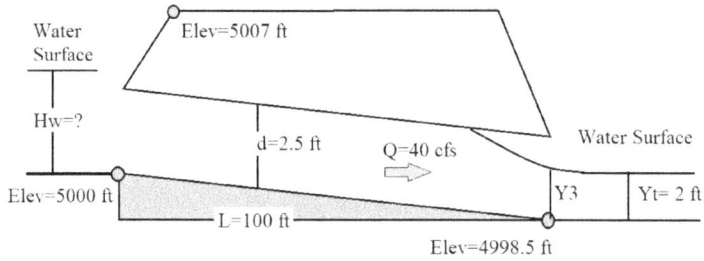

Figure 12.7 Design of a culvert

Solution: Let use try a circular pipe with a diameter of 2.5 ft to carry the design flow of 40 cfs. The detailed design information and calculations of H_w for the given culvert flow are presented in the following:

Design Information (input)		
Pipe Diameter in inches	d= 30.0	inches
Invert Elevation at Entrance	E-in= 5000.0	ft
Invert Elevation at Exit	E-out= 4998.5	ft
Pipe Length	L = 100.0	ft
Pipe Manning's Roughness	N = 0.0140	
Entrance Loss Coeff	Ke = 0.50	
Exist Loss	Kx = 0.00	
Bend Loss Coeff	Kb = 0.00	
Design Discharge		
Total Design Discharge	Qt= 40.00	cfs

Under the inlet-control conditions, the headwater is calculated as:

$$40 = C_o A \sqrt{2g\left(H_w - \frac{d}{2}\right)} = 0.52 \times \frac{3.14 \times 2.5^2}{4} \sqrt{2 \times 64.4\left(H_w - \frac{2.5}{2}\right)}$$

Under the outlet-control and flowing full conditions, the headwater is calculated as:

$$R = \frac{A}{P} = \frac{d}{4} = \frac{2.5}{4} = 0.625 \text{ ft}$$

$K_e = 0.5$

$$K_f = \left[\frac{2g}{k^2} \frac{N^2}{R^{\frac{4}{3}}} L \right] = \left[\frac{2 \times 32.2}{1.486^2} \frac{0.014^2}{0.625^{\frac{4}{3}}} \times 100 \right] = 1.06$$

$$Y_3 = \max \left[Y_t, \frac{Y_c + d}{2} \right] = \max \left[2, \frac{2.13 + 2.5}{2} \right] = 2.31 \text{ ft}$$

$$40 = A \left[\frac{2g(Y_1 - Y_3)}{1 + K_e + K_f} \right]^{0.5} = \frac{3.14 \times 2.5^2}{4} \times \sqrt{ \frac{2 \times 32.2 (H_w + 100 \times 0.015 - 2.31)}{1 + 0.5 + 1.06} }$$

Both inlet- and outlet-control headwater depths are solved by trial and error. The details are summarized as follows:

Headwater Depth by Inlet Control

Orifice Coefficient	Co = 0.52	0.52	
Pipe Diameter	d= 2.50 *ft*	63.50	*mm*
Headwater Depth -Inlet control	Hw-In= 5.06 *ft*	1.54	*m*
Pipe Cross Sectional Area	Ao= 4.91 *sq ft*	0.46	*sq m*
Pipe Flow Velocity	Vo= 8.15 *fps*	2.48	*mps*
	0.00 *ft*	0.00	*m*

Headwater Depth by Outlet Control

If known, enter the Tailwater Depth	Given Yt = 2.00 *ft*	0.00	*m*
Critcal flow depth	Yc= 2.13 *ft*	0.66	*ft*
Yt=Max(Y-tail, 0,5(D+Yc))	Used Yt= 2.31 *ft*	0.71	*m*
Friction Loss Coeff	Kf = 1.06		
Sum of All Loss Coeff's	K's = 1.56		
Headwater Depth - Outlet control	Hw-Out= 3.46 *ft*	1.05	*m*
Design Headwater Depth Ratio			
Headwater for Design	HW= 5.06 *ft*	1.54	*m*
HW/D ratio =	HW/d= 2.03	2.03	

Based on the above computation, this is a case of outlet-control culvert flow. The ratio $H_w/d = 2.03$. Next, check the recommended H_w/d in Table 12.1; the acceptable ratio is $H_w/d < 1.5$. Therefore, we shall increase the pipe size to reduce the headwater.

For instance, let us try a pipe of 3.0 ft in diameter. Repeating the above process, the solution is:

Headwater Depth by Inlet Control

Orifice Coefficient	Co = 0.52	
Pipe Diameter	d= 3.00	*ft*
Headwater Depth -Inlet conrol	Hw-In= 3.34	*ft*
Pipe Cross Sectional Area	Ao= 7.07	*sq ft*
Pipe Flow Velocity	Vo= 5.66	*fps*
	0.00	*ft*

Headwater Depth by Outlet Control

If known, enter the Tailwater Depth	Given Yt = 2.00	*ft*
Critcal flow depth	Yc= 2.06	*ft*
Yt=Max(Y-tail, 0, 5(D+Yc))	Used Yt= 2.53	*ft*
Friction Loss Coeff	K_f = 0.83	
Sum of All Loss Coeff's	K_s = 1.33	
Headwater Depth - Outlet control	Hw-Out= 2.19	*ft*

Design Headwater Depth Ratio

Headwater for Design	HW= **3.34**	*ft*
HW/D ratio =	HW/d= **1.11**	

Example 12.6: Referring to Fig. 12.7, the headwater depth of 4 feet is present at the entrance of the concrete culvert. The culvert pipe has a diameter of 2.5 ft. The water depth at the exit of the culvert is 2.0 feet. Determine the discharge.

Solution: Let us find the inlet-control culvert capacity under H_w = 4 ft. Apply Eq. 12.17 to the 2.5-ft circular pipe with C_o = 0.52. The inlet-control discharge, Q_{IN} = 33.97 cfs. Next, apply Eq. 12.23 to find the outlet-control discharge under H_w = 4 ft. The outlet-control discharge, Q_{OUT} = 43.59 cfs. According to Eq. 12.26, Q = 33.97 cfs for this case. Details of the computations are listed below.

Headwater Depth by Inlet Control

Discharge	Q_{IN}= 33.97	*cfs*
Orifice Coefficient	Co = 0.52	
Pipe Diameter	d= 2.50	*ft*
Headwater Depth -Inlet control	Hw-In= 4.00	*ft*
Pipe Cross Sectional Area	Ao= 4.91	*sq ft*
Pipe Flow Velocity	Vo= 6.92	*fps*

Headwater Depth by Outlet Control

Discharge	Q_{OUT}= 43.59	*cfs*
If known, enter the Tailwater Depth	*Given* Yt = 2.00	*ft*
Critical Flow Depth	Yc = 2.20	*ft*

(a) Trash Rack at Entrance   (b) Need Fence for Safety

Figure 12.8 Safety around a culvert

Yt=Max(Y-tail, 0, 5(d+Yc))	*Used Yt =*	**2.35** *ft*
Friction Loss Coeff	*Kf =*	1.06
Sum of All Loss Coeff's	*K's =*	1.56
Headwater Depth - Outlet control	*Hw-Out=*	3.99 *ft*

12.4 CONCLUSIONS

Culverts are often used as a drainage facility in urban areas. The major concerns associated with culvert design include, but are not limited to, hydraulic efficiency and public safety. On top of the hydraulic analysis, it is recommended that a trash rack, as shown in Fig. 12.8a, be installed at the entrance. The function of such a trash rack is not only for urban debris control, but is also a life saver in case of an accident involving humans or/and animals.

Fences and railing are important elements in the construction of large culvert structures. Fig. 12.8b presents a case that needs right and left fences upstream of the culvert bridge. These fences serve as warning signs and prevent human and animals from mistakenly walking into the flood zone in the dark. As always, sufficient flash-flood-warning signs are posted next to the culvert on the major waterway.

HOMEWORK

Q12-1 A circular culvert of 36-inch in diameter is laid on a slope of 1.5% with Manning's $N = 0.025$. Determine the inlet-control capacity when $H_w/d = 1.0$ and 2.0.

Q12-2 A box culvert of 3-ft by 4-ft (height by width) is laid on a slope of 1.5% with Manning's $N = 0.025$. Determine the inlet-control capacity when $H_w/d = 1.0$ and 2.0, where $d = 3$ ft, i.e. the height of the box culvert.

Q12-3 As illustrated in Fig. 12.6, the crossing culvert under a highway is designed to pass the design flow of 45 cfs through a length of 300 feet. A single concrete barrel of 3-ft in diameter is laid on a slope of 2.0%. The Manning's N for the barrel is 0.020. The entrance and exit loss coefficients are $K_e = 0.5$ and $K_x = 1.0$. The tailwater depth for the design condition is one foot above the crown at the exit.

1. Determine the design headwater depth at the entrance of the culvert.
2. The available water depth at the entrance is 8 feet. Can you apply the headwater of 8 feet to reduce the culvert size? If so, what is the diameter of the culvert? If not, explain.

Q12-4 The double 36-inch circular pipes in Fig. Q12-4 are used as a temporary bridge. The lengths of the culverts are so short that the operation of these two culverts is under inlet control. The discharge coefficient is 0.52 for these two culverts. Estimate the flow depth when the discharge is 143 cfs.

Figure Q12-4 Temporary bridge under high flow

Q12-5 As shown in Fig. Q12-5, a culvert system is designed to pass the design flow of 260 cfs underneath the highway. The upstream invert elevation is: elevation-in = 4987 ft at the culvert entrance. Verify that two concrete pipes of 60 inches in diameter are acceptable for this case.

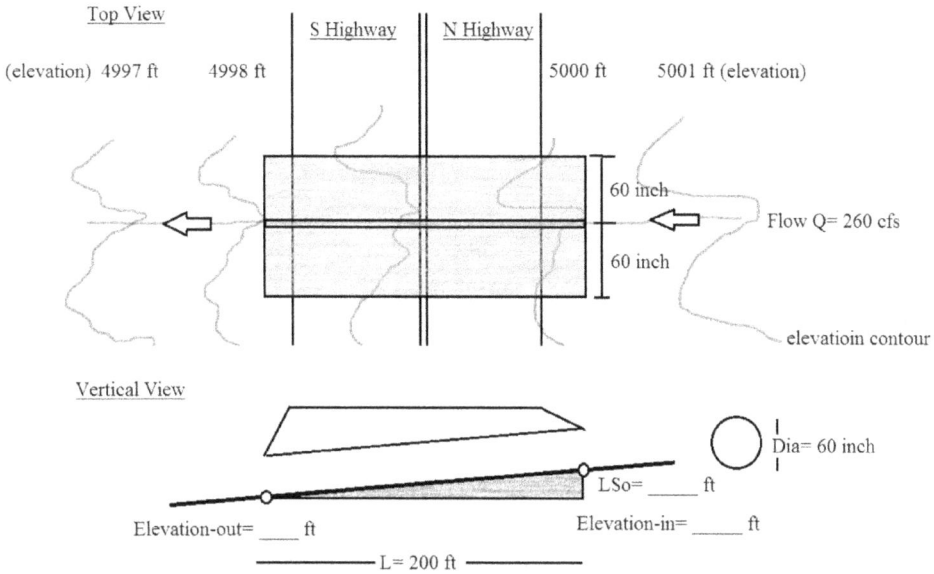

Figure Q12-5 Design of a culvert bridge

REFERENCES

ASCE (1991). *Design and Construction of Urban Stormwater Management Systems*, Manuals and Reports of Engineering Practice No 77 or Water Environmental Federation WEF Manual of Practice FD-20.

CDOT (2015). *Roadway Design Manual*, Colorado Department of Transportation, Denver, CO.

Guo, J.C.Y. (2017). *Urban Flood Mitigation and Stormwater Management*, CSC, Amazon.

HEC14 (1975). *Hydraulic Design of Energy Dissipaters for Culverts and Channels*, US Department of Transportation, Federal Highway Administration, Washington, DC.

USWDCM (2010). *Urban Stormwater Design Criteria Manual*, Urban Drainage and Flood Control District, Denver, CO.

Chapter 13

Storm sewer system design

13.1 LAYOUT OF STORM SEWER SYSTEM

A storm sewer system consists of a series of pipes and manholes. Surface runoff enters a manhole through street inlets. Between two adjacent manholes, the flow is carried by a pipe. Fig. 13.1 shows the plan view of the sewer system. The layout of a storm sewer system is governed by many factors, including:

1. *street alignment for the sewer line,*
2. *street inlet placement,*
3. *existing utility locations,*
4. *sewer system outfall location,* and
5. *watershed topography tributary to the sewer system.*

13.2 VERTICAL PROFILE

The layout of a sewer system is depicted by the *plan view* and *vertical view*, as illustrated in Fig. 13.2. The vertical profile of a sewer line must avoid any conflict with the existing underground utilities and satisfy all design constraints. As a rule of thumb, the ground slopes provide guidance to select the first approximation of the sewer line grades. The vertical profile of a sewer system has to be adjusted several times in order to satisfy the design criteria. For instance, the slope of a sewer line must be repeatedly adjusted with manhole drops several times until the flow velocity in the sewer line is within permissible limits, and also must meet the required minimum soil cover. Sewer flows can be very complicated hydraulically. On a steep slope, the sewer carries an open-channel flow. After a hydraulic jump, the sewer is surcharged. Under continuous backwater effects, the sewer may be pressurized and become flowing full. The energy grade line (EGL) and hydraulic grade line (HGL) are sensitive to the vertical profile selected for the sewer system. It takes several iterations to develop an acceptable vertical profile for the sewer system under design.

As illustrated in Fig. 13.3, all manholes are marked by identification numbers such as M1 and M2. All sewers are marked with their length, slope, and identification number such as S12 from M1 to M2, and S23 from M2 to M3. A manhole is covered

DOI: 10.1201/9781003284239-13

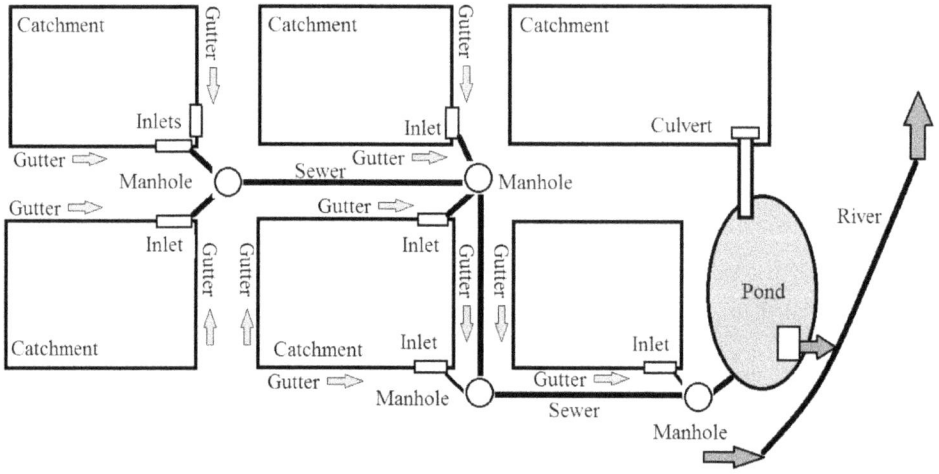

Figure 13.1 Plan view of a sewer system

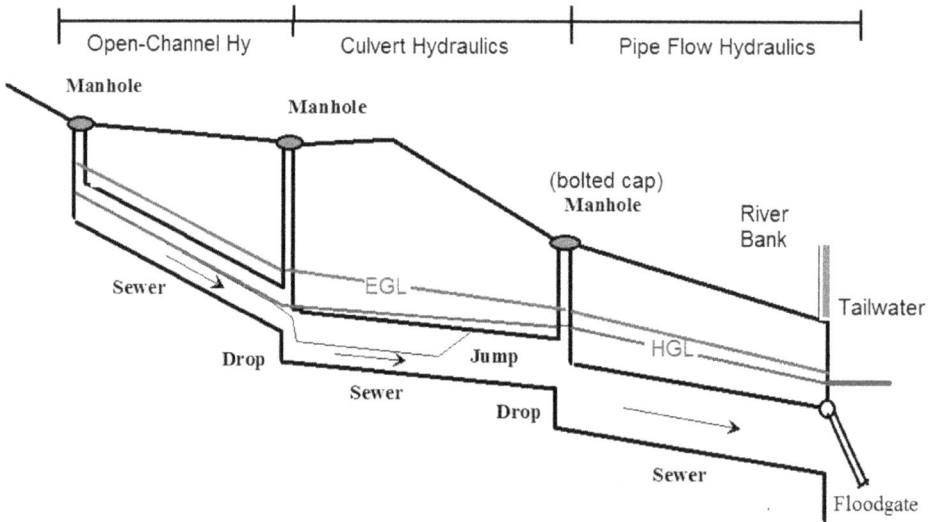

Figure 13.2 Flow conditions in a sewer-manhole system

by a metal cap. The *rim elevation* of the metal cap is set to be at the local ground elevation. Each incoming sewer at the manhole is marked with its *crown* and *invert elevations*, as is the outgoing sewer. The elevation difference between the incoming and outgoing sewer inverts is called the *manhole drop*. As a common practice, it is preferred to add a floor drop of 0.20 feet at a manhole.

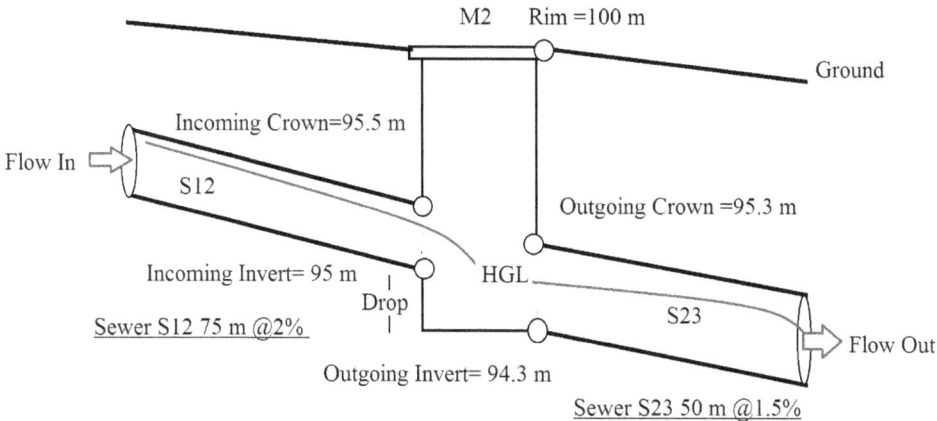

Figure 13.3 Details of vertical profile for a sewer system

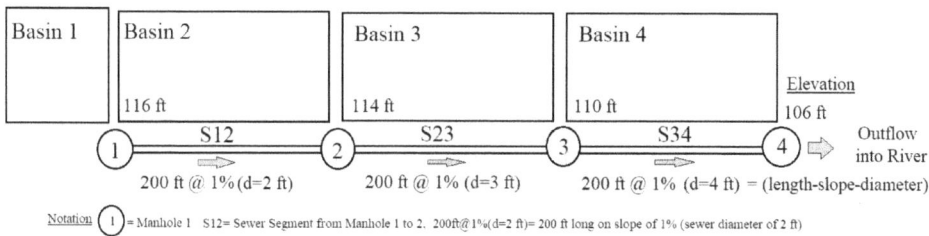

Figure 13.4 Street map for sewer line design

Example 13.1: Given the ground elevations, length, slope, and diameter for the three sewer segments in Fig. 13.4, design the manhole drops to have a sewer vertical profile laid on a constant slope of 1% and a minimum soil cover of 2 ft.

Solution: Fig. 13.4 shows the top view of the given sewer system. The sewer is identified by its upstream manhole and downstream manhole. For instance, S12 is the sewer segment from Manhole 1 to Manhole 2. Each sewer is depicted with its length in feet at a given slope, as a percentage, with its diameter, *d*, in feet. Set the soil cover at Manhole 4 (M4 in Fig. 13.5) to be 2 ft and then plot the vertical sewer profile using a slope of 1%, which is chosen based on the allowable water flow velocity. As illustrated in Fig. 13.5, the downstream side of each manhole has a vertical drop, which is selected to make sure there is a minimum of 2-ft soil cover on the upstream side of the manhole. In this case, a 3-ft drop is sufficient at both Manhole 2 and 3. For the sake of safety, a manhole drop should not exceed 5 feet.

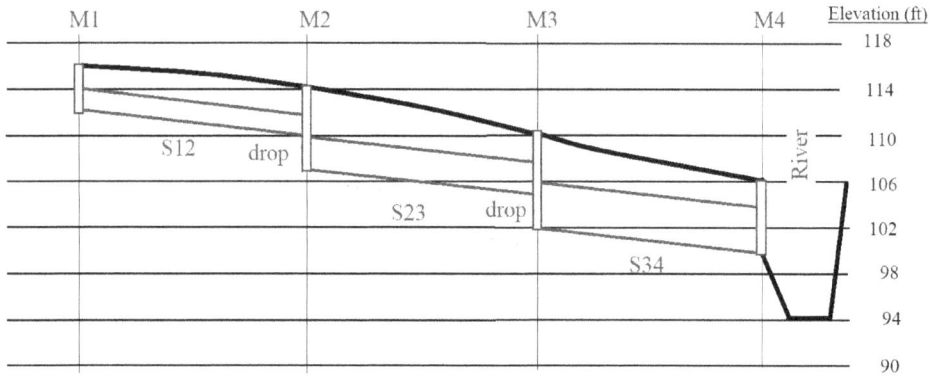

Figure 13.5 Design of a manhole drop

13.3 MANHOLE IN SEWER SYSTEM

A manhole is constructed to provide an efficient transition between incoming and outgoing sewers or to serve as an access to storm sewers for maintenance purposes. A manhole should be placed where:

1. the pipe size changes,
2. the direction of sewer line changes,
3. the invert grades (slopes) along the sewers change,
4. the drops are added to the sewer vertical profiles,
5. the laterals join the trunk line,
6. cleanout access is needed, and
7. the spacing between manholes exceeds 400 feet.

13.4 INCOMING LATERALS

As shown in Fig. 13.6, the angle of confluence between the main line and the lateral should not exceed 45 degrees. A connector pipe from a street inlet box may join the main line at an angle greater than 45 degrees up to a maximum of 90 degrees. Care must be taken when the backwater effects from the main line may severely impact the flow conditions in the lateral. A smooth transition can significantly reduce the surcharge conditions under backwater effects.

13.5 CLASSIFICATION OF SEWERS

The design discharge at a manhole is determined by the cumulative tributary area. The downstream sewer at each manhole is then sized by the assumption of open-channel flow under normal flow conditions. The calculated hydraulic pipe size is often not available commercially, so the next larger commercially available pipe size should be used. For instance, the commercially available circular pipe sizes (in diameter) are: 6,

(a) Manhole and Incoming Lateral   (b) Manholes and Sewer

Figure 13.6 Sewer between manholes and an incoming lateral

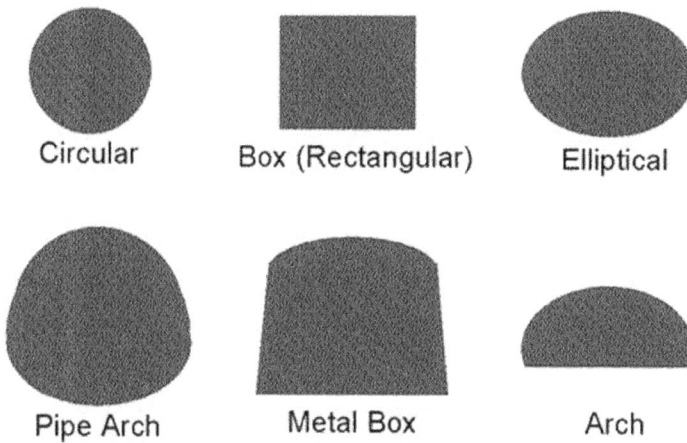

Circular   Box (Rectangular)   Elliptical

Pipe Arch   Metal Box   Arch

Figure 13.7 Various sewers

8, 10, 12, 15, 18, 21, 24, 30, 33, 36, 42, 48, 54, 60, 66, 72, 78, 84-inch etc. Sewers are prefabricated in various shapes (see Fig. 13.7) and materials including concrete or corrugated steel metal. If the detailed hydraulic properties are not readily available for sewers of various shapes, the equivalent circular shape is recommended for hydraulic calculations.

13.6 SEWER SIZING

In practice, all sewers in a drainage system are firstly sized under the assumption of open-channel flow without considering downstream and upstream influences. Manning's formula for open-channel flow is recommended for sewer sizing as:

$$V = \frac{K_n}{n} R^{\frac{2}{3}} \sqrt{S_s} \qquad (13.1)$$

$$R = \frac{A}{P} \qquad (13.2)$$

$$Q = VA \qquad (13.3)$$

$$F_r = \sqrt{\frac{Q^2 T}{gA^3}} \qquad (13.4)$$

where V = cross-sectional normal velocity in [L/T], K_n = 1.486 for foot-second units or 1.0 for meter-second units, R = hydraulic radius in [L], S_s = sewer slope in [L/L], n = Manning's roughness, P = wetted perimeter in [L], A = flow area in [L^2], such as square feet, Q = flow rate in [L^3/T], g = gravitational acceleration in [L/T^2], T = top width of the flow area in [L], and F_r = Froude number in [$L^0 M^0 T^0$]. Eqs 13.1 to 13.4 are further expanded into various cross-sectional geometries.

13.6.1 Circular sewers

As shown in Fig. 13.8, the hydraulic parameters for a partially full flow in a circular pipe can be related to half of the central angle as:

$$y = \frac{d}{2}(1 - \cos\theta) \qquad (13.5)$$

$$A = \frac{d^2}{4}(\theta - \sin\theta\cos\theta) \qquad (13.6)$$

$$P = d\theta \qquad (13.7)$$

$$T = d\sin\theta \qquad (13.8)$$

where θ = half of the central angle in radians, d = diameter of pipe in [L], T = top width in [L], and y = flow depth in [L].

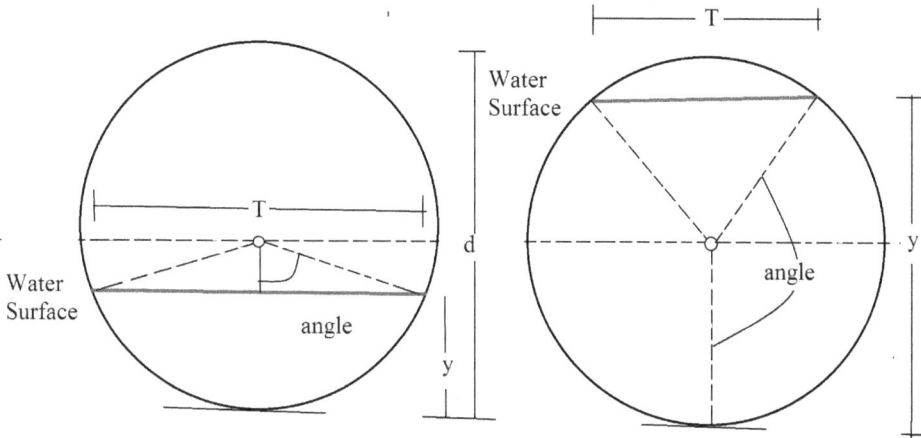

Figure 13.8 Hydraulic parameters in a circular sewer

When a circular pipe is under a flowing full condition, half of the central angle is 180-degree. The hydraulic parameters for flowing full are calculated as:

$$A_f = \frac{\pi d^2}{4} \tag{13.9}$$

$$R_f = \frac{d}{4} \tag{13.10}$$

Substituting Eqs 13.9 and 13.10 into Eq. 13.3 yields:

$$d = \left(\frac{nQ}{K\sqrt{S_0}} \right)^{\frac{3}{8}} \tag{13.11}$$

where Q = design flow in [L³/T], S_o = conduit invert slope in [L/L], d = hydraulically required circular diameter in [L], A_f = full flow area in [L²], R_f = full flow hydraulic depth in [L], and K = 0.462 for foot-second units or 0.311 for meter-second units. Eq. 13.15 is employed to calculate the pipe diameter after the design discharge, Q, and the sewer slope are known. The diameter determined by Eq. 13.11 is called the *hydraulically required sewer diameter*. The calculated pipe size is often not commercially available. As a result, the next larger commercially available pipe size should be used.

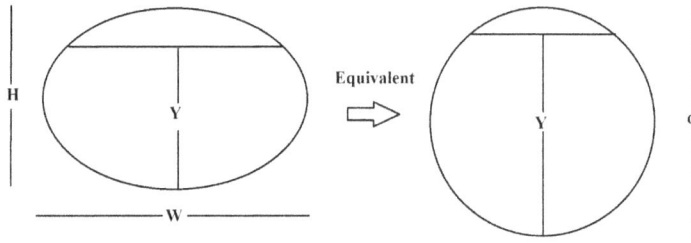

Figure 13.9 Elliptical sewer

13.6.2 Arch (elliptical) sewer hydraulics

Often the narrow clearance due to exiting underground utilities sets constraints on the sewer's vertical profile. Between two manholes, a flat and wide sewer may fit into the narrowed corridor. As a result, an elliptical or arch pipe is selected as a replacement for a circular pipe. The flow through an elliptical or arch pipe in Fig. 13.9 is dictated by its cross-sectional geometry. For simplicity, an equivalent circular pipe may be used as an approximation in hydraulic computations.

The equivalent diameter is approximated by:

$$d = 0.5\sqrt{HW} \tag{13.12}$$

where H = rise of arch or elliptical sewer in [L] and W = span of arch or elliptical sewer in [L].

13.6.3 Box sewer hydraulics

As illustrated in Fig. 13.10, the hydraulic parameters in a box sewer are related to the flow depth as:

$$A = BY \tag{13.13}$$

$$P = 2Y + B \tag{13.14}$$

When a box sewer is selected, the width of the box sewer must be specified first. The hydraulic calculation is to provide the flow depth.

Example 13.2: Referring to Fig. 13.5, Sewer S23 is designed to carry a peak discharge of 60.0 cfs. Knowing that $n = 0.014$, suggest the sewer slope that ensures that the flow velocity does not to exceed 10.0 ft/second.

Solution: Try $S_o = 0.01$ ft/ft.

$$d = \left(\frac{nQ}{K\sqrt{S_0}}\right)^{\frac{3}{8}} = \left(\frac{0.014 \times 60.0}{0.462\sqrt{0.01}}\right)^{\frac{3}{8}} = 2.96 \text{ ft.} \quad \text{Try } d = 3.0 \text{ ft.}$$

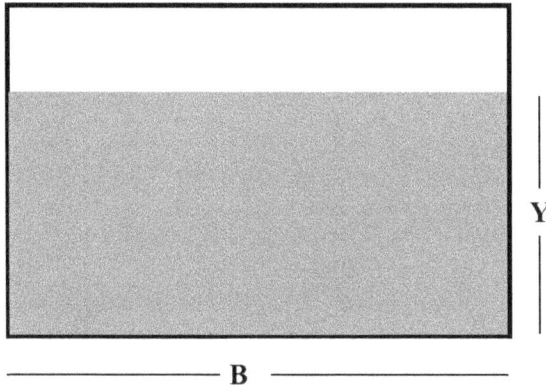

Figure 13.10 Box sewer

Under the assumption of full flow, the full flow velocity, V_f, is calculated as:

$$A_f = \frac{\pi d^2}{4} = 7.06 \text{ sq ft}$$

$$V_f = \frac{Q}{A} = \frac{60.0}{7.06} = 8.5 \text{ fps} < 10 \text{ fps Ok!}$$

Under the assumption of partially full flow (i.e. normal flow), by trial and error, the half of the central angle is set to be:

$\theta = 2.19 \text{ rad} = 125 \text{ degree}$

$$y = \frac{d}{2}(1 - \cos\theta) = \frac{3.0}{2}(1 - \cos(2.19)) = 2.36 \text{ ft}$$

$$A = \frac{d^2}{4}(\theta - \sin\theta\cos\theta) = \frac{3.0^2}{4}(2.19 - \sin 2.19 \times \cos 2.19) = 6.0 \text{ sq ft}$$

$$P = d\theta = 3.0 \times 2.19 = 6.57 \text{ ft}$$

$$R = \frac{A}{P} = \frac{6.0}{6.57} = 0.91 \text{ ft}$$

$$V = \frac{K_n}{n} R^{\frac{2}{3}}\sqrt{S_s} = \frac{1.486}{0.014}0.91^{2/3}\sqrt{0.01} = 9.96 \text{ fps} < 10 \text{ fps Ok!}$$

$Q = VA = 9.96 \times 6.0 = 60 \text{ cfs Ok!}$

In operation, the flowing full condition in the sewer may occur when the downstream tailwater is high. The partially flowing full condition may dominate if an open-channel flow develops without a backwater condition from the downstream manhole.

13.7 DESIGN CONSTRAINTS

As a trial-and-error process, the sewer system design is finalized by continuously adjusting sewer sizes and manhole drops until all the design criteria and constraints are satisfied, including:

1. permissible flow velocity in sewer line between 3 and 20 ft/sec,
2. minimum earth coverage of 2 feet,
3. minimum sewer diameter of 18 inches,
4. minimum manhole drop of 0.20 feet
5. minimum of 2 feet used for sewer trench bottom width
6. earth side slope of 1V:1H used in sewer trench excavation,
7. maximum manhole spacing of 400 feet, and
8. maximum ratio of normal flow depth to sewer diameter or height of 0.8.

After all sewers are sized by open-channel flow under normal flow conditions, the sewer system is further subject to a performance evaluation under the given tailwater at the system exit. The energy and hydraulic grade lines (EGL and HGL) will predict if any manholes in the sewer system are surcharged. Further modifications of the vertical profile may be required until the HGL for the design flow is kept below the ground at all manholes in the system.

13.8 DESIGN PROCEDURES

Design of a sewer system begins with the placement of inlets and manholes. All sewer crowns between two manholes must be set at 2 feet or more below the ground surface. Manhole drops can be introduced to reduce the water flow velocity or to avoid conflicts with utilities. Street inlets and storm sewers are usually designed to pass a 2- to 5-year event. As shown in Fig. 13.11, during a major event, the excess stormwater would be carried by street gutters.

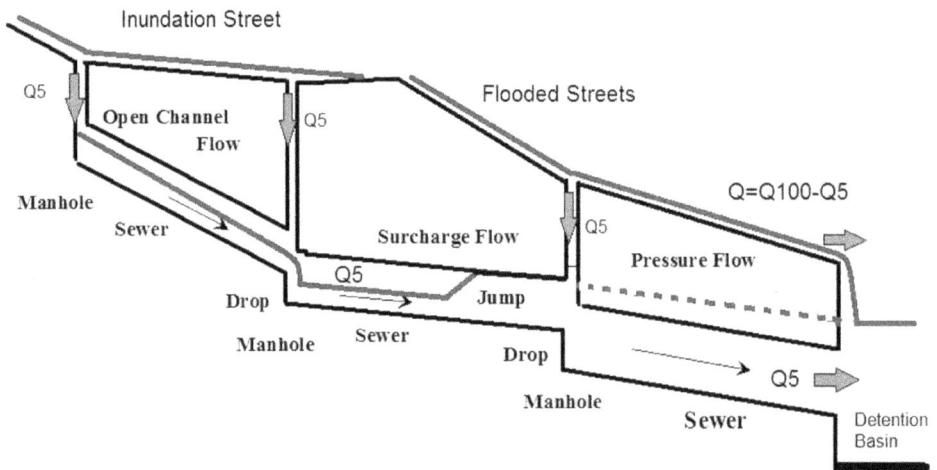

Figure 13.11 Street cutter and underground sewer (double-decker drainage systems)

The sewer system collects the surface storm runoff flows from manholes. The sewer sizing process starts from the most upstream manhole and then moves downstream as the flow rate and the flow time accumulate toward the system exit. At a manhole, the flow times along various flow paths must be first calculated, and then the longest of these should be considered to be the design rainfall duration. The flow time in a sewer can be computed by either the flowing full velocity or the partially full flow velocity determined by Manning's formula. The difference in flow times between these two velocities is nearly negligible. For a flat or negatively sloped sewer, its flowing full velocity is used for hydraulic computations.

13.9 DESIGN DISCHARGE

An urban watershed is divided into small catchments based on the locations of the inlets. Often the street center lines along street crowns serve as catchment boundaries. The watershed in Fig. 13.12 is divided into four catchments for four inlets.

The outgoing sewer from a manhole is sized based on the total design flow at the manhole. At Manhole 1, the tributary area is area *acdeh in* Fig. 13.12. At Manhole 2, the entire watershed is the tributary area. There are two flows coming into Manhole 1, therefore the longer time of concentration dictates the design flow at Manhole 1. For convenience, the Rational method is re-arranged using the effective area, which is defined as the product *CA* for each catchment.

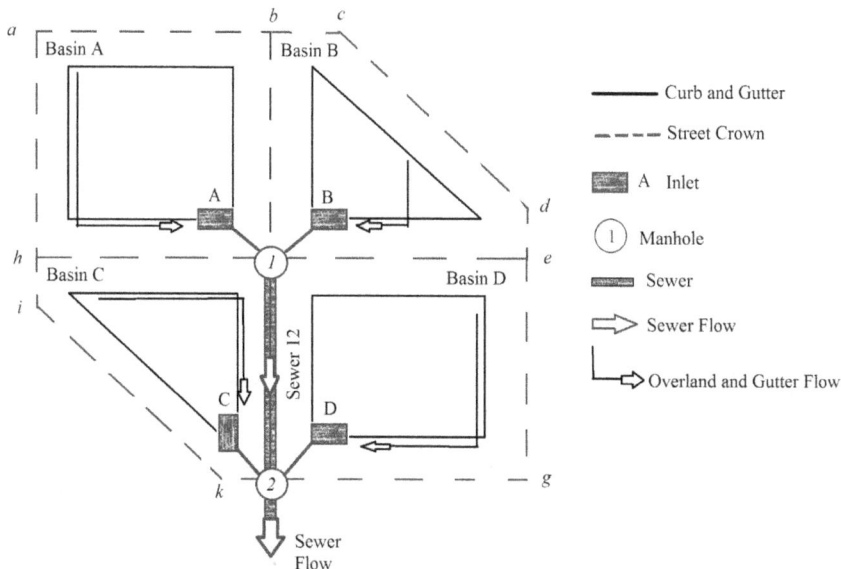

Figure 13.12 Design flows for inlets and manholes

$$A_e = \sum_{k=1}^{k=M} C_i A_1 \tag{13.15}$$

$$T_c = Max(T_{C1}, T_{C2}, T_{C3}, ... T_{CM}) \tag{13.16}$$

$$I = \frac{K_1 P_1}{(K_2 + T_c)^{K_3}} \tag{13.17}$$

$$Q = IA_e \tag{13.18}$$

where C = runoff coefficient, A = catchment area in $[L^2]$, A_e = effective area in $[L^2]$, T_c = time of concentration in $[T]$, I = rainfall intensity in $[L/T]$, Q = peak flow in $[L^3/T]$, i = i-th catchment, k = counter in summation, M = number of catchments, and K_1, K_2, and K_3 = constants in the rainfall intensity–duration–frequency (IDF) curve. Eq. 13.4 is the design flow to size the outgoing sewer. The numerical routine from Eqs 13.15 to 13.18 is repeated at every manhole for sewer sizing.

Example 13.3: The hydrologic parameters for Basins A and B are already given in Fig. 13.13. Consider the rainfall IDF formula with $K_1 P_1 = 38.5$, $K_2 = 10$, and $K_3 = 0.789$. Determine the peak flow at Manhole 1.

Solution:

$$A_e = \sum_{k=1}^{k=2} C_i A_1 = 0.65 \times 2.2 + 0.9 \times 0.67 = 2.03\,acres$$

$$T_c = max(T_{C1}, T_{C2}, T_{C3}, ... T_{CM}) = max(7.5, 5.5) = 7.5\,min$$

$$I = \frac{K_1 P_1}{(K_2 + T_c)^{K_3}} = \frac{38.5}{(10 + 7.5)^{0.789}} = 4.02\,inch/hr$$

$$Q = IA_e = 4.02 \times 2.03 = 8.2\,cfs$$

Knowing the design flow of 8.2 cfs, we are ready to size the outgoing sewer.

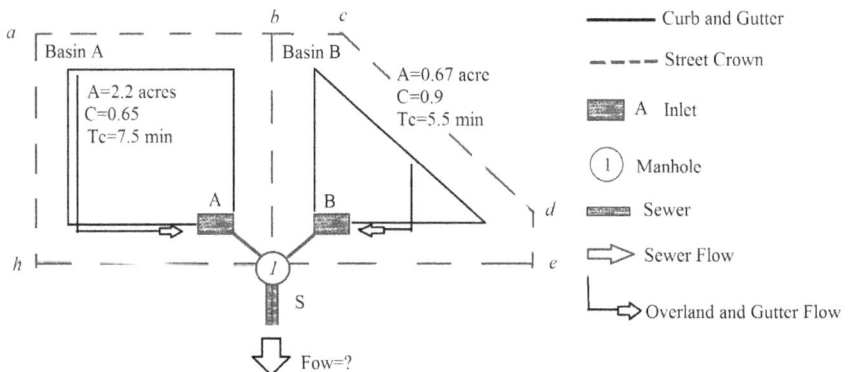

Figure 13.13 Design flow at manholes

Example 13.4: Design a circular sewer to deliver a discharge of 8.2 cfs in Example 13.3 on a slope of 1.0% with a Manning's coefficient of 0.012.

Solution:

(1) Find the hydraulically required pipe size

$$d = \left(\frac{0.012 \times 8.2}{0.462\sqrt{0.01}} \right)^{\frac{3}{8}} \times 12 = 16.5 \text{ inches}$$

(2) Use an 18-inch circular sewer that has a full flow capacity as:

$$Q_f = \frac{1.486}{0.015} \times (1.5/4)^{\frac{2}{3}} \times \left(\frac{3.1416 \times 1.5^2}{4} \right) \times \sqrt{0.01} = 11.5 \text{ cfs}$$

(3) Determine the design flow condition in the 18-inch pipe. The trial-and-error procedure begins with a guessed central angle. Using Eqs 13.9 to 13.12, we will check if the central angle satisfies Manning's equation with a design flow of 8.2 cfs. For instance, try $\theta = 1.83$ radians. Check Manning's equation for the design flow of 8.2 cfs as:

$$Y = \frac{d}{2}(1 - \cos\theta) = \frac{1.5}{2}(1 - \cos 1.83) = 0.94 \text{ ft.}$$

$$A = \frac{d^2}{4}(\theta - \sin\theta\cos\theta) = \frac{1.5^2}{4}(1.83 - \sin 1.83 \times \cos 1.83) = 1.17 \text{ ft}^2$$

$$P = d\theta = 1.5 \times 1.83 = 2.75 \text{ ft}$$

$$Q = VA = \frac{1.486}{0.012}\left(\frac{1.17}{2.75}\right)^{2/3}\sqrt{0.01} \times 1.17 = 8.2 \text{ cfs Ok!}$$

$$V = \frac{Q}{A} = \frac{8.2}{1.17} = 7.0 \text{ fps}$$

$$T = d\sin\theta = 1.5 \times \sin 1.83 = 1.45 \text{ ft}$$

$$Fr = \sqrt{\frac{Q^2 T}{gA^3}} = \sqrt{\frac{8.2^2 \times 1.45}{32.2 \times 1.17^3}} = 1.38 \text{ for normal flow}$$

The above analysis is under normal flow conditions. In fact, a sewer pipe is also subject to downstream backwater effects. Under surcharge conditions, the sewer pipe likely becomes flowing full. Its full flow velocity is:

$$V_f = \frac{8.2}{3.1416 \times 1.5^2 / 4} = 4.64 \text{ fps}$$

Sizing a sewer system starts from the most upstream manhole. Referring to Fig. 13.12, Manhole 2 has three incoming flows, including two *local inlet flows* and one *sewer flow* from Manhole 1. For comparison, the flow times along these three flow paths must be calculated because the longest one is chosen for the design rainfall duration. For instance, from Manhole 1 to Manhole 2, the travel time is:

$$T_s = L_s / (60 \times V_s) \tag{13.19}$$

where T_s = travel time through Sewer 12 in minutes, L_s = length for Sewer 12, and V_s = flowing full velocity through Sewer 12. The sewer flow velocity is assumed to be either full flow or partially full flow velocity. Often, the difference in flow times between these two velocities is numerically negligible. For this case, the time of concentration at Manhole 2 is:

$$T_{M2} = \max(T_{c3}, T_{c4}, T_{M1}+T_s) \tag{13.20}$$

where T_{M2} = time of concentration at Manhole 2, T_{c3} = time of concentration from Basin 3, and T_{c4} = time of concentration from Basin 4. The design rainfall intensity, i_2, derived at Manhole 2 applies to Basins 1, 2, 3, and 4. Therefore, the design discharge at Manhole 2 is:

$$Q = i_2 \, (C_1 A_1 + C_2 A_2 + C_3 A_3 + C_4 A_4) \tag{13.21}$$

Repeat the above procedure to each manhole in the system until the outfall sewer pipe is sized. When all sewers have been individually sized, the sewer system is then ready for performance evaluation under tailwater effects at the system outfall. The EGL and HGL will detect if any manholes and sewers are surcharged. To avoid reverse flows on the street from a surcharged manhole, it is necessary to continue modifying the sewer vertical profile until the HGL for the design flow is kept below the ground at all manholes in the system.

Example 13.5: Continue Example 13.4. The hydrologic parameters for Basins C and D are given in Table 13.1. Determine the design discharge at Manhole 2.
   Sewer 12 has a length of 250 feet with a full flow velocity of 4.64 fps (see Example 13.4). The travel time through Sewer 12 is:

$$T_s = \frac{L_s}{60V_s} = \frac{250}{4.64} \frac{1}{60} = 0.90 \, \text{min}$$

$$T_{M2} = \max(T_{c3}, T_{c4}, T_{M1} + T_s) = \max(5.5, 7.5, 7.5 + 0.90) = 8.4 \, \text{min}$$

$$I = \frac{K_1 P_1}{(K_2 + T_c)^{K_3}} = \frac{38.5}{(10+8.4)^{0.789}} = 3.87 \, \text{inch/hr}$$

$$A_e = \sum_{k=1}^{k=M} C_i A_1 = \sum_{k=1}^{k=4} (0.65 \times 2.2 + 0.9 \times 0.66 + 0.65 \times 0.66 + 0.65 \times 2.2) = 3.89 \text{ acres}$$

$$Q = IA_e = 3.87 \times 3.89 = 15.03 \text{ cfs}$$

The outgoing sewer from Manhole 2 should be sized to carry a peak flow of 15.03 cfs.

Table 13.1 Hydrologic parameters for basins C and D

Catchment Parameters	Basin C	Basin D
Area (acres)	0.66	2.2
T_c to basin outlet (minutes)	5.5	7.5
Flow time through sewer AB	0	0
Runoff C	0.65	0.65

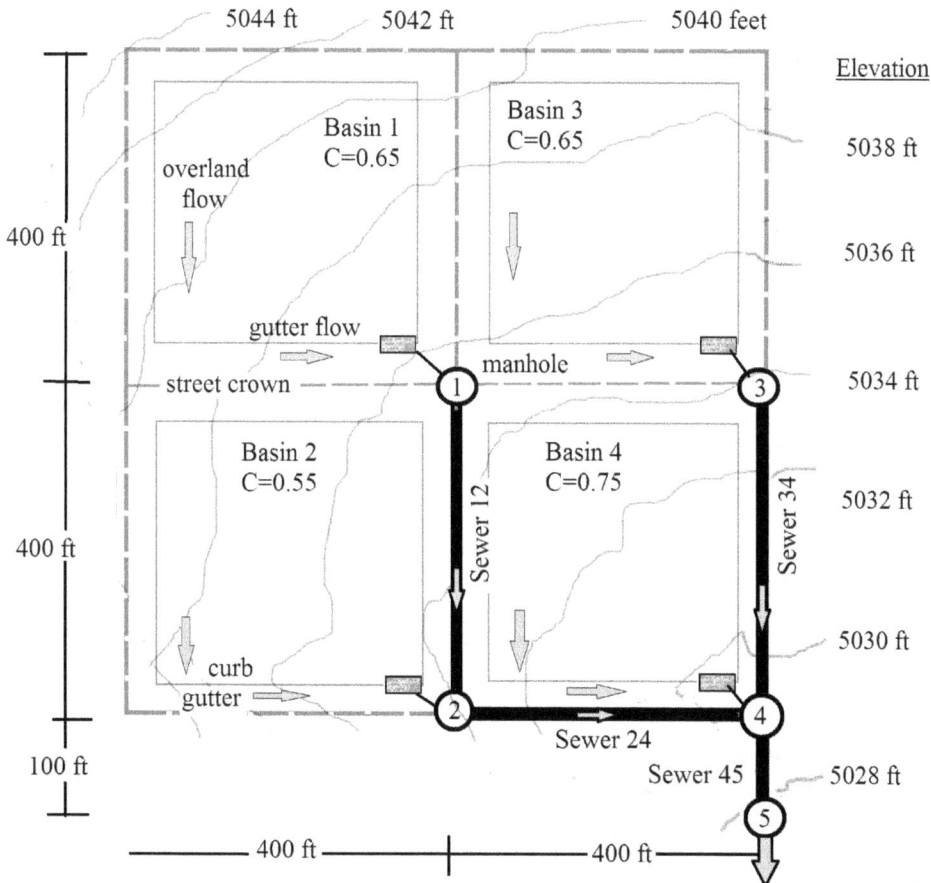

Figure 13.14 Layout of a storm sewer system

13.10 CASE STUDY

Sewer system A is located in the City of Denver, Colorado, and is illustrated in Fig. 13.14. Use the following information to size the sewer pipes.

(a) The 5-yr design rainfall intensity, I (inch/hr), is given as:

$$I = \frac{28.5P_1}{(10 + T_d)^{0.79}}$$

Where $P_1 = 1.35$ inch and T_d = rainfall duration in minutes.

(b) The overland flow time, T_o, should be calculated by the airport formula:

$$T_o = \frac{0.393(1.1 - C)\sqrt{L_o}}{S_o^{0.33}}$$

where L_o = overland flow length ≤ 300 feet, C = runoff coefficient, S_o = overland flow slope in ft/ft. The swale and gutter flow velocities, V_f, can estimated using the SCS upland method with a conveyance coefficient of 20.

$V_f = 20\sqrt{S_f}$, where V_f = flow velocity (fps) and S_f = ground slope in ft/ft.

(c) The tailwater surface at the system exit is at elevation 5022 feet.

Solution:

Basin ID	Area acres A	Runoff Coef C	Imp Ratio Ia	Effective Area acres CA	Overland Flow Slope % So	Length ft Lo	Time min To	Swale Flow Length ft L2	SCS K fps K2	Time min T2	Channel Flow Slope % S2	Length ft L3	SCS K fps K3	Time min T3
1	3.67	0.65	0.70	2.39	1.00	300.00	14.03	100.00	20.00	0.83	1.00	400.00	20.00	3.33
2	3.67	0.55	0.60	2.02	0.50	300.00	21.55	100.00	20.00	1.18	1.00	400.00	20.00	3.33
3	3.67	0.65	0.70	2.39	1.00	300.00	14.03	100.00	20.00	0.83	0.50	400.00	20.00	4.71
4	3.67	0.75	0.80	2.75	0.50	300.00	13.72	100.00	20.00	1.18	1.50	400.00	20.00	2.72

Avg Slope % Sa	Exist Tc Comp min TcComp	Future Tc Check min TcCheck	Storm Duration min Td	Rainfall Intensity in/hr I	Peak Flow cfs Qp
1.00	18.20	12.12	12.12	1.37	3.28
0.75	26.07	15.41	15.41	1.19	2.41
0.75	19.58	13.42	13.42	1.30	3.10
1.00	17.62	10.32	10.32	1.49	4.12

2. Combined Peak Flows at Street Intersection

| | Local Catchment | | | | | Cumulative Parameters at Upst design Pt. | | | | | | | |
| | | | | | | | | | Travel Time Through Street | | | Effective | Storm Duration |
Street Gutter ID	Gutter Length Ls ft	Gutter Slope Ss ft/ft	Basin ID	Effective Area CA acres	Rainfall Duration Td min	Incoming Gutter ID	Effective Area CA acre	Rainfall Duration Td min	length Ls ft	velocity Vs fps	time Ts min	Area CA acre	Td min
12	400	0.01	1.0	2.39	12.12							**2.39**	**12.12**
24	400	0.005	2.0	2.02	15.41		2.02					2.02	15.41
					12.12	12	2.39	12.1	400.0	9.45	0.71	**2.39**	**12.83**
												4.41	**12.83**
34	400	0.01	3.0	2.39	15.41							**2.39**	**15.41**
45	200	0.01	4.0	2.75	13.42							2.75	15.41
						24	4.41	12.8	400.0	7.55	0.88	4.41	13.71
						34	2.39	15.4	400.0	9.07	0.74	2.39	16.15
												9.55	**16.15**

Rainfall Intensity i in/hr	Peak Flow Qp cfs	Pipe Dia Req'd D-req'd inch	Pipe Dia Used D-used inch	Sewer Flow Velocity Vs fps
3.34	7.98	16.70	18.00	9.45
3.26	14.37	23.71	24.00	7.55
3.00	7.15	16.03	18.00	9.07
2.93	27.98	26.73	27.00	6.72

HOMEWORK

Q13-1 A sewer conduit is sized to be circular with a diameter of 42 inches. Due to the limited clearance, the height of the sewer conduit cannot exceed 36 inches. Recommend an elliptic pipe as a replacement.

Q13-2 The proposed sewer vertical profile is shown in Fig. Q13-2. (1) Determine the manhole drops at Manholes M2, M3, and M4, (2) The tailwater elevation is 105 ft at the sewer exit, and all of manholes are aligned in a straight row. With Manning's $N = 0.014$, determine the EGL and HGL for the proposed sewer vertical profile.

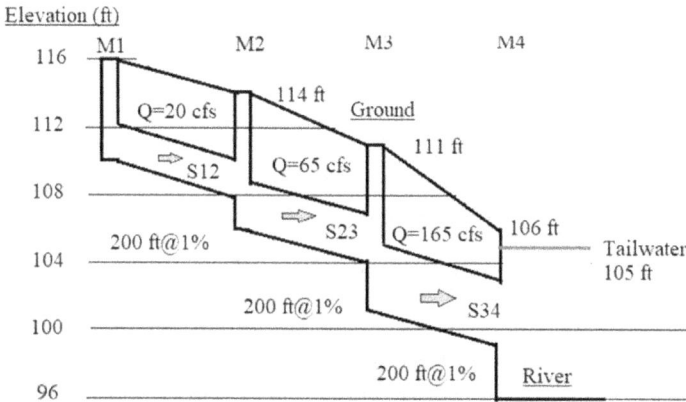

Figure Q13-2 Sewer vertical profile

Solution: Fill in your answers to the blanks.

Sewer ID	Sewer Length ft	Sewer Flow cfs	Sewer Diameter inches	Flow Velocity fps	Friction Slope ft/ft	Friction Loss ft	Manhole Loss Coefficient	Manhole Loss ft	Downstream Invert Elevation ft
	L	Q	d	V	Sf	Hf	Km	Hm	Ee
34	200.00	165.00		13.13	0.01522	3.04	0.00	0.00	
23	200.00	65.00					0.00	0.00	
12	200.00	20.00					0.00	0.00	

Manhole ID	Energy associated with Exit Invert ft	Energy associated with Outgoing Flow ft	Energy Selected at Exit ft	EGL at Manhole ft	HGL at Manhole ft	Ground Elev ft
m	ED2	ED1	ED	EGL(m)	HGL(m)	
4				105.00		106.00
3	105.68	105.00	105.68	108.72	106.04	111.00
2						114.00
1						116.00

Q13-3　A sewer system shown in Fig. Q13-3 consists of three catchments and two manhole-sewer units. Determine the design flows and size circular pipes for Sewer 12 and Sewer 23. Propose the vertical plan for the sewer profile. The design information is given as follows:

a.　Design rainfall intensity formula $I(\text{inch/hr}) = \dfrac{40}{(10 + T_d)^{0.76}}$ where

T_d = duration in minutes and I = intensity in inch/hr.

b.　The design rainfall duration is set to be the time of concentration of the catchment. The overland flow time is estimated by the airport formula, and the gutter flow velocity is estimated by the SCS upland method for the pavement surface.

c.　The design constraints are:

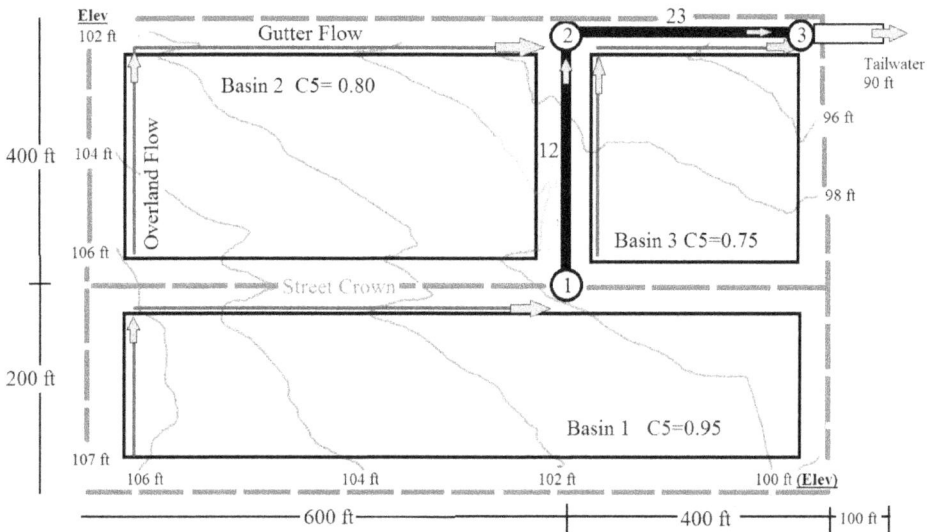

Figure Q13-3 Sewer system design for homework

1. permissible flow velocity in a sewer of between 20 feet/second and 3 feet/second,
2. minimum earth coverage of 2 feet,
3. minimum sewer diameter of 18 inches, and
4. minimum manhole drop of 0.50 feet.

REFERENCES

American Society of Civil Engineers (1979). *Design and Construction of Sanitary and Storm Sewer*, New York.

Guo, J.C.Y. (1985). *Technical Manual for UDSEWER Computer Model*, Research Report, Department of Civil Engineering, University of Colorado at Denver, Denver, CO.

Guo, J.C.Y. (1989a). "Auto Sizing Techniques for Storm Sewer System Design Using UDSWMM", Proceedings for the National EPA Conference, Denver, CO.

Guo, J.C.Y. (1989b). "Energy Dissipation in Storm Sewer System", Proceedings of ASCE Conference on Stormwater Modeling and Management, Denver, CO, May.

Rossman, L. (2005). *EPA SWMM5 User Manual*, www2.epa.gov/

SWMM5 (2005). www2.epa.gov/water-research/storm-water-management-model-swmm

UDFCD Manual (2010). *Storm Water Drainage Design Criteria Manual*, vols 1 and 2, Urban Drainage and Flood Control District, Denver, CO.

Yen, B.C., Wenzel, H.G., Mays, L.W., and Tang, W.H. (1976). "New Models for Optimal Sewer System Design", Proceedings for the EPA Conference on Environmental Modeling and Simulation, Cincinnati, OH, July.

Chapter 14

Detention basin sizing

14.1 TYPES OF DETENTION BASIN

Detention basins are designed to meet different needs and for multiple purposes. Using the functionality as the criterion, detention basins are classified into three major categories. They are: (1) the *flood-control detention basin*, (2) the *stormwater retention basin*, and (3) the *infiltrating basin and trench*. These are discussed in more detail below.

1. *Flood-control detention basin (dry basin)*
    A flood-control detention basin in Fig. 14.1a is placed at the stormwater system outfall to temporarily store excess storm runoff and then to discharge the stored water volume at a rate no more than the allowable (Akan 1990). Between two sequential storm events, a flood control detention basin remains dry and can be accessible as an open space for the public.
2. *Stormwater retention basin (wet basin)*
    A retention basin, shown in Fig. 14.1b, is installed at the low point and operated with a permanent pool. The basin is sized to capture stormwater for the purposes of groundwater recharge, water quality enhancement, stormwater reus, or/and local runoff volume disposal. A retention basin is often mixed with wetland features to settle the solids and pollutants in stormwater
3. *Infiltrating basin and trench (porous basin)*
    Infiltrating basins and trenches are utilized as common low-impact-development (LID) devices to reduce the increased runoff volume. Infiltrating basins, shown in Fig. 14.1c, include rain gardens, infiltration beds, riprap trenches, and vegetation beds. An infiltrating basin consists of an on-surface storage basin, vegetal landscape, porous bottom, and overtopping weir. They are often located at the outlet of an industrial park, business district, or highway intersection as a pollutant-source control device (see Fig. 14.1d).

DOI: 10.1201/9781003284239-14

(a) Detention Basin

(b) Retention Basin

(c) Infiltration Basin

(d) Porous Pool in Detention Basin

Figure 14.1 Examples of storage basins

14.1.1 Classification based on location

Basin location is an important factor in determining the collection of stormwater. Based on the location, stormwater storage basins are classified into the following:

1. *Upstream and downstream basins*
   Upstream basins include small, shallow water-quality porous basins and infiltration basins. As shown in Fig. 14.2, an upstream porous basin shall be placed to target solids removal, while a *downstream basin* shall be installed for the purpose of peak-flow reduction. For instance, urban trash carried in storm runoff from a sports field should be collected into an *upstream basin* before the street inlet. At the exit of a sewer trunk line, a *downstream basin* is needed to control the flow release into the downstream natural water body.
2. *Local and regional basins*
   As shown in Fig. 14.2, a *local detention basin* serves a small, local residential subdivision for flow release control, and a regional detention basin is designed to mitigate flood flows in a major waterway.
3. *On-stream and off-stream basins*
   Widening the floodplain width and constructing an embankment across the floodplain bottom create *on-stream detention* (see Fig. 14.3b). Diverting the

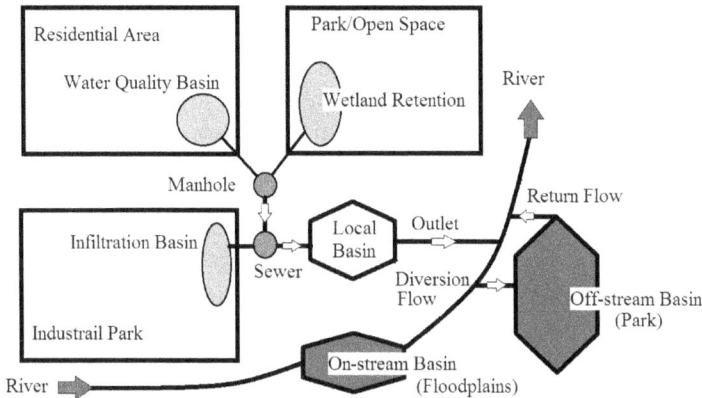

Figure 14.2 Types and locations of detention basins

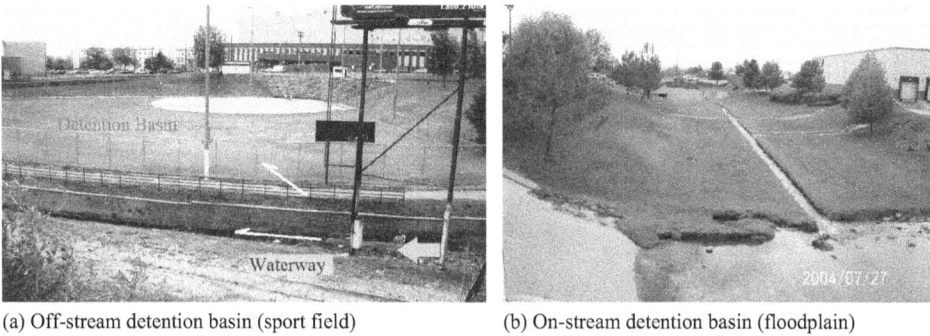

(a) Off-stream detention basin (sport field)   (b) On-stream detention basin (floodplain)

Figure 14.3 Off-stream and on-stream detention basins

excess flood flow from a waterway into adjacent open areas such as depressed parks and sports fields to control stormwater release is termed *off-stream detention* (Fig. 14.3a).

4. *On-site and off-site basins.*

An *on-site detention* is implemented to dispose of the increased runoff volumes on the project site. However, if the easement is available, the flood flows may be transferred to downstream *off-site detention basin* to reduce increased peak flows. Generally, a porous basin is more suitable for on-site infiltration operation, while a detention basin is often recommended for off-site flood control operation.

14.2 DESIGN CONSIDERATIONS

The design of a detention system is an integration of functional integrity, land value, aesthetics, recreation, and safety to merge into the urban setting. From the engineering aspects, the design of a stormwater detention basin should take the following factors into consideration.

14.2.1 Location

In an urban area, parking lot, parks, sport fields, road embankments, and depressed areas provide stormwater storage volume. The selection of a basin site depends on costs, public safety, and maintenance. It is important to impose the concept of multiple land uses to blend a stormwater detention feature into a park, ball field, or/and green belt. A large detention system should also provide recreational functions, such as jogging, walking, bicycling, playground, skating,, or golfing. Different specialists need to work together to develop desirable and acceptable criteria that fit the community's recreational needs as well as flood mitigation purposes.

14.2.2 Basic layout

The basic elements for a detention basin include an *inlet structure* to collect runoff flows, *energy dissipation system* for erosion control at the entrance, *fore bay*, shown in Fig. 14.4a, for sediment settlement, *trickle channel*, shown in Fig. 14.4b, to pass frequent nuisance flows, *storage basin*, shown in in Fig. 14.4c, for the mitigation of design events, and *outlet structure*, shown in Fig. 14.4d, to control flow releases. At the entrance, a proper energy dissipator should be placed for erosion protection. A trickle channel should be installed through the bottom of the basin to pass frequent

(a) Trickle Channel

(b) Forebay at Inlet

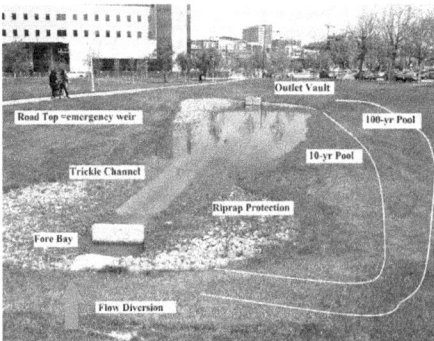

(c) Multiple Layer Detention Basin

(d) Wetland Pool and Outlet Box

Figure 14.4 Detention basins

nuisance flows. In general, the capacity of a trickle channel is 1.0 to 3.0% of the 100-year peak discharge. Proper drops should be placed along the trickle channel to reduce erosion potential. The trickle channel drains into a permanent pool for stormwater quality control. This permanent pool is directly connected to the outlet structure (USWSCM 2010).

An outlet structure is formed from a perforated plate, riser, orifices, and weirs to collect low to high flows into the concrete vault, and outfall pipes discharge the water flow from the concrete vault into the downstream receiving water body. The basin width-to-length ratio must be greater than two so that the flood flows can be suffi-ciently expanded and diffused into the water body to enhance the sedimentation pro-cess. Slopes on embankments have to maintain the stability of the bank slope. As a rule of thumb, slopes on earthen embankments should not be steeper than $1V{:}4H$ and on riprap embankments should not be steeper than $1V{:}2H$.

The cross-sectional geometry of the basin should be designed for multiple events. As shown in Fig. 14.5, the bottom storage volume in a basin is shaped for the water-quality capture volume. The mid layer is shaped to store the 10-year detention volume. From the 10-year water surface up to the weir crest should provide an add-itional storage volume to accommodate a 100-year event. From the weir crest up to the brimful of the basin is the height of the freeboard. To mimic pre-development watershed hydrologic conditions, it is preferred to drain a low storage volume over 6 to 24 hours, whereas a 100-yr storage volume should be emptied within 24 to 72 hours.

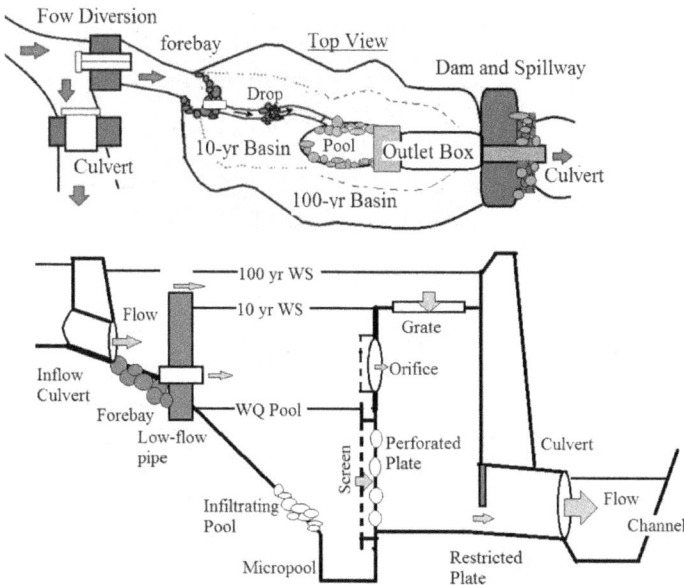

Figure 14.5 Layout of a detention basin for multiple storm events

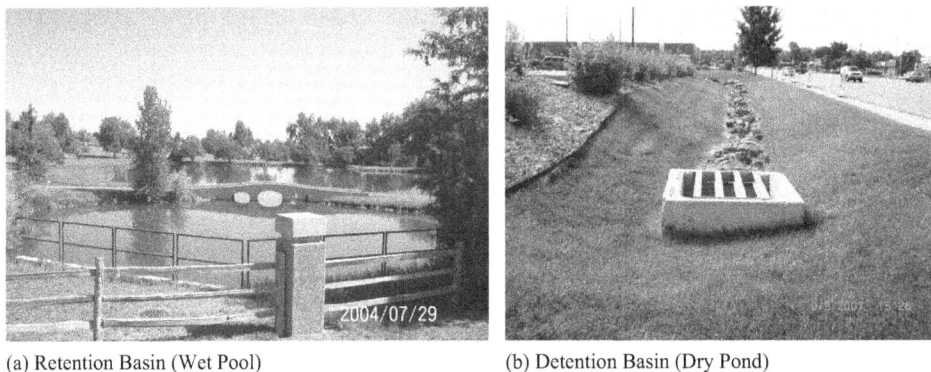

(a) Retention Basin (Wet Pool)   (b) Detention Basin (Dry Pond)

Figure 14.6 Layout of a detention basin for multiple storm events

14.2.3 Groundwater impacts

The detention basin in Fig. 14.6b operates with a dry bottom between two adjacent events. It is necessary to make sure that the drain time of the detention basin does not exceed the average inter-event time between two adjacent storm events. In contrast, a retention basin, as shown in Fig. 14.6a, is designed to be a wet pool. Care must be taken in assessing infiltration to and exfiltration from the local ground-water table. It is necessary to carefully evaluate the water budget among groundwater, surface water, and associated hydrologic losses.

To design a retention basin without an outlet, soil infiltration tests must be carefully conducted. Although vertical drainage wells backfilled with aggregate gravel can be installed to increase the infiltration rate, the subsurface soil's hydraulic conductivity must be carefully examined to make sure that the subsurface geometry sustains the infiltration rate on the land surface. Otherwise, the soil medium below the basin will become saturated, which would cause water mounding to the local groundwater system (Guo 2017).

14.2.4 Inlet and outlet works

Inlets and outlets of a detention basin should be protected from erosion and the deposition of sediment. The concentrated inflow from an inlet pipe shown in Fig. 14.7a has to be diffused into a forebay for energy dissipation and solid settlement. The forebay is a shallow pool that evenly spreads the flow overtopping a level spreader.

The outlet box in Fig. 14.7b is composed of a vertical perforated plate and a horizontal grate on top of the concrete box. The flow capacity through the outlet box is dictated by the operations of the orifices, weirs, and culverts attached on the concrete box. The outlet system for a basin must be designed with a full understanding of downstream tailwater effects. The performance of the outfall culverts must be evaluated for a range of headwater depths at the entrance and tailwater depths at the exit. It is preferable that the outfall pipe is designed under the inlet-control conditions. The trash rack shown in Fig. 14.7c is critically important with regard to public safety. As a rule of thumb, a trash rack is necessary at the entrance of any outfall pipe larger than

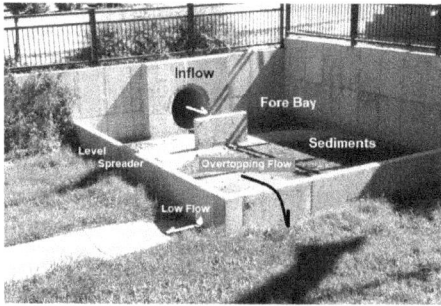

(a) Inlet and Energy Dissipator

(b) Outlet Box for Slow Flow Release

(c) Steel Trash Rack

(d) Notch Weir for Flow Measurement

Figure 14.7 Inlet and outlet works

18 inches (450 mm) in diameter. Often, a monitoring system, such as the one shown in Fig. 14.7d, is installed to measure the inflow and outflow using weirs and orifices. It is important to select locations without vegetal interferences.

14.2.5 Other considerations

The operation of a stormwater detention system also involves many institutional issues, including the infrastructure needed to ensure proper planning, design, construction, operation, and maintenance. A monitoring or regulatory mechanism is required to ensure that the approved design is constructed, operational integrity is implemented, and maintenance is regularly provided. Other considerations include public safety, access facilities, landscaping, and aesthetics.

14.3 DETENTION PROCESS

Urbanization-induced floods reflect post-development stormwater drainage conditions. A *flood control detention basin* is recommended for peak-flow reduction and the delay of the time to peak-flow. As shown in Fig. 14.8, the stormwater detention storage volume is the volume difference between the inflow and outflow hydrographs.

During a major event, the operation of a flood detention basin is composed of a *filling period*, when the inflow rate is greater than the outflow rate, and a *depletion*

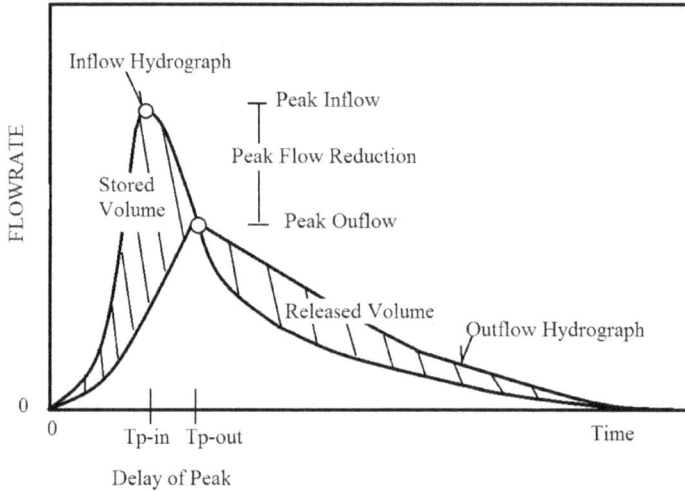

Figure 14.8 Concept of flood detention

period, when the outflow rate is greater than the inflow rate. These two distinct periods are separated by the peak outflow. Fig. 14.8 shows that the higher the peak outflow, the less the storage volume. According to the continuity principle, the total stored (detention) volume during the filling period is equal to the total released volume during the depletion period. As shown in Fig. 14.8, stormwater detention reduces the peak flow and also delays the time to peak.

14.4 ALLOWABLE FLOW RELEASE

Design of a storm water system should not transfer any on-site flooding problems to downstream properties. The allowable stormwater to be released at a design point is often determined by the following:

1. peak discharge under pre-development conditions,
2. critical capacity of the downstream existing drainage facility,
3. allowable flow release published on the regional master drainage plan, and
4. recommended flow release using local design criteria.

In addition to the above, any design constraints associated with the project must also be reviewed. Among all considerations, the smallest release should be adopted for the design. Table 14.1 is a typical example in which the recommended flow releases were defined based on an imperviousness of 5% (USWDCM 2010).

Table 14.1 Allowable flow release rates recommended for Denver area

Design Frequency	Type A Soil (cfs/acre)	Type B Soil (cfs/acre)	Type C or D Soil (cfs/acre)
2-year	0.02	0.03	0.04
5-year	0.07	0.13	0.17
10-year	0.13	0.23	0.30
25-year	0.24	0.41	0.52
50-year	0.33	0.56	0.68
100-year	0.50	0.85	1.00

Example 14.1: A lot of 100 acres is ready for development. The soil texture at the site is classified as a Type B soil. From the regional master drainage study, the flow release from this site was set to be 90 cfs. The downstream existing storm sewer line can take no more than 80 cfs from the site. Determine the 100-yr peak-flow release from the site.

Solution: According to Table 14.1, the 100-yr release for Type B soil is 0.85 cfs/acre or 85 cfs for 100 acres. In comparison with the critical capacity for the existing sewer, the design release for this site is set to be:

Q-allowable = min (85, 90, 80) = 80 cfs.

14.5 DESIGN PROCEDURE

The main objective of stormwater detention is to mitigate the increased storm runoff peak-flow rates. Although the design event for a detention basin is often specified to be a major event such as a 50 to 100-year storm, the operation of a detention basin needs to accommodate events of all kinds. Inflows to the detention basin should be studied for both existing and future conditions. For each project development, it is necessary to identify the changes and mitigation measures between the pre-development and post-development conditions. This effort is an attempt to identify existing and future flood problems. It will serve as a basis for impact evaluations and alternative selections.

Fig. 14.9 outlines the design steps, beginning with the basin site selection. During the preliminary design stage, little information is available. Therefore, it is suggested that the basin geometry be approximated by a triangular, rectangular, or circular shape, and the basin operation be approximated by an inlet-control capacity determined by weir and orifice hydraulics only. Of course, when the project moves to the final stage, the preliminary design can be further refined with more information. For instance, tailwater and backwater effects should be considered to refine the basin characteristic curves, and basin performance must be evaluated by hydrologic routing techniques.

```
┌─────────────────────────────────────────────┐
│  Watershed Studies to identify flooding probems │
└─────────────────────────────────────────────┘
                      │
                      ▼
          ┌────────────────────────┐
          │  Selection of Basin Site │
          └────────────────────────┘
                      │
                      ▼
        ┌─────────────────────────────┐
        │  Pre-sizing Detention Volumes │◄─────────────────┐
        │  for 10- and 100-year Events  │                  │
        └─────────────────────────────┘                  │
                      │                                    │
                      ▼                                    │
       ┌──────────────────────────────────┐              │
       │  Pre-shaping Basin Geometry        │              │
       │  using a triangular or circular shape │           │
       └──────────────────────────────────┘              │
                      │                                    │
                      ▼                                    │
     ┌──────────────────────────────────────┐            │
     │  Size Outfall Struture                 │            │
     │  to control the 10- and 100-yr release rates │      │
     └──────────────────────────────────────┘            │
                      │                                    │
                      ▼                                    │
     ┌──────────────────────────────────────┐            │
     │  Preliminary Design to refine the      │            │
     │  basin geometry and outfall works      │            │
     │  Computer Modeling for Hydrograph Routing │         │
     └──────────────────────────────────────┘            │
                      │                             No     │
                      ▼                                    │
     ╭──────────────────────────────────────╮            │
     │  Are Release Rates < Allowable Rates?   ├───────────┘
     ╰──────────────────────────────────────╯
                      │ Yes
                      ▼
     ┌──────────────────────────────────────────┐
     │  Final Design                               │
     │  to refine tailwater, groundwater, and maintenance │
     └──────────────────────────────────────────┘
```

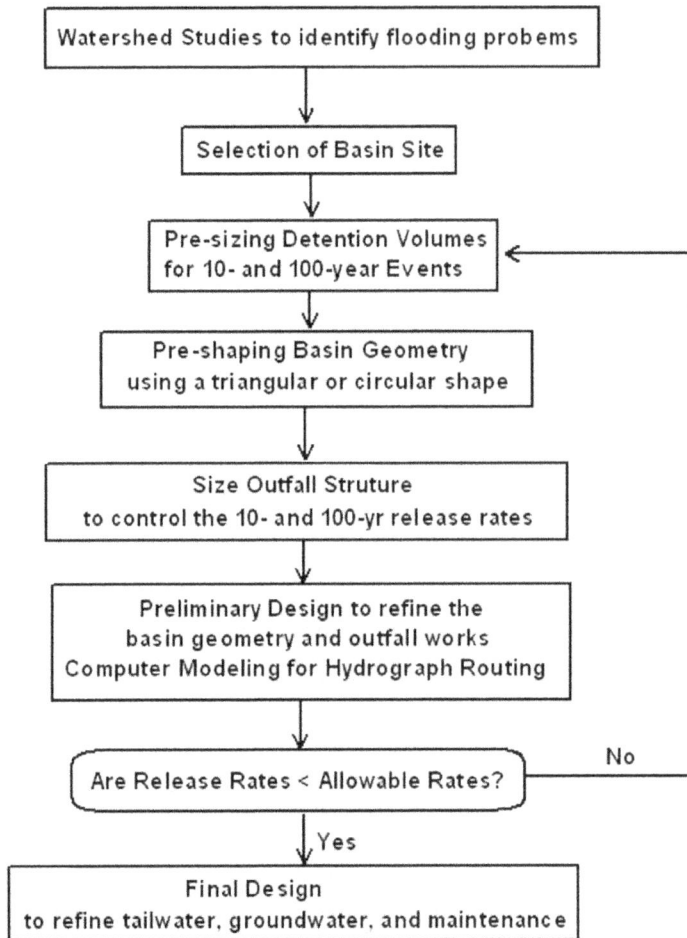

Figure 14.9 Procedure for detention basin design

The above procedure is an iterative process until all design criteria and safety concerns are satisfied.

14.6 DETENTION VOLUME

Detention volume is defined as the difference between the inflow and outflow volumes. The inflow volume is generated from the tributary area under a specified design storm, whereas the outflow volume is determined according to flow release control. The inflow hydrograph to the detention site should be predicted for future developed conditions. The allowable release is determined to either not exceed the existing flow release or the critical capacity of the downstream drainage facilities. At the planning stage, details of the outlet structure are not yet known. As a result, the after-detention hydrograph, as shown in Fig. 14.10, is approximated by the linear rising hydrograph

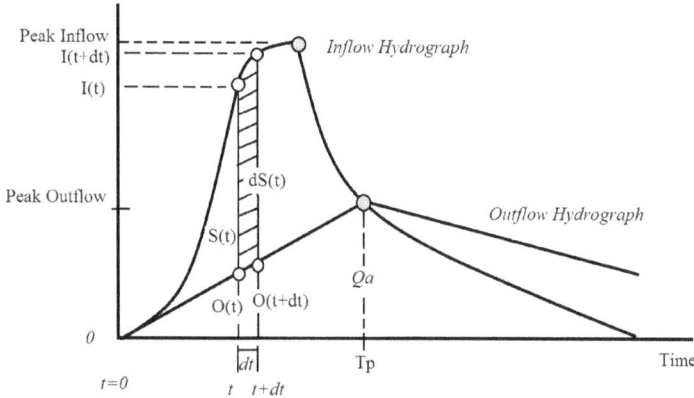

Figure 14.10 Detention volume by the hydrograph method

from the beginning of the event to the allowable release rate on the inflow recession hydrograph (Guo 1999). The required detention storage volume is then calculated by the volume difference between the inflow and outflow hydrographs (McCuen 1998). As shown in Fig. 14.10, the outflow rate, $O(t)$, at time t on the linear rising limb is estimated as:

$$O(t) = \frac{O_a}{T_P} t \quad for \ 0 \le t \le T_P \tag{14.1}$$

where $O(t)$ = outflow rate in [L³/T] at time t, Q_a = allowable flow release in [L³/T], T_p = time to peak on the after-detention hydrograph in [T], and t = elapsed time in [T]. The accumulated storage volume, $S(t)$, is the volume difference between the inflow hydrograph and the rising outflow hydrograph as:

$$S(t) = \sum_{t=0}^{t=T_P} \left[I(t) - O(t)\right] \Delta t \tag{14.2}$$

where $S(t)$ = cumulative storage volume in [L³], $I(t)$ = inflow rate in [L³/T] at time t, and Δt = time increment in [T].

The design storage volume is then calculated by Eq. 14.2 from $t = 0$ to $t = T_p$ as:

$$S_m = S(T_P) \tag{14.3}$$

where S_m = detention storage volume in [L³]. The hydraulic performance of a detention basin is described by its characteristic curve, i.e. the *storage-outflow curve*. For instance, HEC HMS (2005) and US EPA SWMM (2005) computer models define the performance of a detention basin by such a curve. During the preliminary study, adequate design information is not available, so the pairs (S, O) in Eqs 14.1 and

Table 14.2 Preliminary storage-outflow curve by hydrograph method

Time (minutes)	Given Inflow I(t) (cfs)	Linear Outflow O(t) (cfs)	Incremental Vol (acre-ft)	Cumulative Vol S(t) (acre-ft)
0.0	0.00	0.00	0.00	0.00
10.0	50.00	41.67	0.11	0.11
20.0	250.00	83.33	2.30	2.41
30.0	750.00	125.00	8.61	11.02
40.0	500.00	166.67	4.59	15.61
50.0	350.00	208.33	1.95	17.56
60.0	250.00	250.00	0.00	17.56
70.0	200.00	250.00	-----	------

14.2 can serve as the preliminary *storage-outflow curve* for the basin under design. Of course, such a preliminary relationship can be refined after detailed information becomes available (Guo 2004).

Example 14.2: As shown in Table 14.2, the inflow hydrograph to a detention basin has a peak flow of 750 cfs. The allowable flow released from the basin is set to be 250 cfs. On the recession curve of the inflow hydrograph, a flow of 250 cfs occurs at 60 minutes. Under the assumption of a linear rising outflow hydrograph, Eq. 14.1 becomes:

$$O(t) = \frac{250}{60} t \qquad (14.4)$$

With a time increment of 5 minutes, the cumulative storage volume, S(t), is computed as:

$$S(t) = \sum_{t=0}^{t=60} \left[I(t) - \frac{250}{60} t \right] \times (10 \times 60) / 43560 \qquad (14.5)$$

Note that Q(t) is expressed in cfs and S(t) is expressed in acre-ft. As shown in Table 14.2, the cumulative volume, $S(T_p = 60) = 17.56$ acre-ft. The pairs of (O, S) in Table 14.2 can serve as the preliminary storage-outflow curve for the detention basin under design.

Example 14.3: As shown in Fig. 14.11, the Belleview Watershed of 40 acres in Example 6.1 will be developed to have a detention basin placed at the northeast corner. The watershed is covered with Type B soils and will be developed to have an area imperviousness of 60%. According to Table 14.1, the 10-yr and 100-yr allowable flow release rates are: $Q_{10} = 0.23 \times 40 = 9.20$ cfs and $Q_{100} = 0.85 \times 40 = 34.0$ cfs.

The 10- and 100-yr inflow hydrographs are given in Table 14.3. The detention volumes for the 10- and 100-yr events are determined.

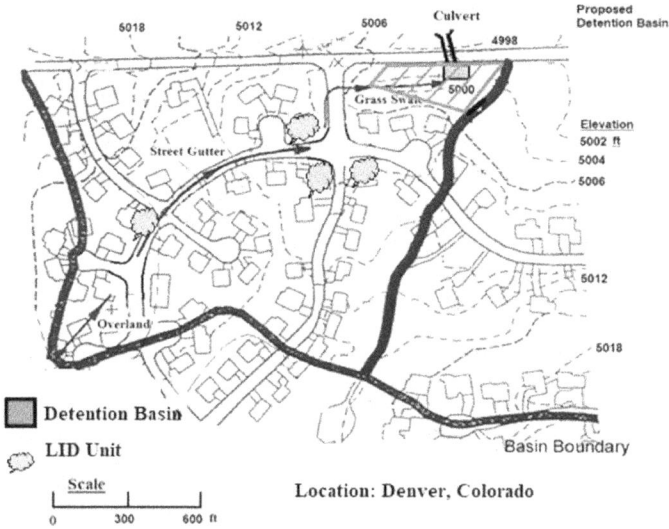

Figure 14.11 Watershed for detention basin design

Table 14.3 Detention volumes for Belleview detention basin

	10-yr Detention Volume				100-yr Detention Volume			
	Peak time = 120.0 min				Peak time = 80.0 min			
Time t	10-yr Hygraph	Linear Outflow	Incremental Volume	Acc. Volume	100-yr Hygraph	Linear Outflow	Incremental Volume	Acc. Volume
min	cfs	cfs	acre-ft	acre-ft	cfs	cfs	acre-ft	acre-ft
0.00	0.00	0.00	0.00	0.00	0.00	0.00	0.00	0.00
10.00	0.00	0.00	0.00	0.00	0.00	0.00	0.00	0.00
20.00	0.00	0.00	0.00	0.00	20.00	8.50	0.16	0.16
30.00	22.69	2.30	0.28	0.28	65.00	12.75	0.72	0.88
40.00	75.58	3.07	1.00	1.28	155.00	17.00	1.90	2.78
50.00	35.65	3.83	0.44	1.72	95.00	21.25	1.02	3.79
60.00	23.14	4.60	0.26	1.97	65.00	25.50	0.54	4.34
70.00	18.48	5.37	0.18	2.15	48.43	29.75	0.26	4.60
80.00	17.01	6.13	0.15	2.30	32.43	32.43	0.00	4.60
90.00	15.71	6.90	0.12	2.43	21.62	0.00	0.00	4.60
100.00	12.00	7.67	0.06	2.48	16.45	0.00	0.00	4.60
110.00	10.26	8.43	0.03	2.51	13.67	0.00	0.00	4.60
120.00	9.40	9.20	0.00	2.51	12.02	0.00	0.00	4.60
130.00	7.00	0.00	0.00	2.51	11.00	0.00	0.00	4.60
140.00	5.00	0.00	0.00	2.51	6.00	0.00	0.00	4.60
150.00	4.00	0.00	0.00	2.51	3.00	0.00	0.00	4.60
	10-yr Storage Volume in ac-ft =			**2.51**	100-yr Storage Volume in ac-ft=			**4.60**

14.7 PRELIMINARY SHAPING

Hydraulic structures are designed to process flows generated from small to extreme events. As a result, a detention basin is built with multiple layers of storage volume, starting from the bottom layer for a 2-yr storage volume, the mid layer to store up to a 10-yr detention volume, and the additional top layer to accommodate a 100-yr storage volume. Having known the detention volumes, the cross-sections of the basin can be approximated by a truncated cone with a circular, triangular, or rectangular base. Refinements to the basin shape can always be added to the future grading plan after the details are finalized.

14.7.1 Rectangular basin

The basic geometric parameters are the width and length of the cross-sectional area for each layer in a rectangular basin, as shown in Fig. 14.12. The cumulative storage volume between two cross-sectional areas is calculated as:

$$L_2 = L_1 + 2zH \tag{14.6}$$

$$B_2 = B_1 + 2zH \tag{14.7}$$

$$A_1 = L_1 B_1 \tag{14.8}$$

$$A_2 = L_2 B_2 \tag{14.9}$$

$$V \frac{1}{3} \left(A_1 + A_2 + \sqrt{A_1 A_2} \right) \approx 0.5 \times \left(A_1 + A_2 \right) H \tag{14.10}$$

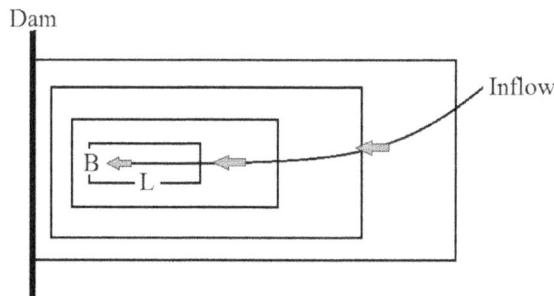

Figure 14.12 Pre-shaping for a rectangular basin

where L = length in [L], B = width in [L], z = average side slope, H = vertical distance in [L], and V = storage volume in [L^3]. The subscript 1 represents the variables for the bottom layer, and 2 represents the variables for the top layer.

Usually, a basin is divided into multiple layers, and the slope of the basin's embankment varies with respect to water depth, starting from as steep as 1V:2H at the bottom to as flat as 1V:10H at the top.

14.7.2 Triangular basin

Repeating the calculations in the previous section, the cross-sectional area in a triangular basin as shown in Fig. 14.13 is calculated by the base width, B_2, and height, L_2, as:

$$A_2 = 0.5B_2L_2 \tag{14.11}$$

The volume between two triangular layers is estimated by Eq. 14.10.

Applying Eqs 14.6 through 14.11 to 2-, 10-, and 100-year storage volumes, the basin is shaped with various stages. Between two adjacent layers, the required side slope can be incorporated into the storage volume calculation. The *stage-storage curve* and *stage-contour area curve* can then be established. Upon completion of pre-shaping and pre-sizing a basin, the engineer can begin to work on the hydrologic and hydraulic modeling and evaluation. During the final design, the detention basin can be further refined to fit the site grading plan.

Example 14.4: The 10- and 100-yr detention volumes for the basin are 2.51 acre and 4.60 ac-ft. Distribute this volume using a triangular basin. The bottom triangle has a width of 200 feet and height of 350 feet. The side slope starts with 1V:4H for water

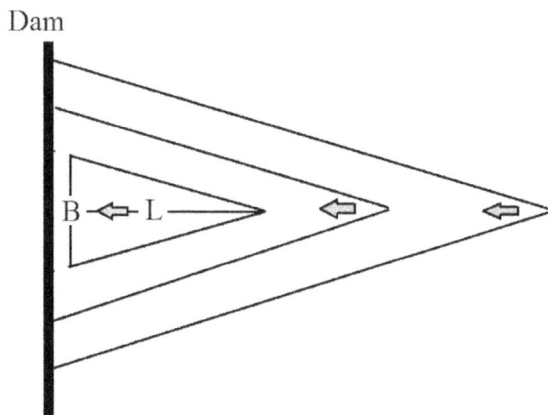

Figure 14.13 Pre-shaping for a triangular basin

Table 14.4 Example for pre-shaping of detention basin

Water Surface Elev feet	Basin Side Slope ft/ft	Water Depth feet	Width of Cross Section feet	Length of Cross Section feet	Cross Sectional Pond Area		Cumulative Storage Pond Volume		Identify Design Water Elev
					sq ft	acres	acre-ft	cubic ft	
5000.00	4.00	0.00	200.00	350.00	35000.0	0.80	0.00	0.0	Inupt
5000.50	4.00	0.50	204.00	354.00	36108.0	0.83	0.41	17777.0	
5001.00	4.00	1.00	208.00	358.00	37232.0	0.85	0.83	36112.0	
5001.50	4.00	1.50	212.00	362.00	38372.0	0.88	1.26	55013.0	
5002.00	4.00	2.00	216.00	366.00	39528.0	0.91	1.71	74488.0	
5002.50	4.00	2.50	220.00	370.00	40700.0	0.93	2.17	94545.0	
5003.00	**4.00**	**3.00**	**224.00**	**374.00**	**41888.0**	**0.96**	**2.64**	**115192.0**	10-yr WS
5003.50	10.00	3.50	234.00	384.00	44928.0	1.03	3.14	136896.0	
5004.00	10.00	4.00	244.00	394.00	48068.0	1.10	3.68	160145.0	
5004.50	10.00	4.50	254.00	404.00	51308.0	1.18	4.25	184989.0	
5005.00	**10.00**	**5.00**	**264.00**	**414.00**	**54648.0**	**1.25**	**4.85**	**211478.0**	100-yr WS
5005.50	15.00	5.50	279.00	429.00	59845.5	1.37	5.51	240101.4	Freeboard
5006.00	15.00	6.00	294.00	444.00	65268.0	1.50	6.23	271379.8	Freeboard

depths less than 3 foot, and then decreases to 1V:10H for recreational bank areas. The pond geometry is summarized in Table 14.4.

Table 14.4 shows that for this case, the 10-yr water surface elevation is 5003.0 ft, while the 100-yr water depth is 5.0 ft. The 1-ft freeboard provides a brim-full capacity of 6.23 acre-ft.

14.8 CONCLUSIONS

Regular geometry is chosen to identify the cross-sectional area in the basin under design. At the final stage, the grading plan at the project site will allocate the space and easement to the proposed detention system. As illustrated in Fig. 14.14, at each elevation, the same area can be stretched into various shapes to fit into the site limits.

The stage-storage curve is constructed based on the contours on the topographic map. The incremental storage volume is estimated as:

$$\Delta S_{t+1} = 0.5 \left(A_{t+1} + A_t \right) \times \left(h_{t+1} - h_t \right) \tag{14.12}$$

$$S(h) = \sum_{h=h_o}^{h=h} \Delta S_i \tag{14.13}$$

where A = topographic contour area in [L^2], ΔS = incremental storage volume in [L^3], $S(h)$ = accumulated storage volume in [L^3] at stage h in [L], and i = i-th contour at stage h in [L].

The stage-storge curve is constructed using Eqs 14.12 and 14.13.

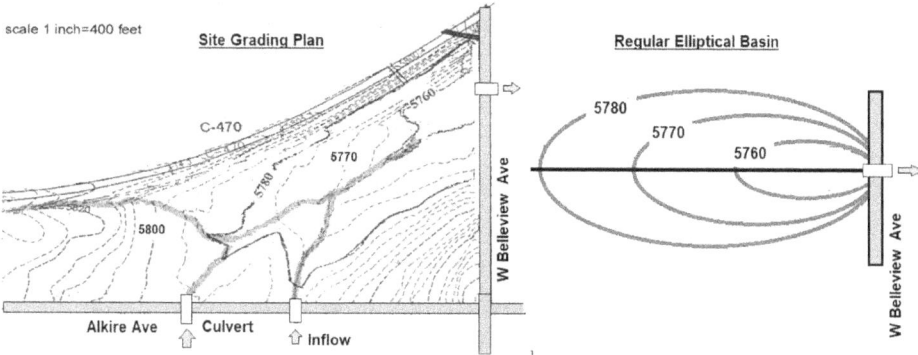

Figure 14.14 Grading plan for a detention basin

Figure 14.15 Safety and maintenance around a basin

For maintenance, a detention basin needs service roads and ramps to access the basin bottom and sediment pool. As a rule of thumb, approximately 20% of the 100-yr water surface area should be dedicated to maintenance operations. For public safety, it is always necessary to install a steel rack, as shown in Fig. 14.15a, at the entrance of a culvert if its diameter is greater than 18 inches, and then to post a warning sign to protect the facility. If a basin is too deep or the side slope is too steep, a fence, as shown in Fig. 14.15b, should be part of the surrounding landscaping for both safety and aesthetics.

HOMEWORK

Q14-1   Three hydrographs are presented in Fig. Q14-1. (1) Identify the post-development, pre-development, and after-detention hydrographs. (2) Did the proposed detention basin preserve the pre-development peak flow? (3) Did the detention process preserve the pre-development hydrograph? (4) The after-detention recession hydrograph carries higher flows than the pre-development hydrograph. What are the major problems associated with high flows over a long duration? (5) How can the detention basin approach be improved so that the pre-development hydrograph can be better preserved?

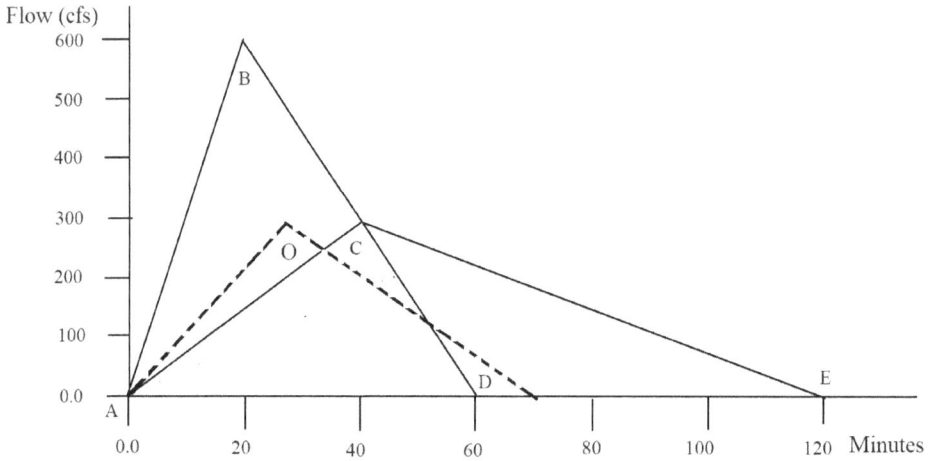

Figure Q14-1 Hydrographs involved in stormwater detention

Q14-2 To design the detention basin shown in Fig. Q14-2, the allowable flow release rates are: 0.23 cfs/acre for a 10-yr event and 0.85 cfs/acre for a 100-yr event. The watershed area is 40 acres. The 10- and 100-yr post-development hydrographs are given. Shape the proposed detention basin.

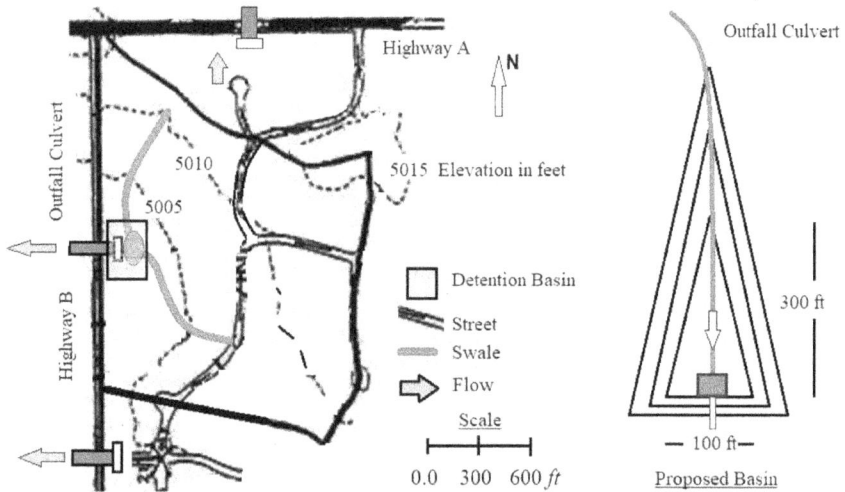

Figure Q14-2 Belleview watershed

1. What are the allowable 10- and 100-year peak flows from this proposed detention basin? (*Answer: Q10-allowable = 9.2 cfs and Q100-allowable = 34 cfs*)
2. From the hydrologic study, the 10- and 100-yr inflow hydrographs to the basin
are given as follows.

Time (min)	10-yr runoff (cfs)	100-yr runoff (cfs)
0	0	0
10	19	17
20	46	64
30	51	108
40	40	103
50	29	78
60	22	56
70	17	37
80	13	22
90	10	13
100	9	10
110	7	7
120	4	4

Apply the hydrograph method to determine the 10- and 100-yr detention
volumes. (*Answer: V10 = 1.4 acre-ft and V100 = 2.3 acre-ft*)

3. Consider a triangular basin with its bottom width = 100 ft and length = 300 ft.
The basin's bottom elevation is set to be 5000 ft, and the freeboard height is set
to be 1.0 ft. Shape this triangular basin using the given side slopes as:

Water Surface Elev (feet)	Basin Side Slope (ft/ft)	Width of Cross Section (feet)	Length of Cross Section (feet)	Cross Pond (acres)	Sectional Area (sq ft)	Cumulative Pond (acre-ft)	Storage Volume C_f	Identify Design Water Elev
5000.00	0.00	100.00	300.00	0.34	15000.0	0.00	0.0	
5000.50	3.00	103.00	303.00	0.36	15604.5	0.18	7651.1	
5001.00	3.00	106.00	306.00	0.37	16218.0	0.36	15606.8	
5002.00	5.00							
5003.00	5.00							
5003.50	5.00	131.00	331.00	0.50	21680.5	1.44	62854.9	W_s = 10 yr
5004.00	7.00							
5004.50	7.00							
5005.00	10.00	155.00	355.00	0.63	27512.5	2.28	99320.4	W_s = 100 yr
5006.00	10.00	175.00	375.00	0.75	32812.5	2.97	129457.9	Freeboard

a. Identify the 10-yr and 100-yr water surface elevations.
b. Establish the stage-cross-sectional area curve.
c. Construct the stage-storage curve.

REFERENCES

EPA SWMM (2005). Stormwater Management Model, www2.epa.gov/water-research/storm-water-management-model-swmm.

HEC HMS (2005). "Hydrologic Analysis System", Hydrologic Engineering Center, Corps of Engineers.

Guo, J.C.Y. (1999). "Detention Storage Volume for Small Urban Catchment", *ASCE Journal of Water Resources Planning and Management*, vol. 125, no. 6, November/December.

Guo, J.C.Y. (2004). "Hydrology-Based Approach to Storm Water Detention Design Using New Routing Schemes", *ASCE Journal of Hydrologic Engineering*, vol. 9, no. 4, July/August.

Guo, J.C.Y. (2012). "Off-stream Detention Design for Stormwater Management", *ASCE Journal of Irrigation and Drainage Engineering*, vol. 138, no. 4, April 1.

Guo, J.C.Y. (2017). *Urban Stormwater Management and Flood Mitigation*, CSC Press, New York.

McCuen, R.H. (1998). *Hydrologic Analysis and Design*, 2nd edition, Prentice Hall, New York.

USWSCM (2010). *Urban Storm Drainage Design and Criteria*, vols 1, 2, and 3, Urban Drainage and Flood Control District, Denver, CO.

Chapter 15

Outlet work

Today, outlet works for a detention basin consist of *extended outlet, low-flow outlet, high-flow outlet*, and *emergency outlet*. As shown in Fig. 15.1a and b, the outlet structure is formed by perforated plates, orifices, weirs, and culverts (ASCE in 1984, Guo in 2017). The concrete outlet box shown in Fig. 15.2 collects flows from the orifices and weirs, and then discharges these flows through the outfall pipes. The capacity of an outfall pipe is determined by culvert hydraulics under headwater and tailwater effects. The operation of the concrete vault reflects its culvert hydraulics under either inlet or outlet control. When a vault collects more water flows than it can release, the water depth in the vault increases. At a given depth, both collection and discharge capacities must be calculated; the smaller of these dominates the operation of the outlet structure.

15.1 PERFORATED PLATE

A perforated plate is often used for extended detention processes, which involve a drain time of 40 to 72 hours. The trickle water release is not only to enhance water quality control, but also to reproduce the base flow in the drainage system. On the vertical plate, a matrix of 1-inch holes, consisting of N columns and M rows, are designed to jointly release the design volume over an extended drain time. At a given water surface elevation, H in Fig. 15.3, the discharge from a row of N submerged holes is the sum of individual orifice flows as:

$$Q_o = \sum_{i=1}^{i=M} C_d \times N \times A_o \sqrt{2g(H - E_o)} \quad \text{or zero if } H \le E_o \text{ for a row of } N \text{ holes} \quad (15.1)$$

where Q_o = discharge in [L³/T] released from a perforated plate at an elevation of E_o, C_d = discharge coefficient such as 0.6 to 0.7, A_o = opening area for a hole, N = number of columns, M = number of rows, g = gravitational acceleration in [L/T²], H = water surface elevation in [L], and E_o = elevation at the center of the hole.

Example 15.1: As shown in Fig. 15.4, a perforated plate has four columns and nine rows of 1-inch holes between elevations of 5001 and 5003 feet. With C_d = 0.60, determine the collection capacity of this plate under water surface elevations from 5001 to 5005 feet.

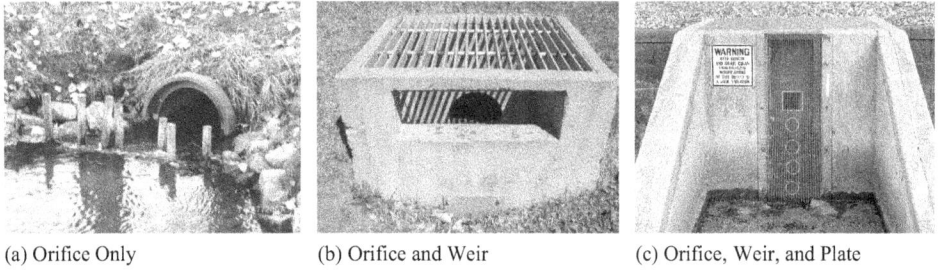

(a) Orifice Only   (b) Orifice and Weir   (c) Orifice, Weir, and Plate

Figure 15.1 Outlet box to control flow releases

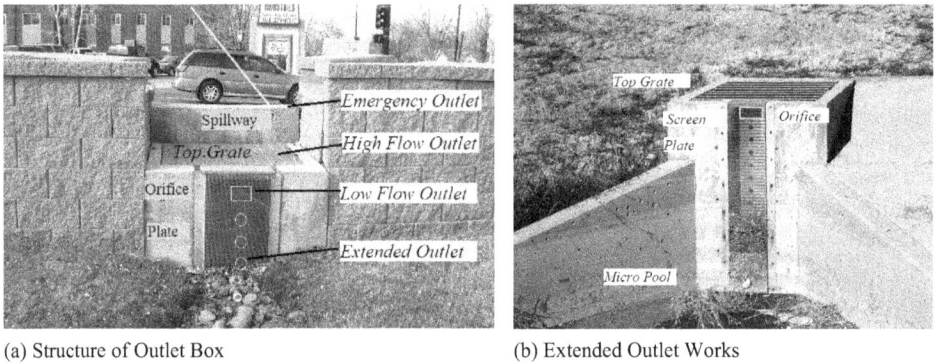

(a) Structure of Outlet Box   (b) Extended Outlet Works

Figure 15.2 Example outlets at a detention basin

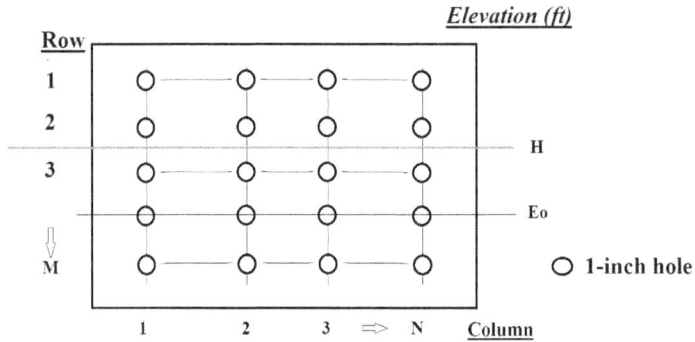

Figure 15.3 Perforated plate

Solution: For $N = 4$ columns, the total opening area of holes at each row is:

$$A = \frac{3.1416 \times \left(\frac{1}{12}\right)^2}{4} = 0.00545 \text{ft}^2 \text{ for a one - inch hole}$$

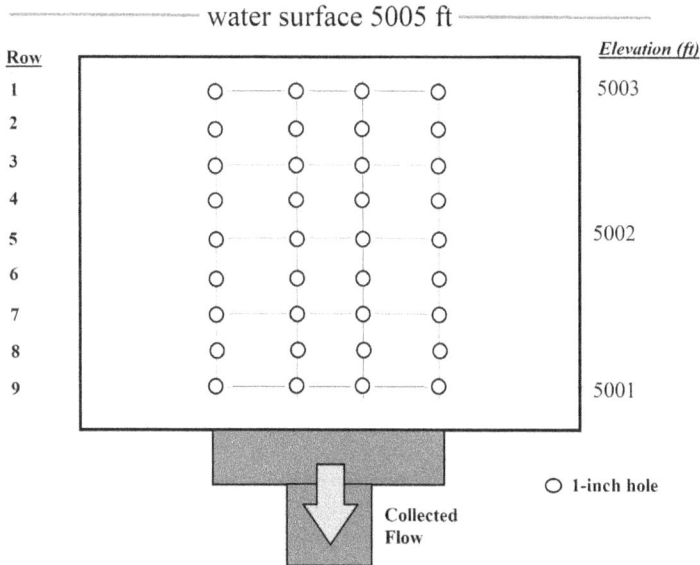

Figure 15.4 Collection capacity of a riser

(a) Detention Basin and Outlet Box   (b) Jet Flow From Perforated Plate

Figure 15.5 Detention basin located at South Knox Ct. Denver, Colorado

There are four holes $(N = 4)$ in a row. The collection discharge for each row is:

$$Q_o = 0.6 \times 4 \times 0.00545 \times \sqrt{2 \times 32.2Y} = 0.105\sqrt{(H - E_o)}$$

As summarized in Table 15.1, the total flow is 6.24 cfs under a water surface elevation of 5005 ft.

Fig. 15.5 presents a detention basin constructed with trickle channels, a micro pool in front of the outlet box and a vertical perforated plate. During a 25-yr storm event,

Table 15.1 Collection capacity of perforated plate

	Collection Capacity in cfs									
	Center elevation for Holes in feet									Flowrate
Stage	5001.00	5001.25	5001.50	5001.75	5002.00	5002.25	5002.50	5002.75	5003.00	per Row
ft				Flow	(cfs)					cfs
5001.00	0.000	0.000	0.000	0.000	0.000	0.000	0.000	0.000	0.000	0.000
5001.25	0.053									0.053
5001.50	0.074	0.053								0.127
5001.75	0.091	0.074	0.053							0.218
5002.00	0.105	0.091	0.074	0.053						0.323
5002.25	0.117	0.105	0.091	0.074	0.053	0.000				0.440
5002.50	0.129	0.117	0.105	0.091	0.074	0.053	0.000			0.569
5002.75	0.139	0.129	0.117	0.105	0.091	0.074	0.053	0.000		0.708
5003.00	0.149	0.139	0.129	0.117	0.105	0.091	0.074	0.053	0.000	0.856
5004.00	0.182	0.174	0.166	0.158	0.149	0.139	0.129	0.117	0.105	1.318
5005.00	0.210	0.203	0.197	0.189	0.182	0.174	0.166	0.158	0.149	1.628
									Total=	6.24

the detention basin was loaded with excess stormwater. The large amount of urban trash floated on the water surface, and the water jets through the holes in the plate continually drained the basin until it was emptied out.

15.2 VERTICAL ORIFICE

A vertical orifice, shown in Fig. 15.6, can be circular or rectangular in shape and is often installed on a concrete box. The hydraulic analysis of a vertical orifice is divided into a submerged zone for $H > E_t$ and an unsubmerged zone for $H \leq E_t$.

$$E_t = E_o + \frac{d}{2} \tag{15.2}$$

$$E_b = E_o - \frac{d}{2} \tag{15.3}$$

$$Q_o = C_d A_o \sqrt{2g(H - E_o)} \text{ for } E_t < H \text{ for a submerged zone.} \tag{15.4}$$

$$Q_o = Q_F \left(\frac{H - E_b}{d}\right)^k = C_o A_o \sqrt{gd} \left(\frac{H - E_b}{d}\right)^k \text{ for } E_b \leq H \leq E_t \text{ for an unsubmerged zone.} \tag{15.5}$$

Where E_t = crown elevation in [L], E_b = bottom elevation in [L], E_o = center elevation in [L] for a vertical orifice, d = orifice height or diameter, A_o = orifice opening

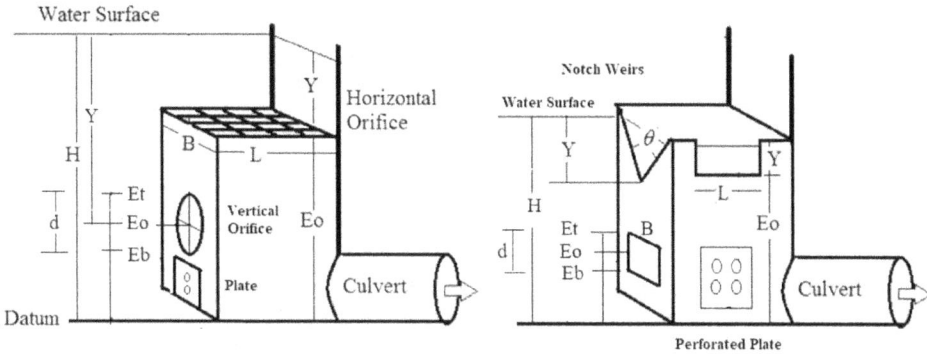

Figure 15.6 Orifice hydraulics

area in [L²], H = water surface elevation in [L], Q_o = vertical orifice flow in [L³/T], Q_F = flowing full capacity in [L³/T] at $H = E_t$, and $k = 2$ for a circular orifice or 1.5 for a rectangular orifice (Guo and Stitt 2017). Eq. 15.5 is a simplified, acceptable approach. More details about unsubmerged orifice hydraulics can be found elsewhere (Guo ad Stitt 2017).

Example 15.2: As shown in Fig. 15.6, a 3-ft wide rectangular orifice is installed between elevations of 5001 and 5003 feet. Considering $C_d = 0.60$, determine the vertical orifice flow under $H = 5005$ ft and $H = 5002$ ft, respectively.

Solution: For this case, $B = 3$ ft, $d = 2$ ft, and $E_o = 5002$ ft.

$$E_b = 5002 - \frac{2}{2} = 5001 \text{ ft}$$

$$Q_o = 0.6 \times (2 \times 3)\sqrt{2 \times 32.2 \times (5005 - 5002)} = 50 \text{ cfs for } H = 3003 \text{ ft}$$

If $k = 1.5$ and $d = 2$ for the rectangular orifice, the partially submerged rectangular weir flow can be approximated as:

$$Q_o = 0.6 \times (2 \times 3)\sqrt{32.2 \times 2} \left(\frac{5002 - 5001}{2}\right)^{1.5} = 10.18 \text{ cfs for } H = 5002 \text{ ft}$$

15.3 HORIZONTAL ORIFICE

The capacity of a horizontal orifice depends on the elevation and size of the orifice opening. The collection rate through an orifice varies with respect to the water surface elevation. When the depth above the orifice opening area is so shallow that the orifice opening area is not completely submerged, the horizontal orifice operates like a weir with a crest length equal to the circumference of a circular orifice. As shown in Fig. 15.6, the horizontal orifice collects flows from three sides of the steel grate. The collection flow into a horizontal orifice under a low head is described as:

$$Q_w = \frac{2}{3}C_d\sqrt{2g}P_eY^{1.5}$$ (15.6)

$$Y = H - E_o \quad \text{for } H > E_o$$ (15.7)

where Q_w = weir flow in [L³/T], H = head in [L] of approach flow or water surface elevation in a reservoir, P_e = effective weir length in [L], and Y = effective headwater depth in [L]. For a horizontal orifice, the collection rate is described by the orifice flow equation as:

$$Q_o = C_dnA_o\sqrt{2gY} = C_dnA_o\sqrt{2g(H-E_o)} \quad \text{for } H > E_o$$ (15.8)

where Q_o = collection capacity of the orifice in [L³/T] and n = area opening ratio of the rack, such as 0.5 to 0.6. In practice, the collection capacity, Q_c, of an orifice under a given water depth should be the smaller of the values from Eqs 15.6 and 15.8, as:

$$Q_c = \min(Q_w, Q_o) \text{ for a horizontal orifice}$$ (15.9)

15.4 WEIR HYDRAULICS

Weirs are classified by their cross-sectional shape, such as rectangular, triangular, or trapezoidal.

15.4.1 Rectangular weir

The capacity of a *rectangular weir* in Fig. 15.6 is determined by its crest width as:

$$Q_w = \frac{2}{3}C_d\sqrt{2g}L_eY^{\frac{3}{2}}$$ (15.10)

$$L_e = L_w - 0.1mY$$ (15.11)

$$Y = H - H_w$$ (15.12)

where Q_w = collection capacity in [L³/T], C_d = discharge coefficient, such as 0.60 to 0.65, L_w = crest width in [L], L_e = effective weir width in [L], E_w = weir crest elevation in [L], m = number of end contractions, and g = gravitational acceleration in [L/T²]. Eq. 15.10 is dimensionally consistent. For English units, Eq. 15.10 is reduced to

$$Q_w = C_wL_eY^{\frac{3}{2}}$$ (15.13)

Table 15.2 Weir coefficients for broad-crested weirs

Breath of weir crest (feet)	Headwater 1.0 ft	Headwater 2.0 ft	Headwater 3.0 ft	Headwater 4.0 ft	Headwater 5.0 ft
5.00	2.68	2.65	2.66	2.70	2.79
10.00	2.68	2.64	2.64	2.64	2.64
15.00	2.63	2.63	2.63	2.63	2.63

Source: Brater and King, 1976.

Table 15.3 Weir coefficients for triangular weirs

Headwater (feet)	H:V 1.0:1.0	H:V 2.0:1.0	H:V 3.0:1.0	H:V 5.0:1.0	H:V 10.0:1.0
0.50	3.85	3.49	3.22	3.05	2.84
1.0	3.85	3.50	3.40	3.13	2.91

Source: Brater and King, 1976.

where C_w = rectangular weir coefficient between 2.6 to 3.8 for English units, as shown in Table 15.2.

15.4.2 Triangular weir

The release rate of a *triangular weir* is governed by its V-notch angle, θ, as illustrated in Fig. 15.6.

$$Q_w = \frac{8}{15} C_d \sqrt{2g} \tan\left(\frac{\theta}{2}\right) Y^{\frac{5}{2}} \qquad (15.14)$$

For English units, Eq. 15.14 is further reduced to:

$$Q_w = C_t \tan\left(\frac{\theta}{2}\right) Y^{\frac{5}{2}} \qquad (15.15)$$

where C_t = triangular weir coefficient. Theoretically speaking, under the same hydraulic conditions, a rectangular weir coefficient, C_t, is approximately 20% less than a triangular weir coefficient, C_w (Table 15.3).

15.4.3 Trapezoidal weir

As illustrated in Fig. 15.7, a *trapezoidal weir* is composed of a rectangular weir with a crest length equal to the trapezoidal bottom width and a triangular weir with a notch angle equal to the trapezoidal side slope, Z, as:

$$\frac{\theta}{2} = \frac{\pi}{2} - \tan^{-1}\left(\frac{1}{Z}\right) \qquad (15.16)$$

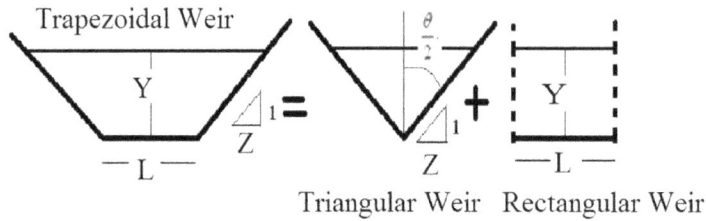

Figure 15.7 Trapezoidal weir hydraulics

Table 15.4 Collection capacity into concrete vault in Belleview detention basin

Water Surface Elev	Low Vertical Orifice Flow	Weir flow	High Grate Orifice Flow	Flow Collected Into Box	Design Identify Design Water	Info Enter Target Release
ft	cfs	cfs	cfs	cfs	Elev	cfs
5000.00	0.00	0.00	0.00	0.00		
5000.50	0.53	0.00	0.00	0.53		
5001.00	2.13	0.00	0.00	2.13		
5001.50	4.79	0.00	0.00	4.79		
5002.00	6.18	0.00	0.00	6.18		
5002.50	7.31	0.00	0.00	7.31		
5003.00	8.29	0.00	0.00	8.29	10-yr WS	92.00
5003.50	Submerged	12.73	18.39	12.73		
5004.00	Submerged	36.00	26.00	26.00		
5004.50	Submerged	66.14	31.84	31.84		
5005.00	Submerged	101.82	36.77	36.77	100-yr WS	34.00
5005.50	Submerged	142.30	41.11	41.11		
5006.00	Submerged	187.06	45.03	45.03		
5006.50	Submerged	235.72	48.64	48.64		
5007.00	Submerged	288.00	52.00	52.00		
Submerged = No Flow because of Hydraulic Equilibrium						

The weir coefficient for various side slopes is summarized in Table 15.4. Typically, a weir discharge coefficient, C_w, for a broad or sharp crested weir ranges between 2.65 and 3.10. Without knowing the weir specifics and downstream tailwater conditions, a value of $C_w = 3.0$ is recommended for sharp weirs and a value of $C_w = 2.65$ is recommended for broad-crested weirs. However, care must be taken in selecting the appropriate discharge coefficient once the downstream tailwater conditions become well understood.

Example 15.3: Continue with the Belleview Watershed in Fig. 15.8, discussed in Examples 14.3 and 14.4. Fig. 15.9 shows the outlet box to be installed in the proposed Belleview detention basin. The 10-yr and 100-yr water depths are 3.0 and 5.0 ft above the basin floor at elevation of 5000 ft. The high-flow orifice is formed by a horizontal

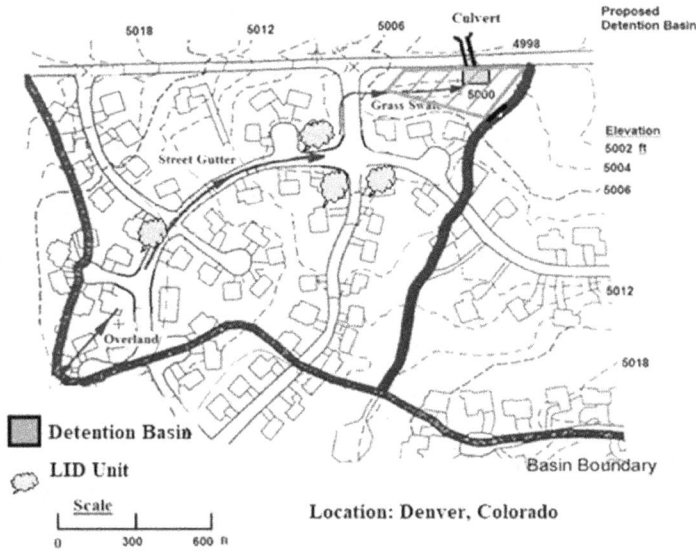

Figure 15.8 Watershed map

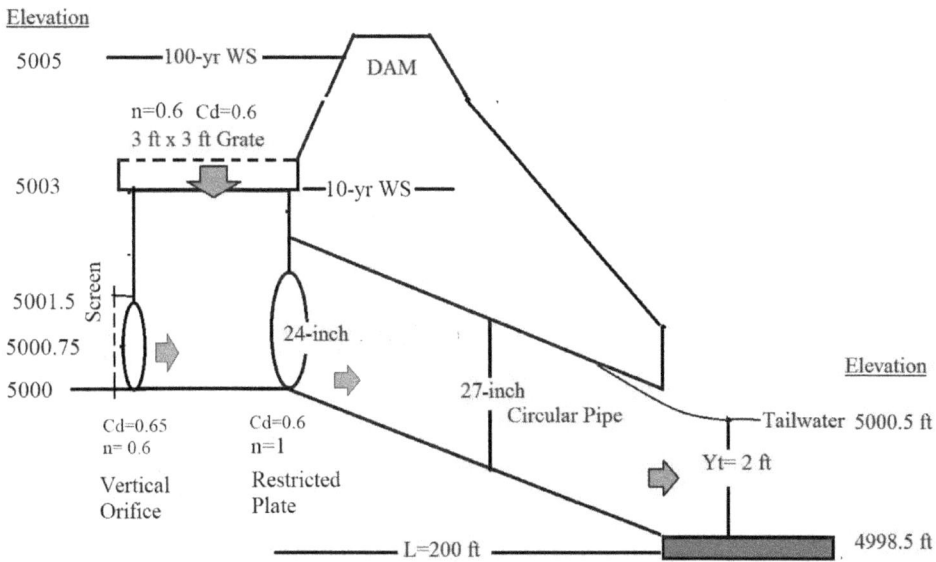

Figure 15.9 Outlet structure for Belleview detention basin

3-ft by 3-ft square grate installed on the top the concrete box at an elevation of 5003 ft. This square orifice is protected by a steel grate with a net area opening ratio of 0.60. The low-flow orifice is a 1.5-ft in-diameter vertical orifice with its center located at 5000.75 feet. The orifice coefficient is 0.60 to 0.65, and the weir coefficient is 3.0. Determine the collection capacity of the low- and high-flow orifices.

Solution: The operation of the vertical orifice is divided into two zones:

(1) If $H > 5001.5$ ft, the water surface is above the orifice crown.

$$Q_o = 0.7 \times 0.7 \times \frac{3.1416 \times 1.5^2}{4} \sqrt{2 \times 32.2 (H - 5000.75)} \text{ if } H \geq 5001.5$$

(2) If $5000 \leq H \leq 5001.5$ ft, the vertical orifice is unsubmerged. The flow rate is proportional to the full flow, Q_F, under the depth to the orifice crown. A weighting factor is recommended to be the flow area ratio, which can be approximated by the squared ratio of flow depth to orifice diameter as:

$$Q_F = 0.7 \times 0.7 \times \frac{3.1416 \times 1.5^2}{4} \sqrt{2 \times 32.2 \times 0.75} \text{ under } H = 5001.5 \, \text{ft}$$

If $k = 3$ for a vertical circular orifice, the unsubmerged flow is estimated as:

$$Q_o = Q_F \left[\frac{H - 5000}{1.5} \right]^2 \text{ if } 5000 \leq H \leq 5001.5$$

The grate will collect water flows from three sides as the water surface elevation is slightly above the elevation of 5003 ft. As the water depth becomes deeper, the operation of the horizontal grate is dictated by the smaller of the weir and orifice flows as:

$Y = H - 5003$
$Q_w = 3.0 \times 2 \times (3 \times 3.0) \times Y^{1.5} = 54.0 \ Y^{1.5}$ cfs

The orifice collection capacity for the top square grate is calculated as:

$$Q_o = 0.60 \times 0.6 \times (3.0 \times 3.0) \times \sqrt{2.0 \times 32.2 \times Y} = 25.92\sqrt{Y} \text{ cfs}$$

At a stage $H \geq 5004$ ft, the collection capacity of the grate is:

$$Q_c = \min\left(54.0 Y^{1.5}, 25.92\sqrt{Y}\right) \text{cfs}$$

As shown in Table 15.4, the flow collection curve is a combination of low- and high-stage flows.

15.5 CULVERT HYDRAULICS

Water enters the concrete vault in Fig. 15.10 through the low- and high-orifices, and then discharges into the downstream receiving system through outlet pipes that are usually short enough, 100 to 200 feet, to act like a culvert. During the preliminary design stage, tailwater information is not yet available. The capacity of the outlet structure in a detention basin is approximated by orifice and weir hydraulics. At the final design stage, the outflow capacity must be refined with culvert hydraulics under tailwater effects.

The conveyance capacity of a culvert is dictated by its inlet and outlet conditions. Under inlet control, the capacity of a culvert is independent of the tailwater at the culvert outlet. For instance, a culvert laid on a steep slope can be operated under inlet control because its capacity is solely determined by the critical depth at the entrance. On the contrary, when a pipe has a submerged exit, the flow capacity in such a pipe may become dictated by the tailwater depth at the exit, namely under outlet control. For a given discharge, one requires a higher headwater depth at the entrance dictates the culvert's operation. Or, for a given headwater depth, the one that passes a smaller discharge dictates the culvert's capacity

In practice, the design tailwater conditions are not always warranted during the design event. As a result, it is preferable to design the outfall pipe under so much inlet control that the flow release from a detention basin is independent of the downstream tailwater conditions. To achieve such an inlet-control operation, it is recommended that a restricted plate be installed at the entrance of the outfall pipe. A restricted plate requires a higher headwater depth and leads to an inlet-control operation. For instance, as shown in Fig. 15.10, a 24-inch plate is inserted at the entrance of the 27-inch pipe.

15.5.1 Outlet-control culvert hydraulics

As illustrated in Fig. 15.10, under outlet control, the culvert capacity is determined by the balance of energy between Sections 1 and 2 as:

$$H = Y + LS_o = \left(K_e + K_x + K_b + K_n\right)\frac{V_c^2}{2g} + Y_t + \frac{V_c^2}{2g}$$

(15.17)

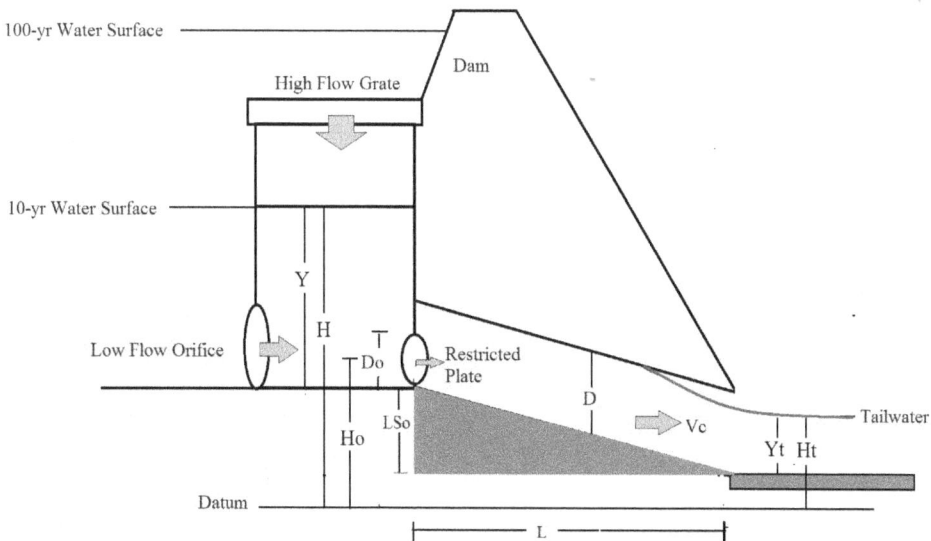

Figure 15.10 Culvert hydraulics

$$K_n = \alpha \frac{N^2 L}{D^{\frac{4}{3}}} \quad \text{for a circular pipe}$$

(15.18)

$$K_n = \beta \frac{N^2 L}{R^{\frac{4}{3}}} \quad \text{for a non-circular pipe}$$

(15.19)

where H = water surface elevation in [L] at the entrance, Y = headwater depth in [L] at the entrance, L = length of the pipe in [L], S_o = pipe slope in [L/L], K_e = entrance loss coefficient (Tables 15.5 and 15.6), K_x = exit loss coefficient between 0.5 and 1.0, K_b = bend loss coefficient as shown in Table 15.7, K_n = friction coefficient, V_c = flowing full velocity in [L/T], N = Manning's roughness coefficient such as 0.025 for metal pipes and 0.015 for concrete pipes, D = diameter of circular pipe in [L], R = hydraulic radius in [L], and Y_t = tailwater depth in [L]. It is noted that $\alpha = 184$ for foot-second units or 124 for meter-second units, and $\beta = 29$ for foot-second units

Table 15.5 Entrance loss coefficients for box culverts

Structure of Box Culvert and Entrance	Coefficient K_e
Headwall parallel to embankment (no wing wall)	
a. square-edged on three edges	0.50
b. three edges rounded	0.20
Headwall with wing walls at 15 to 45 degrees to barrel	
a. square-edge top corner	0.40
b. top corner rounded	0.20

Table 15.6 Entrance loss coefficient for circular culverts

Structure of Circular Culvert and Entrance	Coefficient K_e
Concrete pipe projecting from fill (no headwall)	
a. socket end of pipe	0.20
b. square cut end of pipe	0.50
Concrete pipe with headwall or headwall and wing walls	
a. socket end of pipe	0.10
b. square cut end of pipe	0.50
c. rounded entrance	0.10
Corrugated metal pipe	
a. projecting from fill (no headwall)	0.80
b. headwall or headwall and wing walls	0.50

Table 15.7 Bend loss coefficients

Angle (degrees)	0	20	40	60	80	90
Bend loss coefficient	0	0.1	0.2	0.45	0.80	1.0

or 19.5 for meter-second units. For convenience, let K be the sum of all the loss coefficients as:

$$K = K_e + K_x + K_b + K_n \qquad (15.20)$$

The culvert capacity under outlet control, Q_O in $[L^3/T]$ is calculated using water depths as:

$$Q_O = V_C A_C = A_C \sqrt{\frac{1}{K+1}} \sqrt{2g(Y + LS_o - Y_t)} \qquad (15.21)$$

Or it may be calculated using elevations as:

$$Q_O = V_C A_C = A_C \sqrt{\frac{1}{K+1}} \sqrt{2g(H - H_t)} \qquad (15.22)$$

where A_c = pipe cross-sectional area in $[L^2]$, and H_t = elevation in $[L]$ for a tailwater depth Y_t. Eq. 15.22 is similar to the orifice equation except that the orifice coefficient is computed by the sum of all the loss coefficients.

15.5.2 Inlet-control culvert hydraulics

Under inlet control, the culvert hydraulics are independent of the tailwater effect and energy losses. The culvert capacity under inlet control should operate like an orifice as:

$$Q_I = C_d \frac{\pi D_o^2}{4} \sqrt{2g\left(Y - \frac{D_o}{2}\right)} = C_d A_o \sqrt{2g(H - H_o)} \qquad (15.23)$$

where Q_I = culvert capacity under inlet control in $[L]$, D_o = equivalent diameter in $[L]$ for the restricted plate installed at the culvert entrance, and H_o = elevation in $[L]$ at the center of the restricted plate. Eq. 15.23 is valid if the headwater elevation is above the tailwater elevation.

15.5.3 Discharge capacity of concrete vault

In practice, the range of headwater depths at the culvert entrance needs to be identified first. For a given headwater, H in $[L]$, the discharge capacity, Q_C in $[L^3/T]$, from the concrete vault is dictated by Eqs 15.22 and 15.23, whichever is smaller.

$$Q_C = \min(Q_O, Q_I) \qquad (15.24)$$

Example 15.4: The outfall pipe in Fig. 15.10 has a diameter of 27 inches. The length of pipe is 200 feet, laid on a slope of 0.75%. The loss coefficients are 0.20 at the entrance, 0.50 at the exit, and 0.014 for Manning's roughness. A 24-inch restricted plate is installed at the entrance. The orifice coefficient for this plate is 0.60. Construct the stage-outflow curve for this culvert under the tailwater of $d = 2$ feet or at an elevation of $H_t = 5000.5$ ft.

Solution:
Applying Eqs 15.18 and 15.20 to the friction loss yields:

$$K_n = 184.1 \frac{0.014^2 \times 200}{2.25^{4/3}} = 2.45$$

$$K = 0.2 + 0.5 + 2.45 = 3.15$$

$$LS_o = 200 \times 0.0075 = 1.5 \text{ft}$$

Substituting the variables into Eqs 15.22 and 15.23 yields:

$$Q_o = \sqrt{\frac{1}{3.15+1}} \times \frac{3.14 \times 2.25^2}{4} \times \sqrt{2 \times 32.2 \times (H - 5000.5)} = 15.65\sqrt{H - 5000.5}$$

$$Q_I = 0.60 \times \frac{3.14 \times 2^2}{4} \sqrt{2 \times 32.2 \times (H - 5001)} = 15.12\sqrt{H - 5001}$$

The above equations provide the stage-outflow relationship for the culvert. For a given H, the discharge capacity from the concrete vault is dictated by the smaller value.

$$Q_C = \min(15.65\sqrt{H - 5000.5},\ 15.12\sqrt{H - 5001})$$

As shown in Table 15.8, the restricted plate effectively dominates the culvert hydraulics as the inlet control.

Table 15.8 Discharge capacity from concrete vault in Belleview detention basin

Water Elev Entrance	Water Depth	Outlet Control Flowrate	Inlet Control Flowrate	Discharge from Box	Identify Design Water	Enter Design Release
feet	ft	cfs	cfs	cfs	Elev	cfs
5000.00	0.00	0.00	0.00	0.00		
5000.50	0.50	0.00	6.61	0.00		
5001.00	1.00	11.08	11.34	11.08		
5001.50	1.50	15.67	14.17	14.17		
5002.00	2.00	19.19	15.12	15.12		
5002.50	2.50	22.16	18.52	18.52		
5003.00	**3.00**	**24.77**	**21.38**	**21.38**	10-yr WS	**9.20**
5003.50	3.50	27.13	23.91	23.91		
5004.00	4.00	29.31	26.19	26.19		
5004.50	4.50	31.33	28.29	28.29		
5005.00	**5.00**	**33.23**	**30.24**	**30.24**	100-yr WS	**34.00**
5005.50	5.50	35.03	32.07	32.07		
5006.00	6.00	36.74	33.81	33.81		

15.6 CHARACTERISTIC CURVE

Hydrologic analyses on basin geometry produce a *stage-storage curve* (SS curve), whereas hydraulic analyses on an outlet concrete vault result in a *stage-outflow curve* (SO curve). At a specified stage, H in [L], the outflow in [L^3/T] from the detention basin is determined as:

$$\text{Outflow} = \min\{\text{collection capacity, discharge capacity}\} \text{ for a given } H \qquad (15.25)$$

In practice, the SS and SO curves are then merged into the *stage-storage-outflow* (SSO) curve, which is called the *characteristic curve* for the detention basin under design. For instance, the SO, SS, and SSO curves developed for the specified outlet structure shown in Fig. 15.9 are summarized in Table 15.9.

A reservoir routing scheme requires a SSO curve to perform numerical simulations to confirm that the detention basin under design will produce an allowable flow release and meet the required detention volume. In this example, the SSO curve was developed using a triangular basin, and the tailwater effect was truncated with a restricted plate. As shown in Table 15.9, during a 10-year event, the water surface elevation would be at 5003.0 ft with a storage volume of 2.64 acre-ft and a peak flow of 8.29 cfs, which is less than the allowable value of 9.2 cfs. During a 100-year event, the water surface elevation would be at 5005.0 ft with a storage volume of 4.85 acre-ft and a peak flow of 30.24 cfs, which is less than the allowable value of 34 cfs.

As a preliminary design, this SSO characteristic curve developed for the basin under design has met the goals of peak-flow reduction. Furthermore, a reservoir routing process may be performed to understand the flow movement through the basin.

Table 15.9 Characteristic SSO curve developed for Belleview detention basin

Water Stage (S) feet	Storage Volume (S) acre-ft	Collection into Box cfs	Discharge from Box cfs	Outflow (O) cfs	Target WS elevation ft	Target Release cfs
5000.00	0.00	0.00	0.00	0.00		
5000.50	0.41	0.53	0.00	0.53		
5001.00	0.83	2.13	11.08	2.13		
5001.50	1.26	4.79	14.17	4.79		
5002.00	1.71	6.18	15.12	6.18		
5002.50	2.17	7.31	18.52	7.31		
5003.00	**2.64**	**8.29**	**21.38**	**8.29**	10-yr WS	9.20
5003.50	3.14	12.73	23.91	12.73		
5004.00	3.68	26.00	26.19	26.00		
5004.50	4.25	31.84	28.29	28.29		
5005.00	**4.85**	**36.77**	**30.24**	**30.24**	100-yr WS	34.00
5005.50	5.51	41.11	32.07	32.07	Overflow	
5006.00	6.23	45.03	33.81	33.81	Overflow	
5006.50	7.01	48.64	35.46	35.46	Overflow	
5007 00	7 86	52 00	37 03	37 03	Overflow	

15.7 MAINTENANCE AND SAFETY

Detention basins present an attraction to the public because they are open space and can be designed as neighborhood parks. During a heavy storm event, flash floods can be lethal. It is necessary to have flash-flood-warning signs posted around the basin. All inlets must be protected with a trash rack (Guo, MacKinzie, and Mommandi 2016). For public safety, a fence may be used around the shoreline where potential hazards exist.

Upon the completion of the basin, an inspection should be conducted to make sure that the basin that has been built complies with the design (Jones, Guo, and Urbonas 2006). During the first three years in service, it is necessary to frequently check on vegetation growth in the infiltration bed and on the side slope. Supplemental plantings should be added as needed to ensure good cover.' On an annual basis, the basin needs basic maintenance before the wet months. The outlet structure should be inspected after a severe storm event and debris blockages should be removed. Trash and debris should be removed regularly (Guo, Shih, and MacKenzie 2012).

HOMEWORK

Q15-1 Referring to Fig. Q15-1, there are two vertical orifices installed on the wall with their centers 3.5 ft above the bottom of the wall. With $C_d = 0.60$, estimate the orifice flows under water surface elevations: WS1 = 7 ft and WS2 = 3.5 ft.

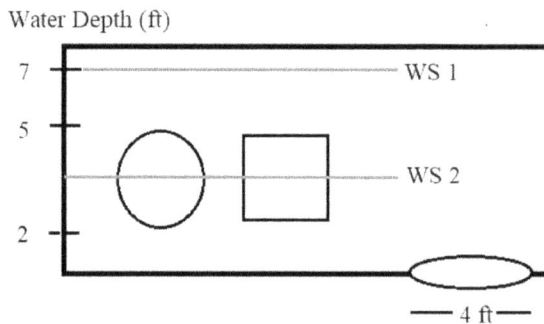

Figure Q15-1 Orifice flows

Q15-2 Referring to Fig. Q15-1, on the horizontal floor there is a 4-ft circular horizontal orifice. With $C_d = 0.6$, estimate the flows under water depths of 2, 4, 6, and 10 ft.

Q15-3 Referring to Fig. Q15-3, with $C_d = 0.60$, (1) estimate the flow released through the trapezoidal weir with $Z_1 = 2$ and $Z_2 = 2$. (2) Estimate the weir flow if $Z_1 = 2$ and $Z_2 = 4$.

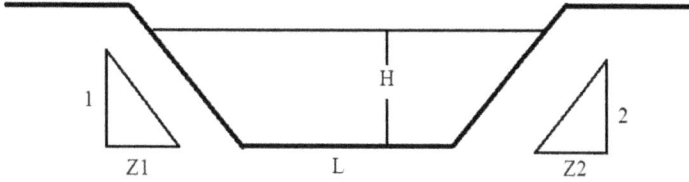

Figure Q15-3 Trapezoidal weir flows

Q15-4 The outlet system shown in Fig. Q15-4 is proposed for the detention basin in Q14-2. The outlet box is composed of low- and high-flow orifices with an outfall culvert. The low-flow outlet is a vertical orifice 18-inch in diameter. The center of this vertical orifice is at an elevation of 5000.75 ft. The high-flow

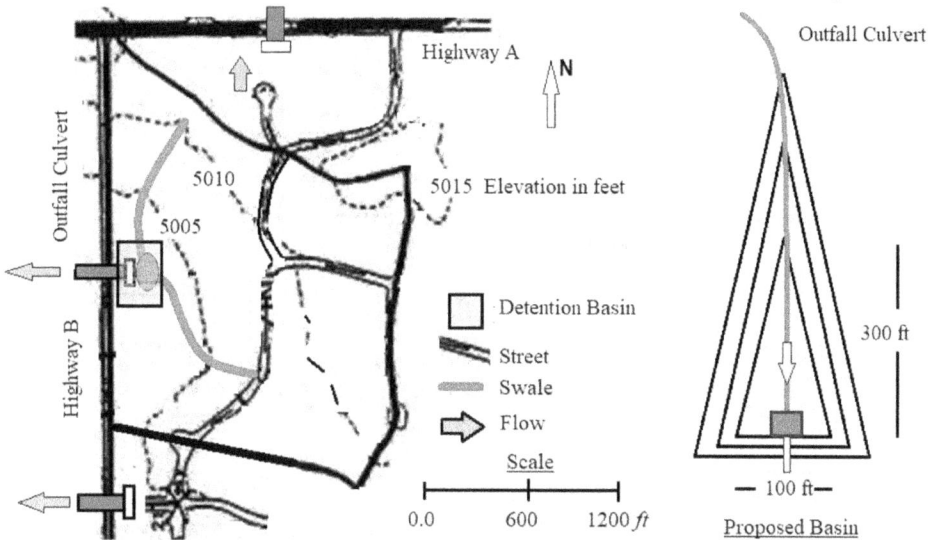

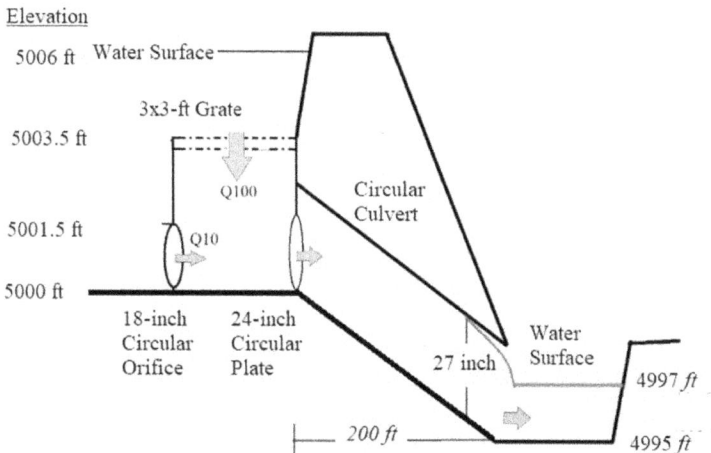

Figure Q15.4 Outlet structure for a proposed detention basin

outlet is another horizontal 3.0 × 3.0-ft grate set at an elevation of 5003.5 feet. The net opening area ratio on the grate is $m = 0.55$. Using foot-sec units, the weir coefficient is 3.0 and the orifice coefficient is 0.65. The tailwater elevation is set to be 4997 feet. A 27-inch circular pipe that is 200-ft long is laid on a slope of 0.025 ft/ft. The pipe roughness is 0.022. The entrance and exit loss coefficients are 0.2 and 0.5. The restricted plate installed at the pipe entrance has a diameter of 24 inches. Construct the stage-outflow curve.

REFERENCES

ASCE (1984). *Final Report of the Task Committee on Stormwater Detention Outlet Structures*, American Society of Civil Engineers, New York.

Brate, E.F. and King, H.W. (1976). *Handbook of Hydraulics*, McGraw Hill, New York.

Guo, J.C.Y. (2017). *Urban Stormwater Management and Flood Mitigation*, CSC Press, New York.

Guo, J.C.Y. and Stitt, R.P. (2017). "Flow Through Partially Submerged Orifice", *ASCE J of Irrigation and Drainage Engineering*, vol. 143, no. 8, March.

Guo, J.C.Y., Shih, H.M., and MacKenzie K. (2012). "Stormwater Quality Control LID Basin with micro pool", *ASCE Journal of Irrigation and Drainage Engineering*, vol. 138, no. 5, May 1.

Guo, J.C.Y., MacKenzie K., and Mommandi, A. (2016). "Hydraulic Interception Capacity for Inclined Grate", *ASCE J of Irrigation and Drainage Engineering*, vol. 142, no. 4, April.

Jones, J., Guo, J.C.Y., and Urbonas, B. (2006). "Safety on Detention and Retention Pond Designs", *Journal of Storm Water*, January/February.

Chapter 16

Performance of detention basin

Following the *3M cascading flow concept*, the detention basin shown in Fig. 16.1 is shaped with a pre-selected micro event for the bottom layer, a minor event for the mid layer, and a major event for the top layer (Guo et. al. 2020). The traditional detention basin design is to target a 100-yr event. As a result, the outfall concrete box can be as simple as a concrete conduit to control a 100-yr flow release. The major problem associated with such a single-event detention basin is the diminishing effectiveness on the flow reduction for frequent events. Often a frequent event trickles through the concrete conduit without an adequate detention process. Under the 3M cascading flow concept, the detention basin should be shaped with three layers to accommodate micro, minor, and major detention volumes. Correspondingly, the outfall concrete box has to be equipped with (a) *perforated plates* (b) a *vertical orifice*, (c) *horizontal top grates*, and (d) *outfall culverts*. A combination of these devices will jointly reduce the micro, minor, and major peak flows to achieve full-spectrum flow release control.

16.1 PERFORMANCE EVALUATION OF BASIN

The detention process is a typical unsteady flow through a storage basin. At each time step, the continuity principle dictates the balance of flow volumes and the storage volume in the storage basin. *Hydrologic routing* is the numerical algorithm used to solve the unsteady finite difference equation of flow movement through the storage basin. The governing equation for hydrologic routing is essentially the general hydrologic continuity principle, as:

$$\frac{I(t)+I(t+\Delta t)}{2} - \frac{O(t)+O(t+\Delta t)}{2} = \frac{S(t+\Delta t)-S(t)}{\Delta t} \tag{16.1}$$

where I = inflow in [L³/T], O = outflow in [L³/T], S = volume storage in [L³], t = at the beginning of time interval, $t+\Delta t$ = the end of time interval, and Δt = time interval in [T]. In practice, the inflows, $I(t)$ and $I(t+\Delta t)$, are specified with the known inflow hydrograph, and $O(t)$ and $S(t)$ are the known initial conditions at the beginning of each time step. For instance, $t = 0$, $O(t)$, and $S(t)$ represent the initial water depth, outflow, and storage volume in the detention basin. To solve the two unknowns, $O(t+\Delta t)$ and $S(t+\Delta t)$, requires a second equation that shall be derived from the detention basin system.

DOI: 10.1201/9781003284239-16

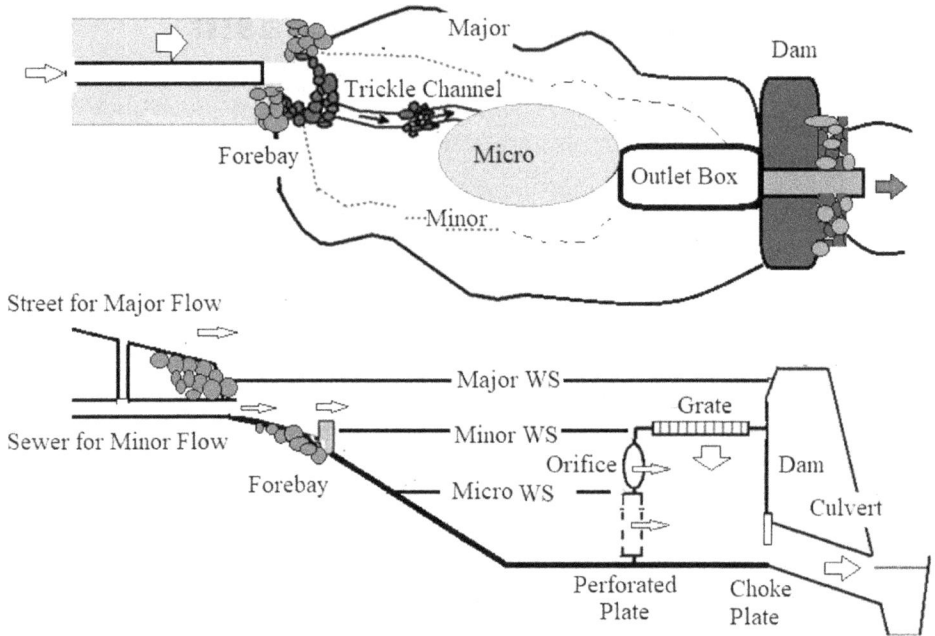

Figure 16.1 Layout of a 3M detention basin

16.1.1 Storage-outflow curve

To solve Eq. 16.2, the second equation is the *storage-outflow curve* that must be developed from the basin and outlet hydraulics. Since both storage volumes and outflows are directly related to the water depth in the basin, the general equation for the storage-outflow curve is written as:

$$S = a_n O^n + a_{n-1} O^{n-1} + \ldots + \text{constant} = \sum_{m=0}^{m=n} a_m O^m \tag{16.2}$$

where a_n = n-th constants derived from regression analysis after plotting the pairs (storage volume, outflow), and m = m-th order of the polynomial equation. In practice, the third-order polynomial equation, $m = 3$, is sufficient for hydrologic routing schemes.

Example 16.1: As summarized in Fig. 16.2, the Belleview Watershed is urbanized to have its imperviousness percentage increased from 5% to 60%. The increased 10- to 100-yr peak runoff flows are attenuated in the proposed Belleview detention basin, which is described by the stage-storage-outflow curve plotted in Fig.16.3. Determine the best-fitted formula for the proposed storage-outflow curve.

Solution: The proposed detention basin is shaped to attenuate 10- and 100-yr peak flows. The outlet box is equipped with a vertical 1.5-ft in-diameter circular orifice

(a) Belleview Watershed

(b) Stage-Storage-Outflow Curve

Water Depth ft	Storage Volume acre-ft	Outflow Cfs	Targeted Water Surface	Pre Development Flows
0.00	0.00	0.00		
0.50	0.41	0.53		
1.00	0.83	2.13		
1.50	1.26	4.79		
2.00	1.71	6.18		
2.50	2.17	7.31		
3.00	2.64	8.29	10-yr WS	9.2 cfs
3.50	3.14	12.73		
4.00	3.68	26.00		
4.50	4.25	28.29		
5.00	4.85	30.24	100-yr WS	34 cfs
5.50	5.51	32.07		
6.00	6.23	33.81		
6.50	7.01	99.54		
7.00	7.86	219.54		

Figure 16.2 Review of Belleview watershed and detention basin

Outflow Rate cfs	Storage Volume acre-ft
0.00	0.00
0.53	0.41
2.13	0.83
4.79	1.26
6.18	1.71
7.31	2.17
8.29	2.64
12.73	3.14
26.00	3.68
28.29	4.25
30.24	4.85
32.07	5.51
33.81	6.23
99.54	7.01

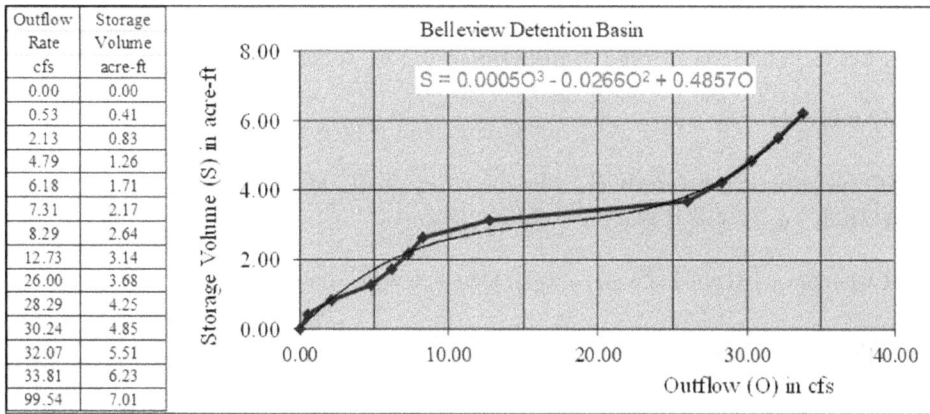

Belleview Detention Basin

$S = 0.0005O^3 - 0.0266O^2 + 0.4857O$

Figure 16.3 Best-fitted formula for the storage-outflow curve for Belleview detention basin

placed at an elevation of 5000 ft, a 3-ft by 3-ft top grate placed at an elevation of 5003 ft, and a 27-inch in-diameter circular culvert.

For this case, the best-fitted formula for the storage-outflow curve is found to be:

$$S = 0.4857O - 0.0266O^2 + 0.0005O^3 \tag{16.3}$$

Eq. 16.3 is sensitive to units, where S = storage volume in acre-ft, O = outflow in cfs. At each time step, Eqs 16.1 and 16.3 are simultaneously solved for $O(t+\Delta t)$ and $S(t+\Delta t)$.

16.2 HYDROLOGIC ROUTING SCHEMES

The numerical algorithm is an iterative relaxation process to progressively approach the solution. To expedite the numerical process, Eq. 16.1 needs to be formulated to have the best computational efficiency. For convenience, Eq. 16.1 is re-arranged as:

$$\frac{2S(t+\Delta t)}{\Delta t}+O(t+\Delta t)=\left[I(t)+I(t+\Delta t)\right]+\left[\frac{2S(t)}{\Delta t}-O(t)\right] \qquad (16.4)$$

There are two unknowns in Eq. 16.4: outflow, $O(t+\Delta t)$ and storage, $S(t+\Delta t)$. At time $t+\Delta t$, both unknowns are directly related to the known value at time t.

16.2.1 Storage routing scheme

The *storage routing scheme* is to seek solutions at time $t+\Delta t$ based on the sum of the storage and the outflow flow volumes at time t (Guo 2006). For convenience, Eq. 16.2 is re-arranged using the unit of flow volumes as:

$$\left[I(t)+I(t+\Delta t)-O(t)\right]\Delta t+2S(t)=O(t+\Delta t)\Delta t+2S(t+\Delta t) \qquad (16.5)$$

Since the storage volumes and outflow rates of a reservoir are related to its stages, the stage-storage curve and stage-outflow curve can be combined into a storage-outflow curve. Let us define the *storage routing function* (*SO* function) as:

$$SO=O\times\Delta t+2S \qquad (16.6)$$

The *SO* function can be established by the pairs (S, O) described in Table 16.1. Aided by Eq. 16.5, Eq. 16.6 is divided into two portions:

$$SO(t+\Delta t)=\left[I(t)+I(t+\Delta t)-2O(t)\right]\Delta t+SO(t) \quad \text{(known volume)} \qquad (16.10)$$

$$SO(t+\Delta t)=O(t+\Delta t)\Delta t+2S(t+\Delta t) \quad \text{(solution at } t+\Delta t) \qquad (16.11)$$

Fig. 16.4 illustrates how solutions can be obtained from the storage-routing function and storage-outflow curve. The value of $SO(t+\Delta t)$ is prescribed by the known

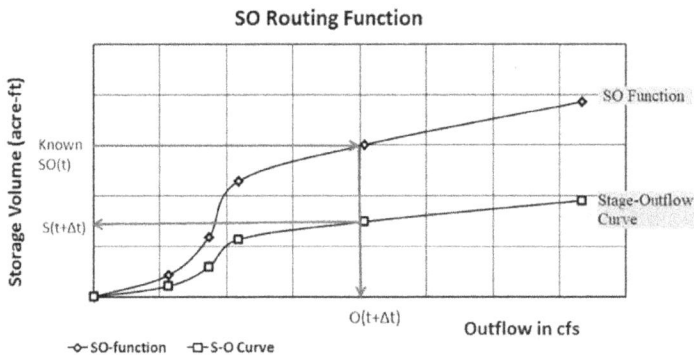

Figure 16.4 Graphic solution using the storage routing function

variables: $I(t)$, $I(t+\Delta t)$, $O(t)$ and $S(t)$. Solutions for the two unknowns, $O(t+\Delta t)$ and $S(t+\Delta t)$, are the pair (S, O) whose SO function satisfies Eq. 16.10. Repeating Eqs 16.10 and 16.11 at each time step, the outflow hydrograph can be generated.

16.2.2 Outflow routing scheme

The *outflow routing scheme* is to provide solutions at time $t+\Delta t$ based on the total outflow rate at time t. Eq. 16.12 for this method is re-arranged as:

$$\frac{2S(t+\Delta t)}{\Delta t}+O(t+\Delta t)=\left[I(t)+I(t+\Delta t)\right]+\left[\frac{2S(t)}{\Delta t}+O(t)\right]-2O(t) \qquad (16.12)$$

Let the *outflow routing function* (*OS* function) be defined as:

$$OS=\frac{2S}{\Delta t}+O \qquad (16.13)$$

It is noted that the outflow routing function has the same unit as the flow rate. Rearranging Eq. 16.12 yields:

$$OS(t+\Delta t)=\left[I(t)+I(t+\Delta t)-2O(t)\right]+OS(t)\,(\text{known outflow at time}\,t) \qquad (16.14)$$

and

$$OS(t+\Delta t)=\frac{S(t+\Delta t)}{dt}+O(t+\Delta t)\,(\text{solution at time}\,t+\Delta t) \qquad (16.15)$$

Fig. 16.5 shows that the outflow routing function and storage-outflow curve are related to the storage volume of the reservoir. At each time step, we firstly compute the value

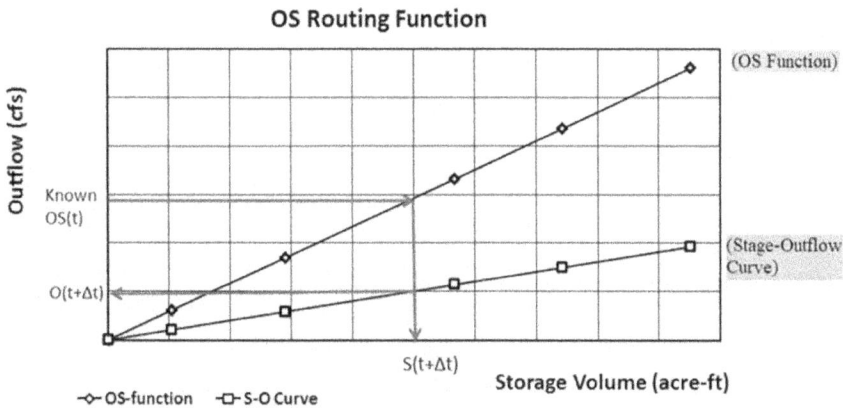

Figure 16.5 Graphic solution using the outflow routing function

of $OS(t+\Delta t)$ in terms of the variables $I(t)$, $I(t+\Delta t)$, $O(t)$, and $OS(t)$. The corresponding $S(t+\Delta t)$ and $O(t+\Delta t)$ can then be determined by the functional relationship described in Fig. 16.4. Repeating this process, we can advance one Δt at a time, until the inflow hydrograph is completely processed.

Example 16.2: Considering the time increment of five minutes, derive the SO function and OS function for Belleview detention basin in Example 16.1.

Solution: Care has to be taken when working on the routing function, because the dimensional units have to be consistent. In this case, the two routing functions are formulated using acre-ft and fps as:

$$OS = \frac{2S \times 43560}{5 \times 60} + O \text{ in cfs} \tag{16.16}$$

$$SO = 2S + O \times 5 \times \frac{60}{43560} \text{ in acre-ft} \tag{16.17}$$

The SO and OS routing functions are developed in Table 16.1.

Example 16.3: Repeat Example 16.2 using the OS routing function to determine the 100-yr after-detention hydrograph. The 100-yr incoming hydrograph is listed in Table 16.2. The initial condition is a dry bed as: $O(t) = 0$ and $S(t) = 0$ at $t = 0.0$.

Solution: Construct the SO routing function curve as shown in Fig. 16.6, which provides the solution at each time step after the known water volume is computed.

Table 16.1 SO and OS functions for Belleview basin in Example 16.2

Time Interval		10.00	min	
Stage H ft	Outflow Rate S acre-ft	Storage Volume O cfs	SO Function SO acre-ft	OS Function OS cfs
5000.00	0.00	0.00	0.000	0.000
5000.50	0.41	0.53	0.824	0.533
5001.00	0.83	2.13	1.687	2.130
5001.50	1.26	4.79	2.592	4.792
5002.00	1.71	6.18	3.505	6.186
5002.50	2.17	7.31	4.442	7.320
5003.00	2.64	8.29	5.403	8.301
5003.50	3.14	12.73	6.461	12.738
5004.00	3.68	26.00	7.711	26.013
5004.50	4.25	28.29	8.883	28.299
5005.00	4.85	30.24	10.126	30.254
5005.50	5.51	32.07	11.466	32.091
5006.00	6.23	33.81	12.926	33.828

Table 16.2 Hydrologic routing using OS function

	Given	KNOWN		SOLUTION from Chart		Calculated	CHECK
Time	Inflow	SO FCT		Outflow	Storage	SO FCT	
t	I(t)	[I(t)+I(t+Δt) −2O(t)]		O(t)	S(t)	O(t)*Δt/43560	Balance
t+Δt	I(t+Δt)	*Δt/43560+SO(t)		O(t+Δt)	S(t+Δt)	+2*S(t)'	in SO FCT
minutes	cfs	acre-ft		cfs	acre-ft	acre-ft	acre-ft
0.00	0.00	0.00		0.00	0.00	0.00	0.00
10.00	0.00	0.00		0.00	0.00	0.00	0.00
20.00	20.00	0.28		0.30	0.14	0.27	0.00
30.00	65.00	1.44		1.64	0.71	1.44	0.00
40.00	155.00	4.42		5.84	2.17	4.42	0.00
50.00	95.00	7.70		13.35	3.76	7.70	0.00
60.00	65.00	9.54		23.48	4.61	9.54	0.00
70.00	48.43	10.46		32.14	5.01	10.46	0.00
80.00	32.43	10.68		33.68	5.11	10.68	0.00
90.00	21.62	10.50		32.46	5.03	10.50	0.00
100.00	16.45	10.13		29.43	4.86	10.13	0.00
110.00	13.67	9.73		25.42	4.69	9.73	0.00
120.00	12.02	9.39		22.10	4.54	9.39	0.00

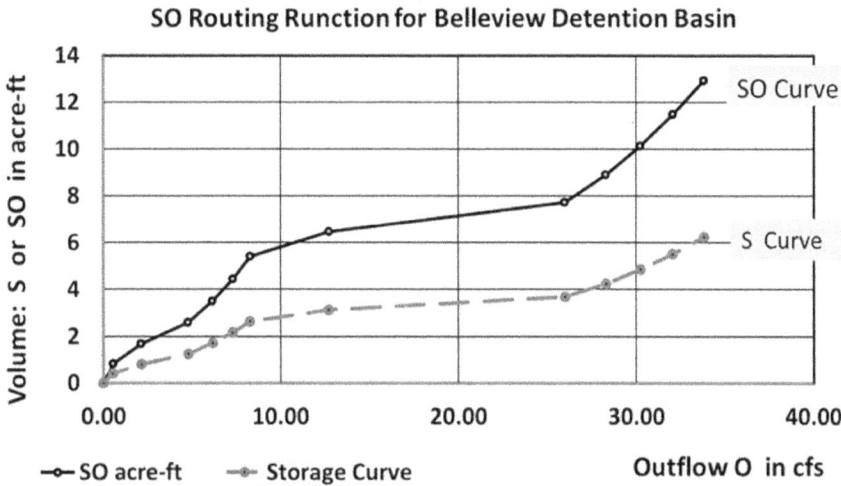

Figure 16.6 SO routing function for Belleview detention basin

At each time step, calculate the value of the SO function. With the known SO, Fig. 16.6 provides solutions for O(t+Δt) and S(t+Δt). The hydrologic routing for the given inflow hydrograph is summarized in Table 16.2.

For example, at t = 70 min, the value of the SO function is 10.46 acre-ft. As illustrated in Fig. 16.7, the corresponding solutions are: O = 32.14 cfs and S = 5.01 acre-ft.

Fig. 16.8 presents the inflow and after-detention outflow hydrographs. In this case, the post-development peak flow of 155 cfs occurs at t = 40 min. After the detention process, the peak flow is reduced to 33.5 cfs at t = 80 min. Repeating Example 16.3, we can confirm the detention effectiveness for 2-, 5-, and 10-yr events.

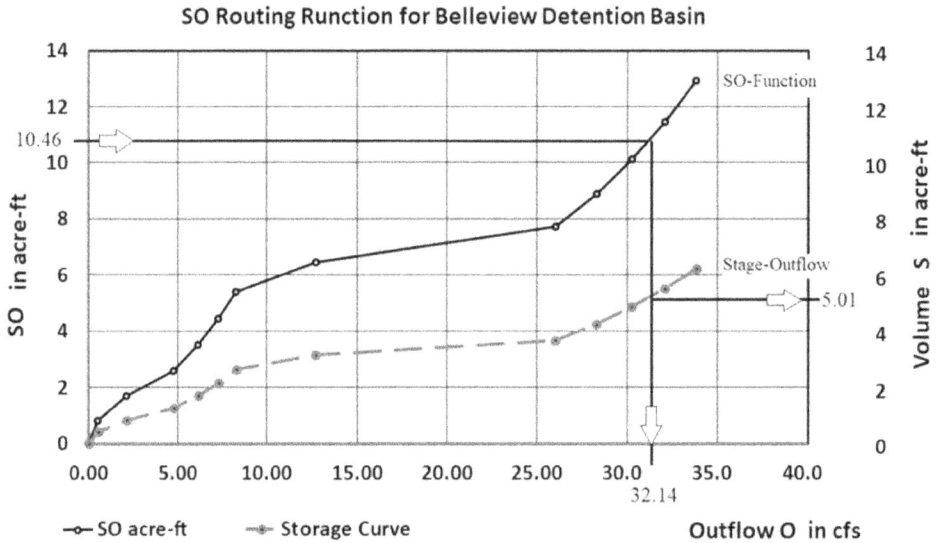

Figure 16.7 Solutions for hydrograph routing using the SO routing function

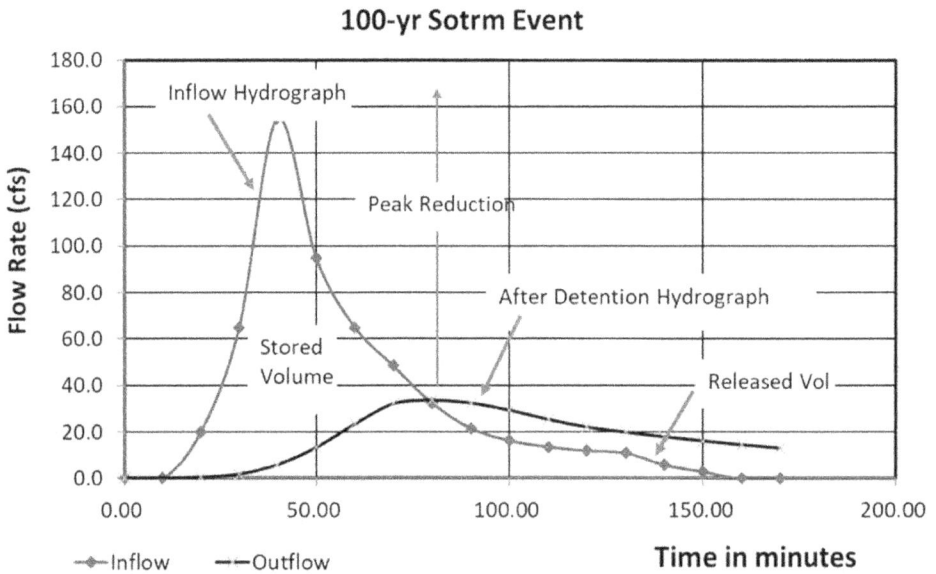

Figure 16.8 Performance of Belleview detention basin for 100-yr peak flow reduction

16.3 EVALUATION OF DETENTION BASIN

The performance of a detention basin is critically sensitive to its outfall box. For comparison, three different outlet boxes were evaluated, including:

1. the *one-layer outlet box* in Fig. 16.9a, using a 27-inch circular culvert to control the 100-yr peak flow,
2. the *two-layer outlet box* in Fig.16.9b, using an 18-inch circular orifice to control the 10-yr peak flow and a 3-ft × 3-ft top grate to reduce the 100-yr peak flow, and
3. the *three-layer outlet box* in Fig. 16.9c, adopted using a perforated plate to control a 2-yr peak flow, a vertical orifice to reduce a 10-yr peak flow, and a top grate to mitigate a 100-yr peak flow.

Repeat the aforementioned procedure to produce the SSO curve for each outlet box, and then predict the 2-, 10-, and 100-yr after-detention hydrographs for comparison. Fig. 16.10 is a plot of the peak-flow-frequency curves for the pre- and post-development conditions. The one-layer outlet box controls the 100-yr peak flow well, but the single outfall culvert fails to provide adequate detention processes to any and all flows that are more frequent than a 100-year storm. Similarly, any more frequent storms than a 10-yr event trickle through the 2-layer outlet box. In comparison, the 3-layer outbox successfully reproduces the pre-development flow-frequency relationship or provides full-spectrum flow release control.

The design example, Belleview detention basin, illustrates how to integrate micro, minor, and major events together to form a three-layer cascading flow system that can successfully mimic the pre-development flow releases and preserve the watershed regime (Guo 2010, 2013).

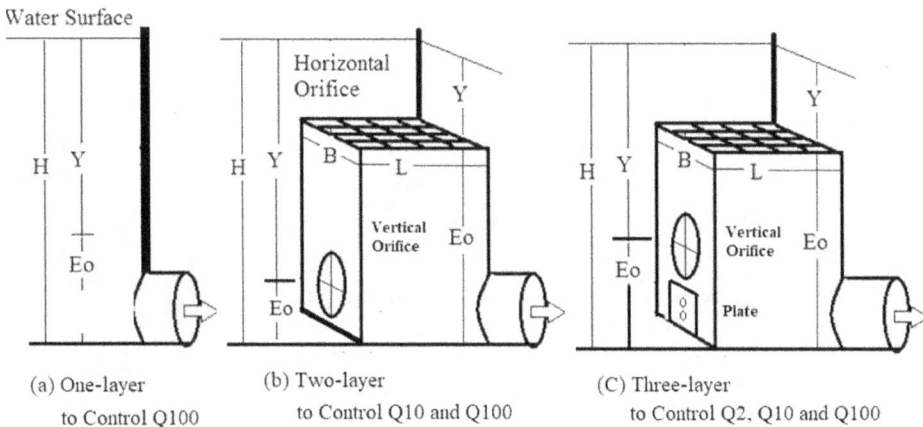

Figure 16.9 Various types of outlet box

16.4 CONCLUSIONS

The practice of stormwater detention was arose from concerns about public safety from flood threats. The general policy of stormwater detention was evolved from a single-layer design to reduce a 100-yr peak flow to a three-layer design for full-spectrum flow release control. The outflow from a detention basin is mainly dominated by the outlet concrete box.

As illustrated in Fig. 16.11, multiple boxes were stacked to provide full-spectrum flow release control. In recent years, such a complicated outlet system has been combined into a simple perforated plate with columns and rows of 1- to 2-inch holes. In the field, a perforated plate can simulate frequent events with a slow release when the water depth in the basin is shallow, then exponentially increase in flow release when the water depth increases. If it is necessary, we can always add a vertical orifice on the perforated plate and another horizontal grate on top of the box to increase the outflow. The outlet

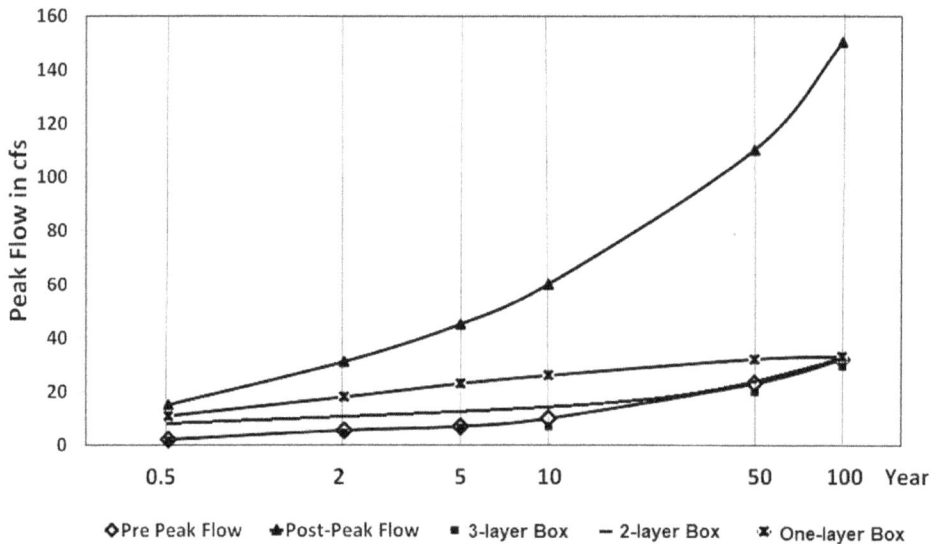

Figure 16.10 Performance evaluation of various outlet boxes

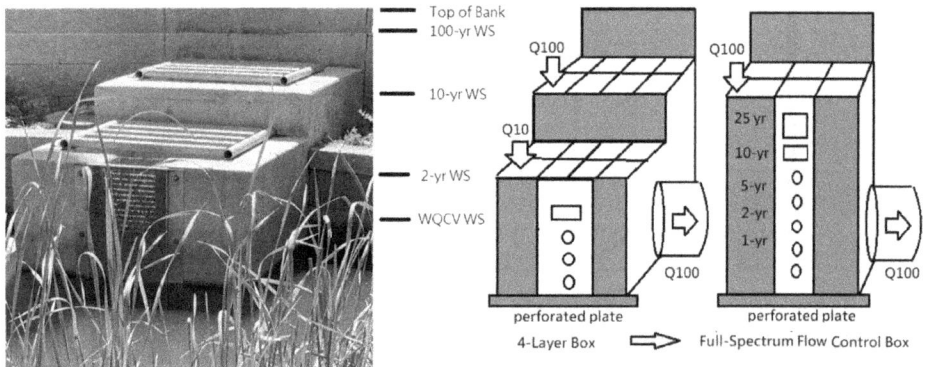

Figure 16.11 Multiple outlet boxes versus a vertical perforated plate

Figure 16.12 Vertical perforated plate for full-spectrum flow releases

box in Fig. 16.12 is the most popular in the field. Behind the grates and screen, we have a vertical perforated plate. In general, a plate is easy to be installed and maintained. Obviously, all small holes on the plate are sensitive to clogging. As a result, a micro pool is preferred to have a reliable suction flow system to reduce the clogging potential.

HOMEWORK

Q16-1 The Belleview detention basin in Fig. Q16.1 is located near Interstate Highway 470 and W. Belleview Avenue in the City of Denver, Colorado. This detention system is designed to have two 24-inch pipes installed on the bottom of the basin at an elevation of 5760 feet, and another 10-ft weir is installed overtopping the road at an elevation of 5,772 feet. The maximum water depth is 16.0 feet. If the orifice coefficient is 0.65 and the weir coefficient is 3.0, determine the stage-storage-outflow curve.

Solution: Using the orifice and weir formulas, the solution is summarized in Table Q16-1.

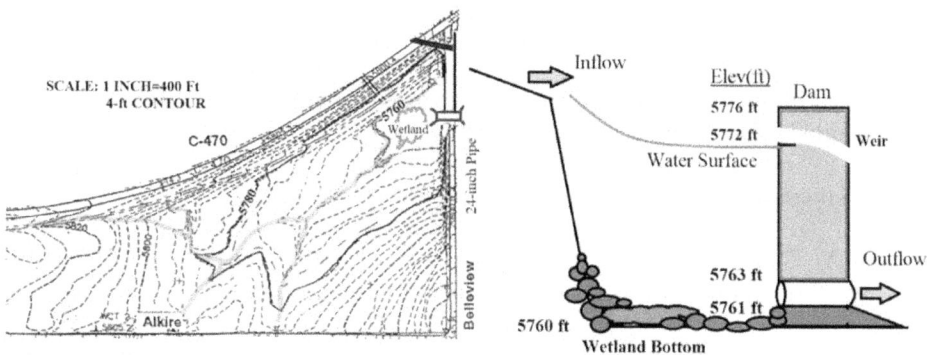

Figure Q16-1 Belleview detention basin

Table Q16-1 Stage-storage-outflow relationship for Belleview basin

Elevation Stage (S) (feet)	Cross-sectional Area (acre)	Cumulative Volume (S) (acre-ft)	Water Depth (feet)	Orifice Flow (cfs)	Weir Flow (cfs)	Total Outflow (O) (cfs)
5760.0	1.80	0.00	0.00	0.00	0.00	0.00
5764.0	3.50	10.40	4.00	56.80	0.00	56.80
5768.0	5.80	29.00	8.00	86.70	0.00	86.70
5772.0	8.00	56.70	12.00	108.70	0.00	108.70
5774.0	9.60	74.30	14.00	118.20	84.90	203.00
5776.0	11.30	95.20	16.00	126.90	240.0	366.90

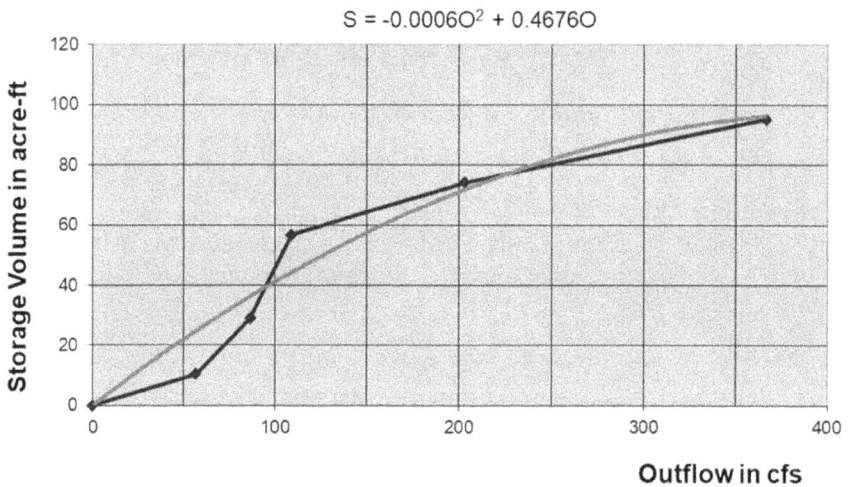

Figure Q16-2 Best-fitted SSO formula for Belleview basin

Conduct a regression analysis to develop an empirical equation for the stage-outflow curve defined in Table Q16-2, as:

$$S = -0.0006\,O^2 + 0.4676\,O \,(\text{correlation coefficient} = 0.94)$$

Q16-2 A stormwater detention basin has a stage-storage-outflow curve as shown in Table Q16-1. The best-fitted equation for the SO relationship is:

$$S \,(\text{acre-ft}) = 0.0002\,O^3 - 0.017\,O^2 + 0.4661\,O \,(O \text{ in cfs})$$

(A) Considering a time step of 10 minutes, derive the SO and OS routing functions for this detention system. (Hint: see Table Q16-2.)

Table Q16-2 Stage-storage-outflow curve

Stage H (ft)	Outflow Rate S (cfs)	Storage Volume O (acre-ft)	SO Function SO (acre-ft)	OS Function OS (cfs)
5000.0	0.00	0.00		
5000.5	0.25	0.41		
5001.0	1.20	0.82		
5001.5	4.09	1.25		
5002.0	5.78	1.70		
5002.5	7.08	2.16		
5003.0	8.17	2.64		
5003.5	9.14	3.13		
5004.0	10.01	3.66		
5004.5	12.73	4.22		
5005.0	28.17	4.81		
5005.5	32.07	5.45		
5006.0	33.81	6.12		

(B) The inflow hydrograph is given in Table Q16-3. The initial condition is an empty basin. For each time step $t+\Delta t$, you may guess the outflow, $O(t+\Delta t)$, and then calculate the corresponding storage volume, $S(t+\Delta t)$, using the best-fitted equation. Next, check the values of the SO routing function calculated with the known variables at time t, and to-be-found variables at $t+\Delta t$. Iterate this trial-and-error procedure until the value of the SO function is converged.

Table Q16-3 Hydrograph routing using SO function

	Given	KNOWN		SOLUTION from Chart		Calculated	CHECK
Time t $t+\Delta t$ minutes	Inflow $I(t)$ $I(t+\Delta t)$ cfs	SO FCT $[I(t)+I\Delta t)-2O(t)]$ $*\Delta t/43560+SO(t)$ acre-ft		Outflow $O(t)$ $O(t+\Delta t)$ cfs	Storage $S(t)$ $s(t+\Delta t)$ acre-ft	SO FCT $O(t)*\Delta t/43560$ $+2*S(t)'$ acre-ft	Balance in SO FCT acre-ft
0.00	0.00	0.00		0.00	0.00	0.00	0.00
10.00	17.00	17.00		0.25	0.12	17.00	0.00
20.00	64.00	97.50		1.50	0.66	97.50	0.00
20.00	64.00	222.50		3.71	1.51	222.50	0.00
30.00	108.00	387.07					
40.00	103.00	598.07					
50.00	78.00	779.07					
60.00	56.00	913.07					
70.00	37.00	1006.07					
80.00	22.00	1065.07					
90.00	13.00	1100.07					
100.00	10.00	1123.07					
110.00	7.00	1140.07					
120.00	4.00	1151.07					

REFERENCES

Bedient, P.B. and Huber, W.C. (1992). *Hydrology and Floodplain Analysis*, 2nd Edition, Addison-Wesley, New York.

Guo, J.C.Y. (2004). "Hydrology-Based Approach to Storm Water Detention Design Using New Routing Schemes", *ASCE Journal of Hydrologic Engineering*, vol. 9, no. 4, July/August.

Guo, J.C.Y.(2009). "Retrofitting Detention Basin for LID Design with a Water Quality Control Pool", *ASCE Journal of Irrigation and Drainage Engineering*, vol. 135, no. 6, October.

Guo, J.C.Y. (2013). "Green Concept in Stormwater Management", *Journal of Irrigation and Drainage Systems Engineering*, vol. 2, no. 3.

Guo, J.C.Y. (2017). *Urban Stormwater Management and Flood Mitigation*, CSC Press, New York.

Guo, J.C.Y. and Cheng, J.Y.C. (2008). "Retrofit Stormwater Retention Volume for Low Impact Development (LID)", *ASCE Journal of Irrigation and Drainage Engineering*, vol. 134, no. 6, December.

McCuen, R.H. (1998). *Hydrologic Analysis and Design*, 2nd edition, Prentice Hall, New York.

Puls, L.G. (1928). "Construction of Flood Routing Curves", House Document 185, US 70th Congress, First Session, Washington, DC.

Chapter 17

Energy dissipation basin

In practice, a *stilling basin* is commonly used to sustain the required tailwater depth. As illustrated in Fig. 17.1, the jump is initiated by a supercritical flow on the spillway and ends with a subcritical flow in the stilling basin. As always, a jump dissipates a tremendous energy by internal friction within the roller, which can be extended over a long distance in the stilling basin. In contrast, a hydraulic drop is gentler in terms of energy dissipation. In practice, a drop is triggered with a discontinuity on the channel floor. As illustrated in Fig. 17.1, a drop is submerged into the *plunging pool* and the flow regime is transferred from a subcritical flow to a supercritical flow.

In this chapter, the principles of energy and force for open-channel flows are reviewed and applied to various types of drop and jump basins as shown in Fig. 17.2. The concepts of specific energy and force are employed to outline the design procedures for sizing the *stilling basin, plunging pool,* and *fan-shaped basin.*

17.1 SPECIFIC ENERGY

The basic principles for open-channel flow were derived using a depth-integration method, where the flow properties are directly correlated to the cross-sectional average velocity. The Bernoulli's sum at a cross-section is defined as:

$$H = Y + Z + \frac{\alpha U^2}{2g} \tag{17.1}$$

$$W_S = Y + Z \tag{17.2}$$

where H = Bernoulli's sum in [L], Y = flow depth in [L] from the water surface to the lowest point in the cross-section, Z = invert elevation in [L], U = cross-sectional average flow velocity in [L/T], α = energy correction factor, and W_S = hydraulic head or water surface elevation in [L]. The Bernoulli's sum at a cross-section consists of *hydraulic* and *dynamic heads*. In practice, the value of α varies from 1.0 to 1.20. For simplicity, $\alpha = 1.0$ is acceptable for practice. At a given cross-section, the invert elevation is specified. The specific energy, E_S, is defined as the sum of the flow depth and velocity head as:

$$E_S = Y + \frac{\alpha U^2}{2g} \approx Y + \frac{U^2}{2g} \tag{17.3}$$

$$U = \frac{Q}{A} \tag{17.4}$$

DOI: 10.1201/9781003284239-17

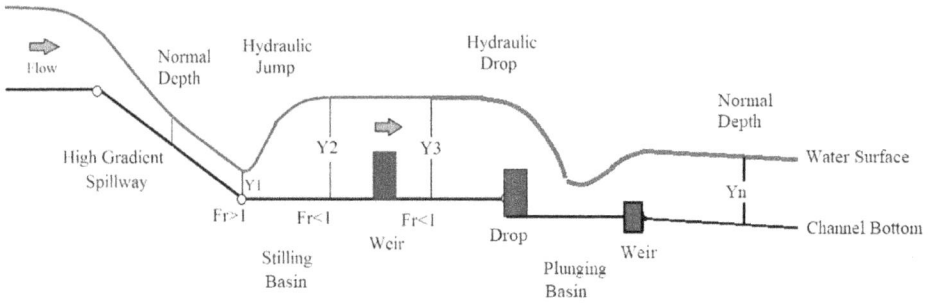

Figure 17.1 Hydraulic jump and drop in basins

(a) Jump Basin in Water Park

(b) Jump on Sloping Chute

(c) Vertical Drop across River

(d) Drop and Plunging Pool

Figure 17.2 Examples of jumps and drops

where E_s = specific energy in [L], Q = flow rate in [L³/T] channel, A = flow area in [L²], and g = gravitational acceleration in [L/T²]. Substituting Eq. 17.4 into Eq. 17.3 yields:

$$E_S = Y + \frac{Q^2}{2gA^2} \tag{17.5}$$

Eq. 17.5 represents the *specific energy curve*, which is the relationship between flow depth and velocity head for a given discharge, Q, in a specified cross-section, A. Although Eq. 17.5 is a cubic equation, only the two roots of real numbers are meaningful for engineering practice. It implies that for a given specific energy, the water flow can have either a high stage with a low velocity head or a low stage with a high velocity head.

Example 17.1: A water flow of 450 cfs is delivered in a trapezoidal channel with a bottom width, B = 10 feet, and side slopes, $Z_1 = Z_2 = 1.0$. Construct a specific energy curve for water depths ranging from 1.0 to 15.0 feet.

Solution: Direct solutions for Eq. 17.5 are listed in Table 17.1. As the flow depth is increased, the top width and flow area are increased, while the flow velocity and dynamic head (or kinetic energy) are decreased. As shown in Fig. 17.3, the specific energy, E_s, reaches its minimum when F_r = 1 or the critical flow with $Y = Y_c = 3.52$ feet.

Fig. 17.3 presents the specific energy curve for the given discharge in the specified channel section. The specific energy curve consists of three flow regimes identified by Froude numbers, including *subcritical*, *critical*, and *supercritical*. As illustrated in Fig. 17.4, the specified energy curve consists of two limbs because there are two roots (Y_1 and Y_2) for a given specific energy, E_S. The high stage, Y_1, has a Froude number less than unity and the low stage, Y_2, has a Froude number greater than unity. The depth of Y_1 is the *alternate depth* of Y_2, and vice versa. The high stage limb represents a *subcritical flow regime*, while the low stage limb represents a *supercritical flow regime*. The interception of these two limbs is the *critical flow*. It is noted that the specific energy at the critical flow is a minimum. The application of the specific energy curve is best explained by a subcritical flow overtopping a submerged dam. The energy loss for such a mild drop is negligible. As a result, both Y_1 and Y_2 in Fig. 17.4 are alternative depths that share the same specific energy as:

$$E_1 = E_2 + \Delta E \approx E_2 \qquad (17.6)$$

For the design flow, the critical depth occurs on the top of the dam.

$$E_1 = E_2 = E_c + H_m \qquad (17.7)$$

where E_1 = specific energy in [L] at Section 1, E_2 = specific energy in [L] at Section 2, ΔE = energy loss in [L], E_c = specific energy in [L] for critical flow, and H_m = height in [L] of the submerged dam.

Table 17.1 Specific energy curve

Flow Depth Y ft	Top Width T ft	Flow Area A sq ft	Flow Velocity u fps	Kinetic Energy u^2/2g ft	Sp. Energy Es ft	Froude Numer Fr
1.00	12.00	11.00	40.91	25.99	26.99	7.53
3.00	16.00	39.00	11.54	2.07	5.07	1.30
5.00	20.00	75.00	6.00	0.56	5.56	0.55
7.00	24.00	119.00	3.78	0.22	7.22	0.30
9.00	28.00	171.00	2.63	0.11	9.11	0.19
11.00	32.00	231.00	1.95	0.06	11.06	0.13
13.00	36.00	299.00	1.51	0.04	13.04	0.09
15.00	40.00	375.00	1.20	0.02	15.02	0.07

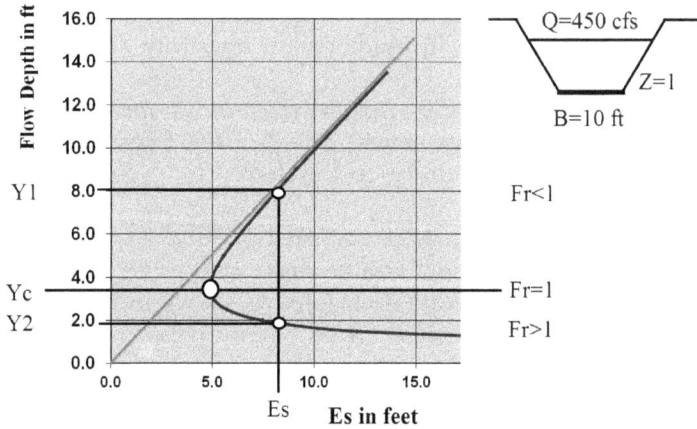

Figure 17.3 Specific energy curve

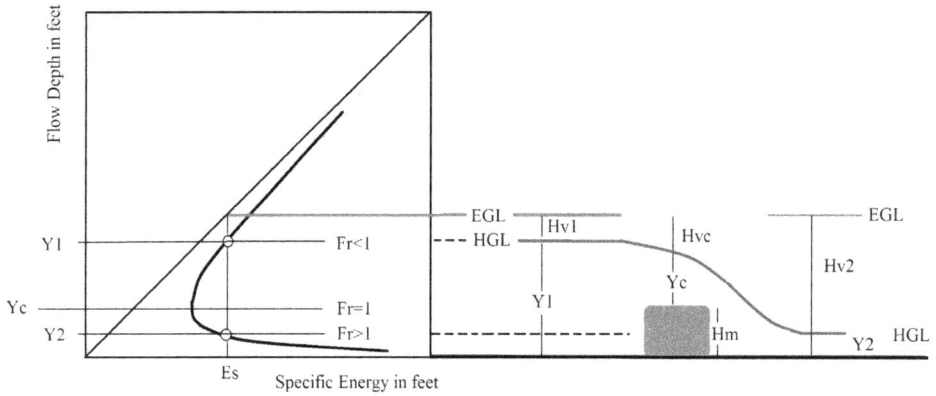

Figure 17.4 Application of the specific energy curve

17.1.1 Critical flow on specific energy curve

Channel flows are classified by the Froude number, F_r, which is defined as the ratio of flow velocity to wave celerity as:

$$F_r = \frac{U}{\sqrt{gH_d}} = \sqrt{\frac{U^2 T}{gA}} = \sqrt{\frac{Q^2 T}{gA^3}} \tag{17.8}$$

As illustrated in Fig. 17.3, the specific energy becomes a minimum at the critical depth. To prove it, taking the first derivative of Eq. 17.3 with respect to depth, Y, yields:

$$\frac{dE_s}{dY} = 1 - \frac{U^2 T}{gA} = 1 - \frac{U^2}{gH_d} = 1 - F_r^2 = 0 \quad \text{(the critical condition)} \tag{17.9}$$

Eq. 17.9 satisfies that $F_r = 1$ at the critical depth. Using Eq. 17.9, the critical hydraulic depth is defined as:

$$\frac{(H_d)_c}{2} = \frac{U_c^2}{2g} \tag{17.10}$$

where $(H_d)_c$ = critical hydraulic depth and U_c = critical flow velocity. For a rectangular channel, $(H_d)_c = Y_c$, which is the critical depth.

17.1.2 Special cases for specific energy

With $Z = 0$, a trapezoidal channel is reduced to a rectangular channel. When $Z = 0$, then $H_d = Y$. Eq. 17.10 is reduced to:

$$\frac{Y_c}{2} = \frac{U_c^2}{2g} \text{ for a rectangular channel} \tag{17.11}$$

Substituting the continuity principle into Eq. 17.11 yields:

$$Y_c = \left(\frac{q^2}{g}\right)^{1/3} \text{ for a rectangular channel} \tag{17.12}$$

$$q = \frac{Q}{B} \quad \text{for a rectangular channel} \tag{17.13}$$

where q = discharge per unit width in $[L^2/T]$. Using Eq. 17.11, the critical depth in a rectangular channel is two thirds of the specific energy:

$$E_c = Y_c + \frac{U_c^2}{2g} = Y_c + \frac{Y_c}{2} = \frac{3Y_c}{2} \text{ for a rectangular channel} \tag{17.14}$$

Example 17.2: Find the critical flow depth for the channel flow in Example 17.1.

Solution: The critical depth is solved by setting the flow Froude number equal to one, as:

$$T_c = B + 2ZY_c = 10.0 + 2.0Y_c$$

$$A_c = (B + ZY_c)Y_c = (10.0 + 1.0Y_c)Y_c$$

$$F_r^2 = \frac{Q^2 T_c}{gA_c^3} = \frac{450^2 (10.0 + 2.0Y_c)}{32.2[(10.0 + 1.0 \times Y_c)Y_c]^3} = 1$$

By trial and error, we have $Y_c = 3.52$ ft, $A_c = 47.5$ ft^2, and $U_c = 9.47$ fps.

Example 17.3: The flow in Example 17.1 is carried on a slope of 0.5% with Manning's N = 0.045. Referring to Fig. 17.3, the incoming flow is set to be the normal flow, $Y_1 = Y_n$.

(1) Determine the alternative depth, Y_2, downstream of the submerged dam.
(2) What is the height of the dam?

Solution: The normal flow condition is solved as:

Flow Depth Yn ft	Flow Area An sq ft	Wetted P-meter Pn ft	Hy- Radius Rn ft	Flow Velocity Un fps	Flow rate Q cfs	Froude Numer Fr
5.53	85.8	25.6	3.35	5.25	450.0	0.46

Referring to Fig. 17.3, set $Y_1 = Y_n$ = 5.53 ft to determine E_s = 5.95 feet and incoming flow F_r = 0.46. It is confirmed that the incoming flow is subcritical. After the drop, the flow becomes supercritical or $Y_2 < Y_C$ (Yc = 3.52 ft). The trial-and-error procedure begins with a value <3.52 feet for Y_2. For this case, Y_2 = 2.40 feet and F_r = 1.88, which is the alternate depth defined by the specific energy: E_s = 5.95 feet.

Station	Flow Depth Y ft	Flow Area A sq ft	Top Width T ft	Flow Velocity u fps	Kinetic Energy u^2/2g ft	Sp. Energy Es ft	Froude Numer Fr
Y=Y1=Yn	5.53	85.8	21.05	5.25	0.43	5.95	0.46
Y=Y2	2.40	29.8	14.80	15.12	3.55	5.95	1.88
Y=Yc	3.52	47.5	17.03	9.47	1.39	4.91	1.00

The balance of flow energy between Section 1 and the critical flow section is:

$$E_{s1} = E_{sc} + H_m \tag{17.15}$$

$$5.96 = 4.91 + H_m$$

So, H_m = 1.05 ft, which is the height of the submerged dam.

17.2 SPECIFIC FORCE

The momentum force associated with a channel flow consists of static force and dynamic force. The hydrostatic force results from the weight of the water body, whereas the dynamic force is equal to the flow momentum or the force needed to bring the flow to a stop.

$$F_{HS} = \gamma Y_o A \tag{17.16}$$

$$F_{HD} = \beta\rho QU \tag{17.17}$$

$$F_S = \gamma Y_o A\beta + \rho QU \tag{17.18}$$

where F_{HS} = hydrostatic force, F_{HD} = dynamic force, F_S = specific force, γ = specific weight of water, Y_o = vertical distance in [L] to the centroid of the flow area measured from the water surface, A = flow area in [L²], i = i-th area element, ρ = density of water in [M/L³], β = momentum correction factor varying between 1.0 and 1.1. All forces and weights are expressed in pounds or newtons. The value of β was developed to account for the non-uniformity of turbulent velocity distribution. In practice, let β be 1.0. The flow area can be divided into regular area elements. The location of the centroid of the flow area is determined by the area-weighted method as:

$$Y_o = \frac{\sum Y_{oi} A_i}{A} \tag{17.19}$$

Example 17.4: The trapezoidal channel in Example 17.1 carries a discharge of 450 cfs with $Y = 4.5$ ft, $A = 65.25$ sq feet, and $U = 6.90$ ft/second. The flow area is divided into two triangles and one rectangle, as shown in Fig. 17.5. Determine the momentum force in pounds.

Solution: Apply the area-weighted method to the three area elements to determine the centroid for the flow area as:

$$Y_o = \frac{\sum Y_{oi} A_i}{A} = \frac{1.5 \times \frac{1}{2}(4.5 \times 4.5) + 2.25 \times (4.5 \times 10.0) + 1.5 \times \frac{1}{2}(4.5 \times 4.5)}{65.25} = 2.01 \text{ ft}$$

The specific force carried by the flow is:

$$F_S = \gamma Y_o A + \rho QU = \frac{62.4 \times 2.01 \times 65.25 + 1.92 \times 450 \times 6.90}{1000} A = 14.22 \, klb$$

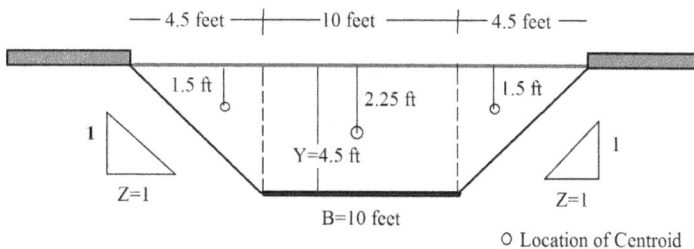

Figure 17.5 Calculating the specific force

Repeat the above procedure for flow depths ranging from 1.0 to 6.5 feet. Table 17.2 is the specific force curve for this case. It is noted that the critical depth is 3.52 feet, which agrees with the specific energy curve in Fig. 17.6.

Similar to the specific energy curve, the specific force curve is developed for a given discharge in a specified channel cross-section. A specific curve consists of three flow regimes: subcritical flow, critical flow, and supercritical flow. As illustrated in Fig. 17.7, the upper limb represents a subcritical flow regime and the lower limb represents a supercritical regime. Both limbs intercept each other at the critical depth. The application of a specific force curve is best explained by the water flow released from a

Table 17.2 Example for specific force curve

Flow Depth Y ft	Top Width T ft	Flow Area A sq ft	Flow Velocity u fps	Central Depth Yo ft	Sp. Force Fs Klbs	Froude Numer Fr	SP Energy Es ft
1.00	12.00	11.00	40.91	0.48	36.05	7.53	26.99
1.50	13.00	17.25	26.09	0.72	23.55	3.99	12.07
2.00	14.00	24.00	18.75	0.94	17.78	2.52	7.46
2.50	15.00	31.25	14.40	1.17	14.84	1.76	5.72
3.00	16.00	39.00	11.54	1.38	13.44	1.30	5.07
3.50	17.00	47.25	9.52	1.60	13.02	1.01	4.91
4.00	18.00	56.00	8.04	1.81	13.33	0.80	5.00
4.50	19.00	65.25	6.90	2.01	14.22	0.66	5.24
5.00	20.00	75.00	6.00	2.22	15.61	0.55	5.56
5.50	21.00	85.25	5.28	2.42	17.47	0.46	5.93
6.00	22.00	96.00	4.69	2.62	19.77	0.40	6.34
6.50	23.00	107.25	4.20	2.81	22.50	0.34	6.77

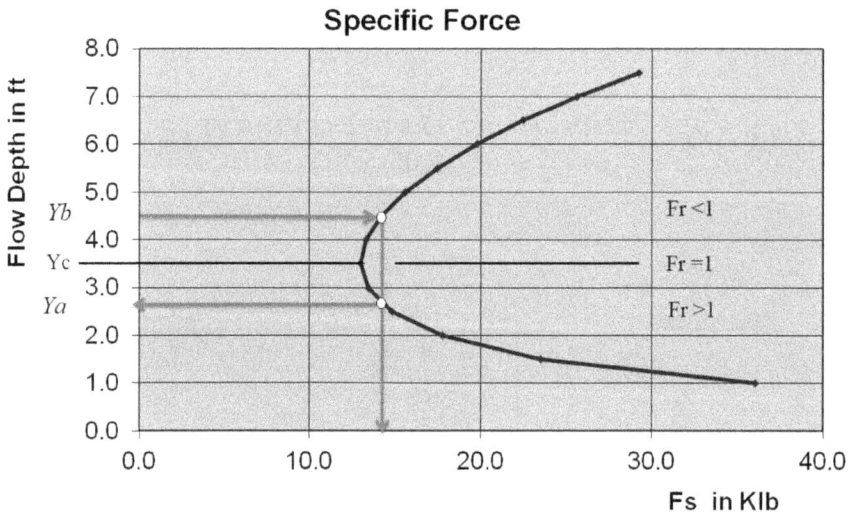

Figure 17.6 Specific force curve

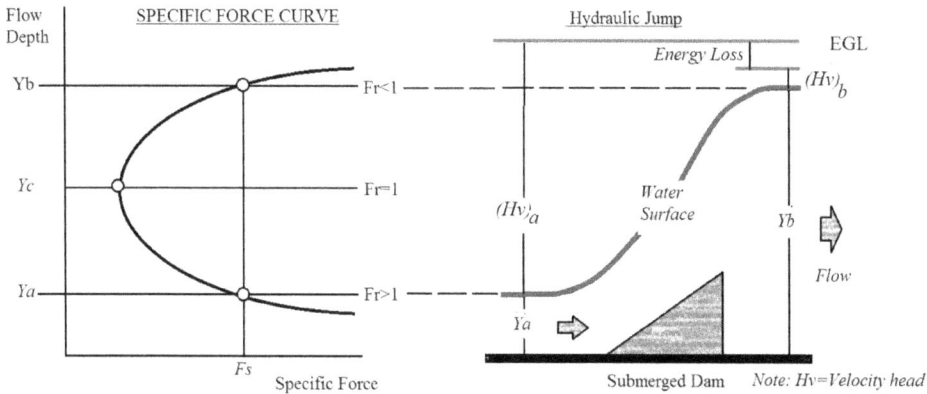

Figure 17.7 Application of the specific force curve to a jump

steep spillway that results in a hydraulic jump in the stilling basin. Although the huge dissipation through a hydraulic jump is not fully understood, it is recommended to use momentum principles to analyze a hydraulic jump on a flat floor. The flow depths before and after a jump, Y_a and Y_b, share the same specific force. Such a pair of depths is termed *sequent depths*.

$$F_a = F_b + \Delta F \approx F_b \tag{17.20}$$

$$\Delta E = E_a - E_b \tag{17.21}$$

where F_a = specific force at Section a, F_b = specific force at Section b, ΔF = external flow, which is negligible, ΔE = energy dissipation in jump, E_a = specific energy at Section a, and E_b = specific energy at Section b.

It is noted that the specific force at the critical flow is again a minimum. This fact can be proved by taking the first derivative of Eq. 17.18 with respect to depth Y and setting the resultant equation equal to zero.

Example 17.5: The spillway in Fig. 17.8 carries a flow of 450 cfs on a slope of 3.5% with Manning's $N = 0.015$. The incoming flow upstream of the jump is set to be the normal flow, $Y_a = Y_n$. A stilling pool backed up with a weir provides a sufficient tail-water water depth, Y_t, because of flow overtops the weir. (1) Determine the sequent depth, Y_b, downstream of the hydraulic jump. (2) Estimate the energy dissipation.

Solution: The normal flow condition for the given condition is computed as:

Flow Depth Y_n ft	Flow Area A_n sq ft	Wetted P-meter P_n ft	Hy-Radius R_n ft	Flow Velocity U_n fps	Flow rate Q cfs	Froude Numer Fr	Sp Energy E_s ft	Depth to Centroid Y_b ft	Sp Force F_s klb
1.70	19.9	14.8	1.34	22.60	450.0	3.27	9.61	0.81	20.73

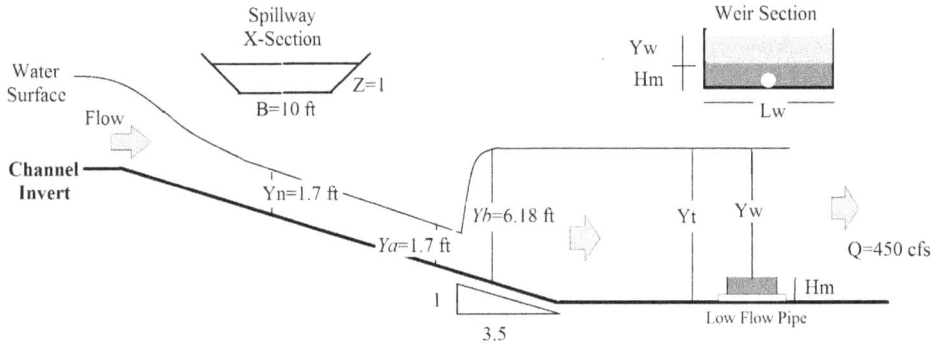

Figure 17.8 Analysis of a hydraulic jump

It is confirmed that the incoming flow is a supercritical flow with $F_r = 3.27$. It will produce a hydraulic jump when there is sufficient tailwater. Let $Y_a = i\,Y_n = 1.7$ feet. It is found that $Y_b = 6.18$ feet and $F_r = 0.37$ can produce the same specific force as 20.73 kilo-pounds.

Station Depth Y ft	Flow Depth A sq ft	Flow Area	Top Width T ft	Flow Velocity u fps	Kinetic Energy u^2 /2g ft	Sp. Energy Es ft	Froude Numer Fr	Depth to Centroid Yo ft	Sp. Force Fs Klbs
Y=Ya=Yn	1.70	19.9	13.40	22.60	7.93	9.63	3.27	0.81	20.73
Y=Yb	6.18	100.1	22.37	4.50	0.31	6.50	0.37	2.69	20.73
Y=Yc	3.51	47.5	17.03	9.47	1.39	4.91	1.00	1.60	13.02

The difference in specific energy before and after the jump is calculated as:

$$\Delta E_s = 9.63 - 6.50 = 3.13 \text{ feet or pound-ft per pound of water.}$$

$$\text{Energy dissipation} = \Delta E_s \, \gamma Q = 3.13 \times 62.4 \times 450 = 87.89 \text{ kilo-pound-ft}$$

A hydraulic jump creates a tremendous energy loss through the plunging pool. Knowing the geometry of the plunging pool, the location and length of the jump can further be determined.

17.2.1 Critical flow on specific force curve

Both specific energy and force curves are centered on the critical flow for the given conditions. In summary, the critical flow is characterized by:

1. a minimum specific energy,
2. a velocity head equal to a half of the critical hydraulic depth,
3. a minimum specific force, and
4. a Froude number equal to unity.

As indicated in Eq. 17.9, the critical flow is defined by $F_r = 1$. For a given cross-section, the flow regime varies with respect to flow depth. However, the critical flow depth is independent of channel roughness and slope.

17.2.2 Special cases for specific force curve

Assuming that the external force is negligible and the value of β is unity, Eq. 17.20 is expanded for a hydraulic jump as:

$$\gamma Y_a + \rho Q A_a = \gamma Y_b + \rho Q A_b + \Delta F (\Delta F = 0) \tag{17.22}$$

The subscript a indicates variables associated with Section a upstream of the jump, and the subscript b indicates variables associated with Section b. For *a rectangular channel*, we have:

$$Y_o = 0.5Y \tag{17.23}$$

$$A = BY \tag{17.24}$$

$$U = Q/BY \tag{17.25}$$

$$F_r = \frac{U}{\sqrt{gY}} \tag{17.26}$$

Substituting the above relationships into Eq. 17.22 results in:

$$\frac{Y_b}{Y_a} = \frac{1}{2}\left[\sqrt{1 + 8(F_r)_a^2} - 1\right] \tag{17.27}$$

Eq. 17.27 depicts the relationship between sequent depths in a rectangular channel. If either the downstream or upstream depth is known, its sequent depth in a rectangular channel can be computed using Eq. 17.27.

For a symmetric *triangular channel* in Fig. 17.9 with a side slope of $1V{:}ZH$, we have:

$$Y_h = Y/3 \tag{17.28}$$

$$A = ZY^2 \tag{17.29}$$

$$U = Q/(ZY^2) \tag{17.30}$$

Substituting the above equations into Eq. 17.22 yields:

$$\frac{1}{3}\left(Y_b^3 - Y_a^3\right) = \frac{Q}{Zg}\left(\frac{1}{Y_b^3} - \frac{1}{Y_a^3}\right) \tag{17.31}$$

Eq. 17.30 can be solved by trial and error if either Y_a or Y_b is given.

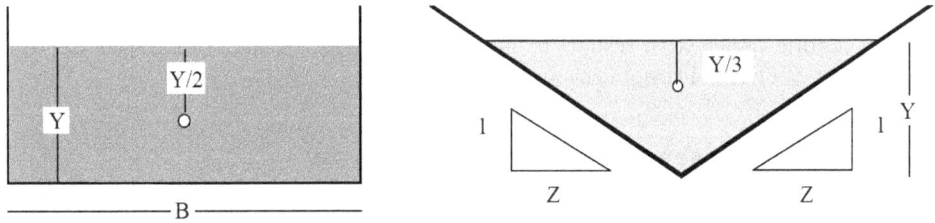

Figure 17.9 Rectangular and triangular channels

(a) Spillway and Stilling Basin

(b) Hydraulic Jump on Spillway

Figure 17.10 Example of a spillway and a stilling basin

17.3 STILLING BASIN

As illustrated in Fig. 17.10a and 17.10b, a stilling basin is located at the downstream end of a steep spillway. The water flow is released from the upstream reservoir into the spillway. As expected, a hydraulic jump is triggered in the basin.

The stilling basin, as illustrated in Fig. 17.11, is designed to provide adequate tail-water depth to trigger the hydraulic jump. The water pool in the basin is sustained by the backwater effect from the downstream weir. The height of the weir is chosen to balance the momentum force associated with the incoming flow. The water surface profile is extended from the basin into the spillway. The water depth on the spillway changes due to the spillway's slope. An inflow will develop normal flow on the spillway, and then will jump at the location where the water depth provides the required specific force associated with the incoming flow.

17.3.1 Stilling basin under design

The design flow is selected based on economic studies or a pre-selected level of protection. The system under design should be sized to accommodate the design flow. The design procedure should begin with the upstream incoming flow conditions. For the given design flow, the incoming flow on the spillway is assumed to have normal flow as:

$$Y_a = Y_n \tag{17.32}$$

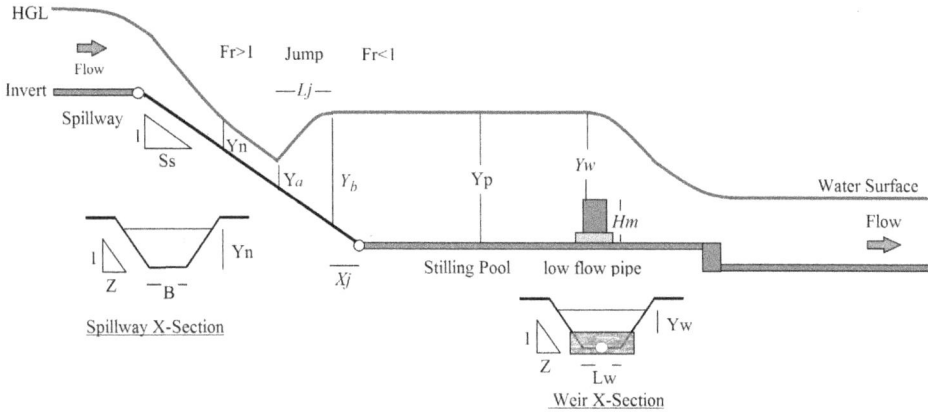

Figure 17.11 Hydraulic jump in a stilling basin

Based on the momentum principle, the sequent depth, Y_b, is determined by:

$$F_a = F_b + \Delta F, \text{with } \Delta F \approx 0 \tag{17.33}$$

Adding the freeboard depth, Fx, to Y_b yields the water depth in the basin as:

$$Y_p = Y_b + Fx \tag{17.34}$$

The location of the sequent depth, Y_b, is at the pre-selection location, X_j, on the spillway as:

$$X_j = \frac{E_p - E_b}{S_s} \tag{17.35}$$

where E_p = specific energy in [L] in the basin, E_b = specific energy in [L] associated with Y_b, X_j = distance from the toe of the spillway in [L], and S_s = slope on the spillway in [L/L]. Next, we shall investigate the geometry of the downstream weir. Under the assumption that the critical depth should be developed on top of the weir, the height of the weir is:

$$Y_w = Y_C \tag{17.36}$$

$$H_m = E_P - E_w = E_P - E_C \tag{17.37}$$

where Y_w = depth on top of the downstream weir, E_p = specific energy for flow depth, Y_p, in the basin, E_w = specific energy at the weir location, and E_c = specific energy of critical flow. Eq. 17.36 is the energy balance along the energy grade line (EGL). As shown in Fig. 17.11, the weir is designed to overtop the design flow and also to pass low flows through the low-flow pipes placed on the floor. The length of the basin mainly depends on the length of the hydraulic jump, which is highly empirical. Based on USBR studies in Fig. 17.12, the jump length is approximated as:

$$\frac{L_j}{Y_b} = 10(Fr_b - 1) \tag{17.38}$$

where L_j = length of jump in [L]. Eq. 17.38 only represents the jump length (USBR 2005). In practice, a safety factor of 1.25 should be considered.

Example 17.6: Referring to Fig. 17.13, the concrete spillway has B = 10 feet, Z = 1, $N = 0.015$, and $S_s = 5\%$. The design flow is 600 cfs. The downstream weir has $L_w = 18$ ft and $Z = 1$. Design the stilling basin to have a jump on the spillway.

Solution:

Step 1. Determine the normal flow on the spillway.

Assuming that the incoming flow is normal, the flow condition is determined with Manning's formula as:

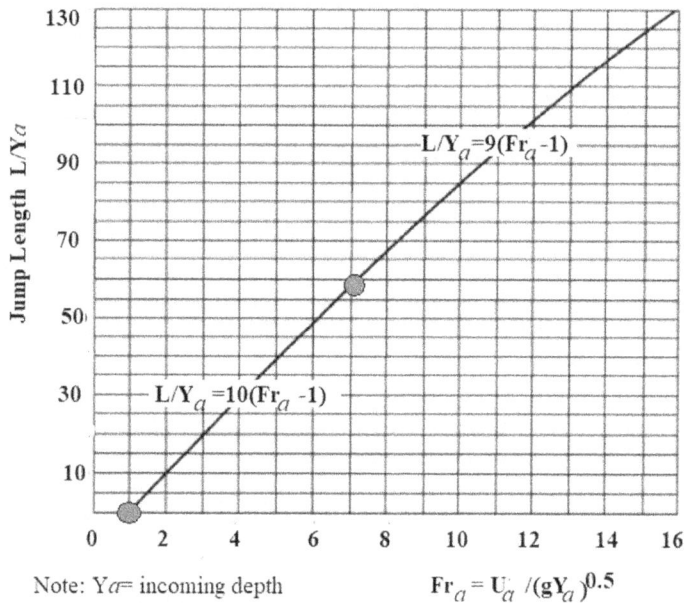

Note: Ya= incoming depth

$Fr_a = U_a / (gY_a)^{0.5}$

Figure 17.12 Length of a hydraulic jump

Bottom Width	B= 10.00	feet
Left Side Slope	Z1= 1.00	ft/ft
Right Side Slope	Z2= 1.00	ft/ft
Manning's N	N= 0.01500	
Longitudinal Slope	Ss= 0.0500	ft/ft
Design Flow	Q100= 600.0	cfs

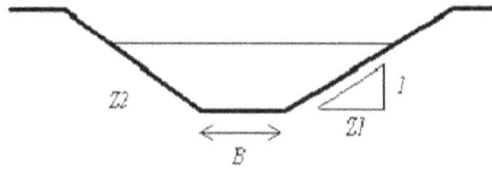

Flow Depth Y ft	Flow Area A sq ft	Wetted P-meter P ft	Hy- Radius R ft	Flow Velocity U fps	Flow rate Q cfs	Froude Numer Fr	Specific Energy Es ft	Depth to Centroid Yh ft	Specific Force Fs klb
1.82	21.4	15.1	1.42	27.98	600.0	3.93	13.93	0.86	33.72

Step 2. Set $Y_a = 1.82$ *feet to find* Y_b.

Solving the force equation $F_a = F_b = 33.72$ kilo-pounds and setting the downstream weir with a crest width, $L_w = 30$ ft, we have:

Enter Downstream Weir Width			Lw= 18 ft							
X-Section Flow Depth	Flow Depth Y ft	X-Section Width B or Lw ft	Flow Area A sq ft	Top Width T ft	Flow Velocity U fps	Specific Energy Es ft	Froude Numer Fr	Depth to Centroid Yo ft	Specific Force Fs Klbs
Y=Ya	1.82	10.00	21.4	13.63	27.98	13.97	3.93	0.86	33.72
Y=Yb	7.79	10.00	138.6	25.58	4.33	8.08	0.33	3.32	33.72
Y=Yw=Yc	3.07	18.00	64.6	24.13	9.29	4.41	1.00	1.46	16.69
Y=Yp	8.79	18.00	235.5	35.58	2.55	8.89	0.17	3.90	60.34
Hm=	4.48	feet	Hm=Ep−Ec (specific energy differnece)						
X100=	19.97	ft	X100=(Yp−Yb)/Ss						

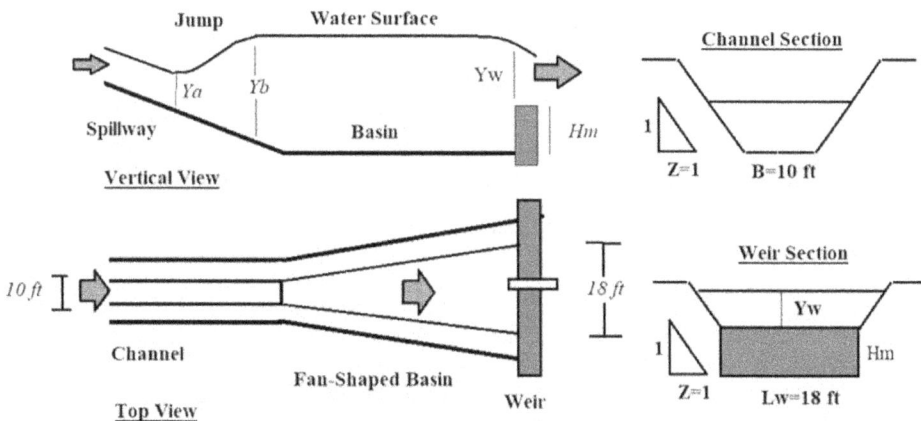

Figure 17.13 Fan-shaped basin for a hydraulic jump

Step 3. Determine the height of the downstream weir based on the critical flow conditions.

Considering that the freeboard is 1 ft, the water surface elevation in the basin is:

$Y_p = Y_b + 1 = 8.79$ ft

The location of Y_b from the toe of the spillway is calculated as:

$$X_j = \frac{\left(Y_p - Y_b\right)}{S_s} = \frac{8.79 - 7.79}{0.05} = 20 \text{ feet}$$

$Y_w = Y_C = 3.07$ ft and $E_w = E_c = 4.41$ ft.

For this case, one 6-inch circular pipe shall be installed as the low-flow pipe to pass frequent flows.

Step 4. Estimate the length of the jump.

$$\frac{L_j}{Y_a} = 10(Fr_a - 1) = 10(3.93 - 1) = 29.3 \quad \text{or} \quad L_j = 53.3 \text{ feet}$$

For this case, the stilling basin should have a length 60 ft > 53.3 ft.

17.3.2 Low flow through stilling basin

After the system is built, a wide spectrum of water flows will run through the system. As mentioned above, the incoming flow conditions are set to be $Y_a = Y_n$ on the spillway for the given flow. Based on the principle of momentum, the alternate depth, Y_b, can be determined. Next, where the jump is located on the spillway depends on the tail-water depth in the pool. The analysis of the tailwater depth should start from the downstream control section, which is the weir section. As illustrated in Fig. 17.14, the downstream trapezoidal weir is built to provide a sufficient tailwater depth to trigger the jump. This weir cross-section can be decomposed into a rectangular weir with a

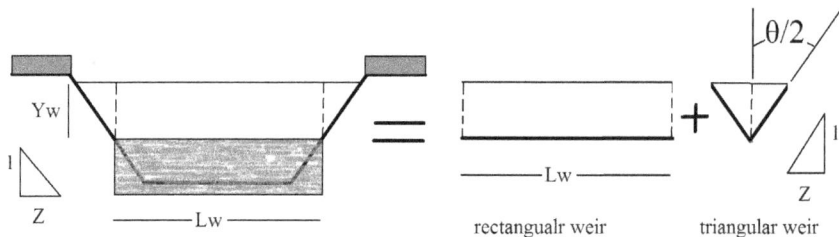

Figure 17.14 Trapezoidal weir at the downstream end of a still basin

crest length equal to the trapezoidal bottom width and a triangular weir with a notch angle equal to the trapezoidal side slope, Z, as:

$$\frac{\theta}{2} = \frac{\pi}{2} - \tan^{-1}\left(\frac{1}{Z}\right)$$

(17.39)

where θ = central angle in radians for a triangular weir and Z = side slope in [L/L]. Ignoring the minor adjustment on the end contraction, the capacity of a trapezoidal weir is derived as:

$$Q_w = \frac{2}{3}\sqrt{2g}C_d\left\{L_w Y_w^{1.5} + \frac{4}{5}\tan\left[\frac{\theta}{2}\right]Y_w^{2.5}\right\}$$

(17.40)

where Q_w = discharge in [L/T³] overtopping the weir, C_d = discharge coefficient, L_w = bottom width in [L] of a trapezoidal weir, and Y_w = water depth in [L] on top of the weir. Eq. 17.40 does not take internal friction into consideration. As a result, it is advisable to use a smaller value of 0.5 to 0.6 for the discharge coefficient.

Next, the water depth in the stilling basin is:

$$E_p = E_w + H_m \approx Y_w + H_m$$

(17.41)

The location of the jump on the spillway is determined as:

$$X_j = \frac{E_p - E_b}{S_S} \approx \frac{Y_p - Y_b}{S_S} \quad \text{under the assumption that the velocity head} \approx 0 \quad (17.42)$$

Example 17.7: Continued from Example 17.6. Recognizing that the system is built as described in Example 17.6. Analyze the hydraulic jump for a flow of 300 cfs.

Solution:

Step 1. Determine the normal flow on the spillway.

Bottom Width	B=	10.00 feet
Left Side Slope	Z1=	1.00 ft/ft
Right Side Slope	Z2=	1.00 ft/ft
Manning's n	N=	0.01500
Longitudinal Slope	S=	0.0500 ft/ft
Low Flow	Q2=	300.0 cfs

Flow Depth Y ft	Flow Area A sq ft	Wetted P-meter P ft	Hy-Radius R ft	Flow Velocity U fps	Flow rate Q cfs	Froude Numer Fr	Sp Energy Es ft	Depth to Centroid Yh ft	Sp Force Fs klb
1.20	13.5	13.4	1.01	22.24	300.0	3.76	8.86	0.58	13.43

Step 2: Set $Y_a = Y_n$ to find Y_b based on the specific force.

Flow Depth	Flow Depth Y ft	Flow Area A sq	Top Width T ft	Flow Velocity U ft	Kinetic Energy U2/2g fps	Specific Energy Es4 ft	Froude Numer Fr ft	Depth to Centroid Yo ft	Specific Force Fs Klbs
Y=Ya	1.20	13.5	12.41	22.24	7.68	8.89	3.76	0.58	13.43
Y=Yb	5.17	78.5	20.35	3.82	0.23	5.40	0.34	2.29	13.43

Step 3. Determine the flow depth on the downstream as-built weir.

For this case, the flow capacity overtopping the weir is calculated as:

$$\frac{\theta}{2} = \frac{\pi}{2} - \tan^{-1}\left(\frac{1}{Z}\right) = \frac{3.1416}{2} - \tan^{-1}\left(\frac{1}{1}\right) = 0.785 \text{ radians} (45 \text{ degrees})$$

Set L_w = 18 feet. The flow over the trapezoidal weir is calculated as:

$$300 = \frac{2}{3}\sqrt{2 \times 32.2} \times 0.55 \times \left\{18.0 \times Y_w^{1.5} + \frac{4}{5}\tan\left[45^\circ\right]Y_w^{2.5}\right\}$$

By trial and error, Y_w = 2.93 ft

Step 4. Determine the tailwater depth in the pool.

Knowing that the height of the as-built weir is 4.48 feet from Example 17.6, the water depth in the basin is:

$$Y_p = Y_w + H_m = 2.93 + 4.48 = 7.41 \text{ feet}$$

Step 5. Locate the hydraulic jump on the spillway.

$$X_j = \frac{Y_o - Y_b}{S_S} = \frac{7.41 - 5.17}{0.05} = 44.8 \text{ feet up stream of the spillway toe.}$$

Step 6. Determine the length of the jump.

$$\frac{L_j}{Y_a} = 10(Fr_a - 1) = 10(3.76 - 1) = 27.6 \text{ ft or } Lj = 33.1 \text{ ft} < 60 \text{ ft}$$

It is important to understand that the stilling system is designed for the design flow, but it has to be tested for a range of frequent flows to make sure the system can pass all flows safely.

17.4 PLUNGING POOL

A drop is often employed to stabilize the streambed in a waterway. A drop structure is composed of a submerged dam (weir), which raises the water depth to create back-water effects in the upstream direction and also to lower the channel floor by 3 to 5 feet immediately downstream of the drop structure. Under backwater effects, the incoming flow slows down. The potential for upstream bed erosion and bank scours is alleviated. However, downstream of the drop needs to be protected from the impingement of the jet flow on the channel floor. Often, a plunging basin, as shown in Figs 17.15a and b, is placed at the downstream end of a drop structure.

A drop is characterized with its drop height, h, as shown in Fig. 17.16. The unit-width flow overtops the drop structure and then plunges into the pool. The water jet produces a parabolic trajectory to impinge on the pool floor.

(a) Plunging Pool in Channel   (b) Jump in Plunging Pool

Figure 17.15 Examples of a plunging pool

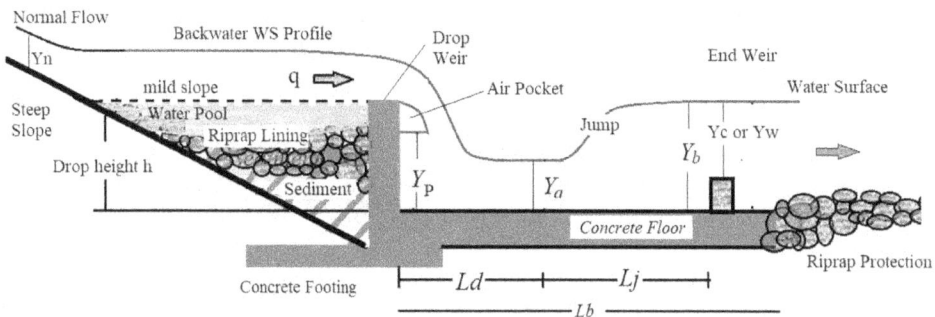

Figure 17.16 Plunging pool

The geometry of this jet flow is described by the drop height and drop number, which are defined as:

$$q = \frac{Q}{T} \qquad (17.43)$$

$$D_n = \frac{q^2}{gh^3} \qquad (17.44)$$

where Q = design flow in [L³/T], q = unit-width discharge in [L²/T], T = top width in [L] of the weir crest, D_n = drop number in [L⁰], g = gravitational acceleration in [L/T²], and h = drop height in [L]. The drop-pool section is designed to have a rectangular cross-section, with its width equal to the top width, T, of the drop structure.

$$\frac{L_d}{h} = 4.30 D_n{}^{0.27} \qquad (17.45)$$

$$\frac{Y_p}{h} = 1.00 D_n{}^{0.22} \quad \text{(drop length)} \qquad (17.46)$$

$$\frac{Y_a}{h} = 0.54 D_n{}^{0.425} \qquad (17.47)$$

$$\frac{Y_b}{h} = 1.66 D_n{}^{0.27} \qquad (17.48)$$

$$\frac{L_j}{Y_a} = 10\left(\frac{U_a}{\sqrt{gY_a}} - 1\right) \quad \text{(jump length)} \qquad (17.49)$$

$$L_b \geq \left(L_j + L_d\right) \quad \text{(basin length)} \qquad (17.50)$$

where L_d = drop length, which is the distance in [L] from the drop wall to depth Y_a, Y_p = pool depth in [L] near the drop wall, Y_a = depth in [L] before the jump, Y_b = depth in [L], U_a = flow velocity in [L/T] for depth Y_a, L_j = jump length in [L], and L_b = length of basin in [L].

Example 17.8: A grass-lined channel carries a flow of Q = 3000 cfs with a top width of T = 84.15 ft. A drop structure is proposed to have a vertical fall, h = 5.0 feet, for flow energy dissipation. Design the dimensions of the plunging pool downstream of the drop.

Solution: Referring to Fig. 17.16, the dimensions of the pool are calculated as:

Drop Height per Reach	$h =$	5.00	ft
Width of Drop Top	$W2$	84.15	ft
Unit Flowrate per foot	$q=$	35.65	cfs/ft
Drop number	$Dn =$	0.32	
Water Depth in Pool	$Yp=$	3.88	ft
Water Depth before Jump	$Ya=$	1.65	ft
Water Depth after Jump	$Yb =$	6.08	ft
Location of Jet Impingement	$Ld=$	15.75	ft
Length of jump	$Lj> =$	32.30	ft
Total length of pool	$Lb> =$	48.05	ft

17.4.1 Composite weir section on top of drop structure

From the point view of safety, a drop structure in Fig. 17.17a is always sized to withstand an extreme event. However, its daily operation is to pass frequent, small flows. As a result, the top weir should be designed with a low-flow notch.

The low flow rate is approximately 1 to 3% of the design flow, such as a 100-yr peak flow. Applying the weir flow formula to the low-flow notch in Fig. 17.17b for the low flow rate yields:

$$Q_L = \frac{2}{3}C_d\sqrt{2g}W_1H_1^{\frac{3}{2}}$$

(17.51)

where Q_L = low flow in [L^3/T], W_1 = notch width in [L], and H_1 = low head depth in [L] on the notch weir. In practice, the low head depth is pre-selected from a range of 1 to 2 feet. The only unknown, W_1, in Eq. 17.50 can be determined. Under the design flow, the high head depth on top of the drop structure applies to both the notch weir in the

(a) Front View of Composite Weir   (b) Cross Section of Composite Weir

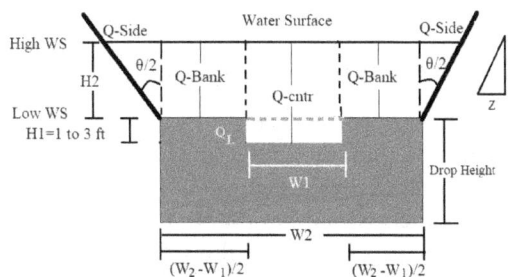

Figure 17.17 Weir section on top of a drop structure

center and two overflow weirs on both sides. The weir flow formula for the composite weir section under the high head depth is written as:

$$Q_H = \frac{2}{3}C_d\sqrt{2g}[W_1\left(H_1+H_2\right)^{1.5}+\left(W_2-W_1\right)H_2^{1.5}+\frac{4}{5}\tan\left[\frac{\theta}{2}\right]H_2^{2.5}] \qquad (17.52)$$

where Q_H = design flow [L³/T] under high head depth, W_2 = top width in [L] of the drop structure, and H_2 = water depth in [L] on top of the overflow weir. In practice, W_2 in [L] is set to be close to the top width of the approaching channel flow upstream of the drop. As a result, the only unknown, H_2, in Eq. 17.51 can be solved for the design flow.

Example 17.9: The rectangular channel in Fig. 17.17b is designed with Q_L = 20 cfs, Q_H = 1000 cfs, H_1 = 1.0 ft and C_o = 0.6 or C_w = 3.21 using feet-second units. The top width of the channel flow is 60.83 ft. Size the dimensions of the weir section on top of the drop structure.

Solution: The low flow, Q_L = 20 cfs, is released from the lower notch weir. Setting H_1 = 1.0 ft, the width of the low notch weir is:

$$Q_L = \frac{2}{3}C_d\sqrt{2g}W_1H_1^{\frac{3}{2}} = \frac{2}{3}\times0.6\times\sqrt{2\times32.2}\ W_1\left(1\right)^{3/2} = 20$$

So, W_1 = 6.23 ft. We know that the upstream channel flow has a top width of 60.83 ft. To have a smooth transition from the channel flow to weir flow, let W_2 = 60 ft. The headwater on the top weir is determined as:

$$Q_H = \frac{2}{3}\times0.60\sqrt{2\times32.2}[6.23\times\left(1+H_2\right)^{\frac{3}{2}}+\left(60.0-6.23\right)H_2^{\frac{3}{2}}]$$

So, H_2 = 2.89 ft.

17.5 CONCLUSIONS

An energy dissipation basin is the most commonly used method of stabilizing waterways, banks, levees, and hydraulic structures from streambed erosion and embankment scours. Typical examples are the plunging pool in Fig. 17.18a downstream of a drop structure in a waterway, and the stilling basin in Fig.17.18b at a conduit outfall that is tied into a natural floodplain.

Under the assumption that external forces are negligible, a *hydraulic jump* can be modeled with a balance of flow momentum and hydrostatic forces.

$$\Delta F \approx 0 \qquad \text{for a hydraulic jump} \qquad (17.53)$$

(a) Plunging Pool in Waterway   (b) Fan-shaped Stilling Basin at Outfall

Figure 17.18 Examples of energy dissipation basins

When a hydraulic jump is triggered by the tailwater depth on a flat basin, it is reasonable to assume that the surface friction and body force of the water volume are negligible. In practice, the weight of water volume may become significantly important when the jump occurs on a steep slope (>10%).

A plunging pool starts with an upstream weir flow that drops into the pool. The plunging pool ends with an overtopping weir flow at the downstream check dam. Care must be taken when choosing the height of the check dam. A properly sized check dam would produce a sufficient tailwater depth to create a submerged condition. Under the assumption that friction losses within a short distance are negligible, the inflow and outflow in a *hydraulic drop* can be reasonably solved by the balance of energy principle.

$$\Delta E \approx K \frac{U^2}{2g} \quad \text{and } K = 0 \quad \text{for a hydraulic drop} \tag{17.54}$$

Eq. 17.54 is acceptable when a drop is triggered under smooth and submerged conditions. Otherwise, a minor loss with $K = 0.29$ of the incoming flow kinetic energy should be taken into consideration when a drop involves a vertical impingement. As a common practice, a fan-shaped pool is preferred for energy dissipation at the confluence when a man-made channel drains into the natural stream. It is important to select the flare angle, i.e. a width ratio of 2 to 4, to produce a non-erodible sheet flow onto the floodplain.

HOMEWORK

Q17-1 The riprap trapezoidal channel shown in Fig. Q17-1 has a bottom width of $B = 10$ feet, side slope of $Z = 1$, and Manning's roughness, $N = 0.045$. This channel runs on a bottom slope of 3%. To reduce the erosion potential, a drop of $h = 5$ ft is proposed to reduce the energy slope from 3% to 0.74%. (1) With an energy slope of 0.74%, determine the normal flow coming to the drop. (2) Considering that the drop is as wide as the top width associated with the incoming channel flow, shape the plunging pool. (3) Size the composite weir on top of the drop to pass 20 cfs as the low flow and 777 cfs as the high flow.

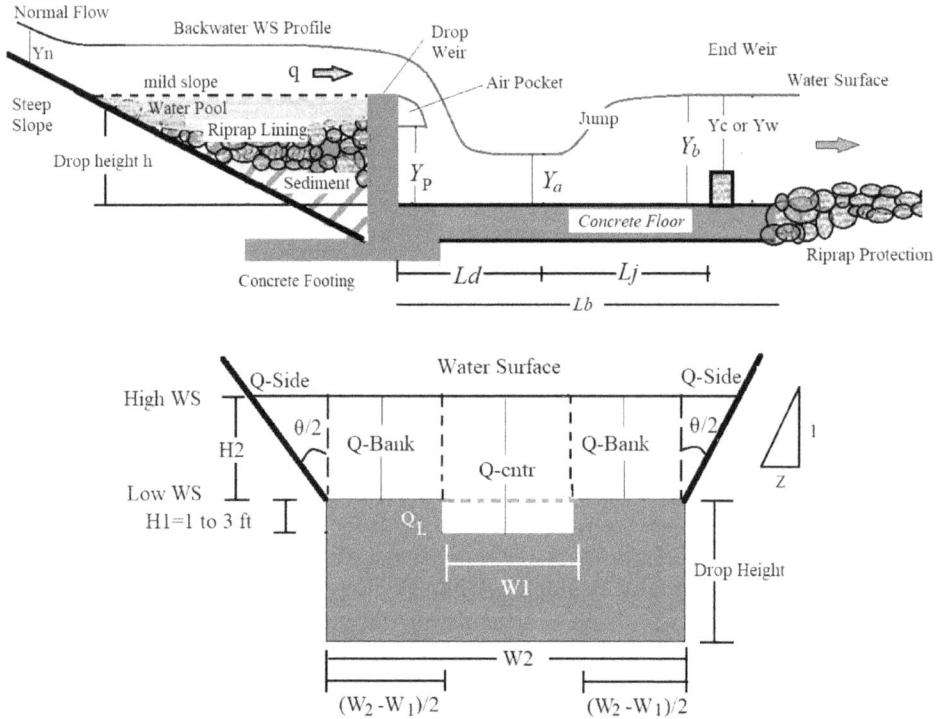

Figure Q17-1 Plunging pool and composite weir

Solution:

INCOMING CHANNEL FLOW

Bottom Width	B= 10.00	feet
Left Side Slope	Z1= 1.00	ft/ft
Right Side Slope	Z2= 1.00	ft/ft
Manning's N	N= 0.045	
Channel Bottom Slope	So= 0.0074	ft/ft
Design Flow	Q= 777.00	cfs
Normal Flow Depth	Y= 6.66	feet
Wetted Perimeter	P= 28.84	ft/ft
Top Width	T= 23.32	ft
Normal Flow Area	A= 111.00	sq ft
Hydraulic Radius	R= 3.85	feet
Flow Velocity	U= 7.00	fps
Froude Number	Fr= 0.57	

PLUNGING BASIN GEOMETRY

Drop Height per Reach	h= 5.00	ft
Width of Drop Top	$W2$= 23.32	ft
Unit Flowrate per foot	q= 33.31	cfs/ft
Drop number	Dn = 0.28	
Water Depth in Pool	Yp = 3.77	ft
Water Depth before Jump	Ya = 1.56	ft
Water Depth after Jump	Yb = 5.86	ft
Location of Jet Impingement	Ld= 15.18	ft
Length of jump	Lj> = 31.36	ft
Total length of pool	Lb> = 46.55	ft

DESIGN OF TOP WEIR

Discharge Coefficient	Cd= 0.55
Weir Coefficient	Cw= 2.94

Low flow Weir

Low Flow	Qlow= 20	cfs
Weir Side Slope	Z= 0.00	0.000
Headwater Depth	H1= 1.00	ft
Weir Width (guess)	W1= 6.80	ft

Overbank Weir

High Flow	QH=Q= 777.00	cfs	Summary of Weir Flow		
Trapezoidal Weir Width	W2= 23.32	ft	Side Angle	45.02	degree
Trapezoidal Weir Side Slope	Z= 1.00	0.785	Weir Width	23.32	ft
Headwater Depth	H2= 4.33	ft	Flow area	126.64	sq ft
Qoverbank in Rectangular Weir	Qbank= 438.66	cfs	Flow V	6.14	fps
Qoverbank in Side Wier	Qside= 92.01	cfs	Flow Depth	5.33	ft
Q-Center in Central (Low) Weir	Qlow= 246.33	cfs	Es =	5.92	ft
Total Computed Flow Q=	777.00	cfs			

Q17-2 Continued from Q17-1. Referring to Fig. Q17-1, the flow condition at the drop structure includes a headwater depth of 5.33 ft and a cross-sectional velocity of 6.14 fps. Balance the flow energy between the drop section and Y_a section to find the flow depth, Y_a, as:

$$E_w = Y_a + \frac{U_a^2}{2g} + h + \Delta E = Y_a + \frac{U_a^2}{2g} + 5 + 0.3 \times \frac{U_a^2}{2g}$$

$$U_a = Q / A_a$$

The Y_a section has the same cross-section as the incoming flow. Compare your answer with the empirical formula in Eq. 17.47.

Q17-3 A concrete trapezoidal chute is designed to pass a discharge of 500 cfs on a slope of 5.0% with a bottom width of 20 feet, side slope of 1*V*:1*H*, and Manning's roughness of *N* = 0.014. A stilling basin with a crossing weir is designed for energy dissipation. (a) Determine the location of the jump, or the distance, *X*, on the ramp, tailwater depth, *Y*$_t$, in the poll, and height, *H*$_m$, of the end weir for a design flow of 500 cfs. (b) Having built the system, analyze a flow of 50 cfs flowing through the system.

Solution: All depths in Fig. Q17-3 are in feet.

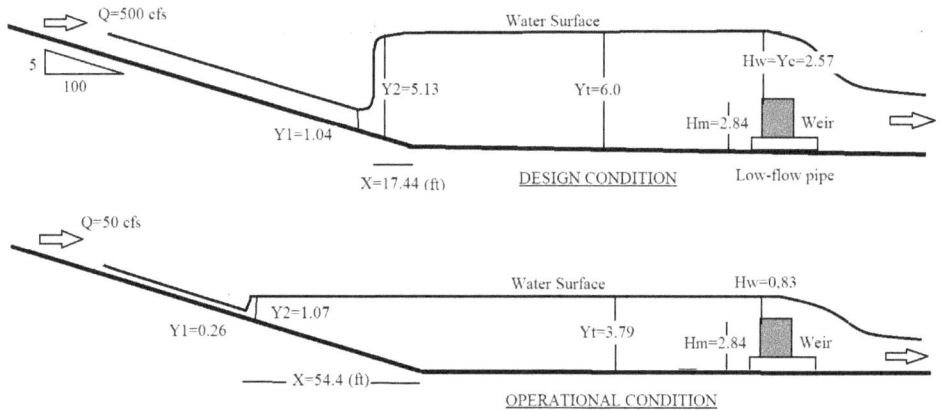

Figure Q17-3 Design of a stilling basin

REFERENCES

Chow, V.T. (1959). *Open-Channel Hydraulics*, McGraw Hill, New York.

Granata, F., de Marinis, G., Gargano, R., and Hager, W.H. (2009). "Energy Loss in Circular Drop Manholes", *Research Gate*, p. 6.

Guo, J.C.Y. (2004). "Design of Urban Channel Drop Structure", *Journal of Flood Hazards News*, December.

Guo, J.C.Y. (2009). "Grade Control for Urban Channel Design", *Journal of Hydro-Environmental Research*, 10.1016/j.jher.2009.01.001, February/March, pp. 1–4.

HEC14 (1975). *Hydraulic Design of Energy Dissipators for Culverts and Channels*, US Department of Transportation, Federal Highway Administration, Washington, DC.

Rand, W. (1959). "Flow geometry at straight drop spillways", *American Society of Civil Engineers*, vol. 8, pp. 1–13.

USBR (2005). "Hydraulic Design of Stilling Basins and Energy Dissipators", *Water Resources*, Tech no. 85.

USWDCM (2010). *Urban Stormwater Design Criteria Manual*, Urban Drainage and Flood Control District, vol. 2, Denver, CO, pp. 9–50.

Chapter 18

Stormwater quality capture volume

18.1 REVIEW OF URBAN DRAINAGE SYSTEMS

Under conventional concepts, urban drainage systems were designed with major concerns about traffic safety and public health. Conventional stormwater systems were designed mainly for the purpose of quickly removing *minor runoff flows* (2- to 5-yr storms) using street inlets and sewers, and *major runoff flows* (10- to 100-yr storms) using street gutters and channels (ASCE 1994). After the US EPA urban runoff study (EPA Report 1983), public concern began to shift from stormwater removal to urban water environmental preservation. Therefore, conventional conveyance systems have been modified to include storage facilities. Many innovative stormwater methods have been developed with the aim of reducing peak flows using stormwater detention facilities (FAA 1977, Aron and Kibler 1990, Guo 1999). After the mandate of *Federal Clean Water Act* (Ryan 2011), the goals in urban drainage design have been widened to include stormwater quality enhancement using on-site runoff volume disposal approaches. As a learning process, the concept of *best management practices* (BMPs) have gradually converged on *low-impact-development* (LID) approaches using infiltration and filtering processes to dispose of the increased runoff volume into vegetation beds, porous basins, and pervious pavements (Booth and Jackson 1997, Davis 2008). Stormwater modeling techniques have been shifted from *event-based approaches* to *long-term continuous rainfall-runoff simulations*. The ultimate goal for innovative stormwater management is to achieve a full spectrum of runoff release control (Guo 2013).

Storm runoff is sensitive to watershed imperviousness, which induces increases in both runoff flows and volumes after development. Without on-site mitigations, increased runoff flows will transfer negative impacts to downstream water bodies. To adhere to the goals of the *Federal Clean Water Act* and related policies, a *cascading flow system*, as shown in Fig. 18.1, is recommended as a three-layer drainage system. It comprises: (1) LID units to intercept 3- to 6-month *micro events*, (2) conveyance systems to transport 2- to 5-year *minor storms* in sewer systems and 10- to 100-yr *major storms* in street gutters, and (3) detention systems to control the flow releases from 2- to 100-yr full-spectrum storm events (Guo, Wang and Li in 2020). As illustrated in Figs 18.1 and 18.2, a waterway starts with an upstream porous basin, which includes a trash catcher and a WQCB. The overtopping flow will drain into the street gutter, where minor stormwater will fall into underground sewer lines and excess stormwater up to a 100-yr storm event will be carried through street gutters to the watershed outlet. At the outfall point, a detention basin is installed to reduce peak outflows.

DOI: 10.1201/9781003284239-18

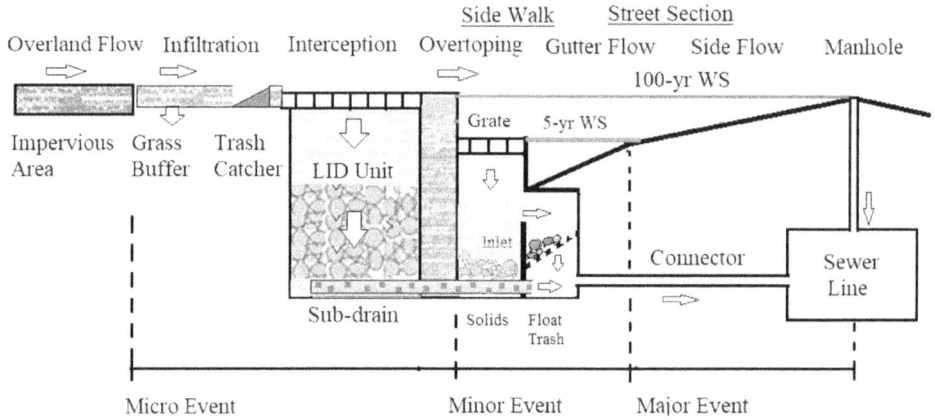

Figure 18.1 Stormwater mitigation on increased runoff volume and flow

(a) Rain Garden as WQCB   (b) Trash Catcher in front of WQCB

Figure 18.2 LID unit and trash catcher

In contrast, a WQCB implemented upstream of the conveyance system is to intercept the *early runoff volume up to the WQCV*, whereas a flood detention basin is often placed at the outfall to reduce post-development *peak flows* to pre-development conditions. In comparison, a LID porous basin is an overtopping flow system built with a shallow and flat pool, such as rain gardens and bioretention basins. In contrast, a flood detention basin is shaped with a deep, large storage pool, which is equipped with inflow dissipation systems and outflow control systems.

Design methods for stormwater detention focus on how to define extreme events with a pre-selected risk level, whereas LID porous basins are sized to optimize the runoff capture rate (Guo and Urbonas 1996). The settlement process in a porous basin depends on the basin's size and resident time. The resident time is often set to be the drain time, which ranges from 12 to 72 hours, depending on the sediment removal rate and water rights policies (UDFCD 2010). Obviously, the higher the sediment removal rate, the longer the basin's drain time. The higher the runoff capture rate, the bigger the basin's storage.

(a) Conveyance Type --Porous Pavement

(b) Storage Type --Infiltration Basin

(c) Rain Garden

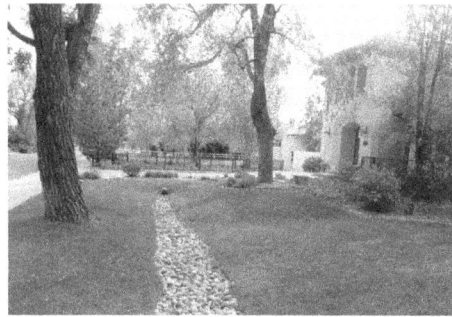

(d) Bio Swale

Figure 18.3 Conveyance and storage low-impact designs

18.2 FILTERING AND INFILTRATING DEVICES

In an urban catchment, impervious areas are the source of storm runoff. Before over-land flows become concentrated, the increased runoff volume per unit area is the cause of water-quality problems and is called the *V-problem*. After overland flows are collected into street gutters, sewers, and channels, the increased runoff flow is the cause of flooding problems and is called the *Q-problem*. Conventional stormwater designs have been focused on how to reduce peak flows using stormwater detention, whereas the latest development under the LID concept is to integrate various infiltrating and filtering facilities to promote on-site runoff volume disposal. A LID design will apply a filtering process to enhance stormwater quality and an infiltration process to reduce stormwater volume. Because LID devices are designed to aim at runoff source control, they are applicable to small tributary areas (up to 5 acres or 2 hectares). As shown in Fig. 18.3, the latest developments on LID designs, including infiltration beds, rain gardens, bio-swales, and porous pavements, can be classified into:

1. *flow-over conveyance types*, such as porous pavements using flat infiltration beds, and
2. *flow-in storage types*, such as rain gardens using shallow infiltration basins.

Obviously, the effectiveness of a LID design depends on how much the surface runoff volume can be intercepted. The storage volume used to size a WQCB is called the *water-quality capture volume* (WQCV). A WQCV should be of the same magnitude as the natural depression volume that was filled and leveled during the urbanization process.

18.3 RAINFALL AND RUNOFF DISTRIBUTIONS

It is important to realize that the conventional criteria used to design stormwater detention basins for extreme events can no longer serve the design of WQCB, simply because the goal of a WQCB is to capture frequent runoff events, not extreme events. Frequent runoff flows must be identified from a continuous runoff record, which can be observed in the field or simulated from a continuous long-term rainfall record. Between two adjacent rainfall events in a continuous record, as shown in Fig. 18.4, the *inter-event time* represents the period of time of no rain. Analyses of a continuous record begins with delimiting individual rainfall events using a pre-selected *minimum inter-event time* (MIET) or *event separation time*, which is defined as the period of no rain between two adjacent events (Guo and Urbonas 1996). As demonstrated in Fig. 18.4, considering a MIET of 6 hours, three individual events are identified because Groups A and B are lumped into a single event, as are Groups D and E. After each individual event has been identified, the *event rainfall depth* and *duration* can be further calculated for statistical analyses.

Fig. 18.5 presents the distribution of the rainfall event-depths recorded at the City of Denver, Colorado, after a continuous rainfall record has been divided into individual events. Although a two-year storm event is often considered a small storm for flood control projects, a 2-yr event, in fact, has a rainfall depth greater than 95% of the rainfall population. Although the skewness of the event rainfall depth distribution varies with respect to the meteorological region, it is generally true that the number of smaller rainfall events absolutely dominate the rainfall population.

As recommended (EPA Report 1986, Discoll et. al. 1989), a MIET of 6 hours should be used to divide a continuous rainfall record into N individual events. Each individual event is constructed with the attributes of event depth, event duration, and inter-event time to the next event. Most of rain gauges are installed above the ground. After subtracting the interception loss, which depends on wind speeds and surrounding bushes, trees, and buildings, the on-ground rainfall depth is determined as:

$$d(i) = D(i) - I_s \qquad (18.1)$$

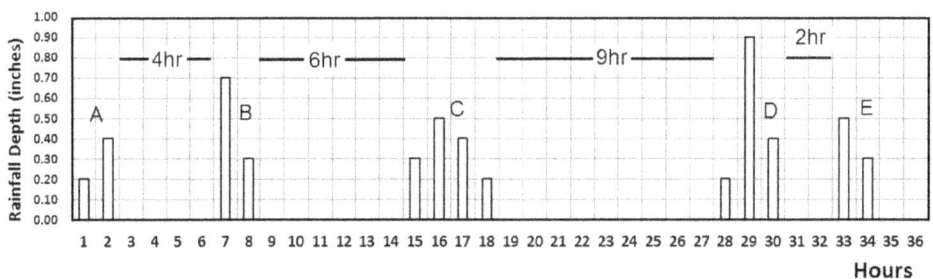

Figure 18.4 Event separation by inter-event time

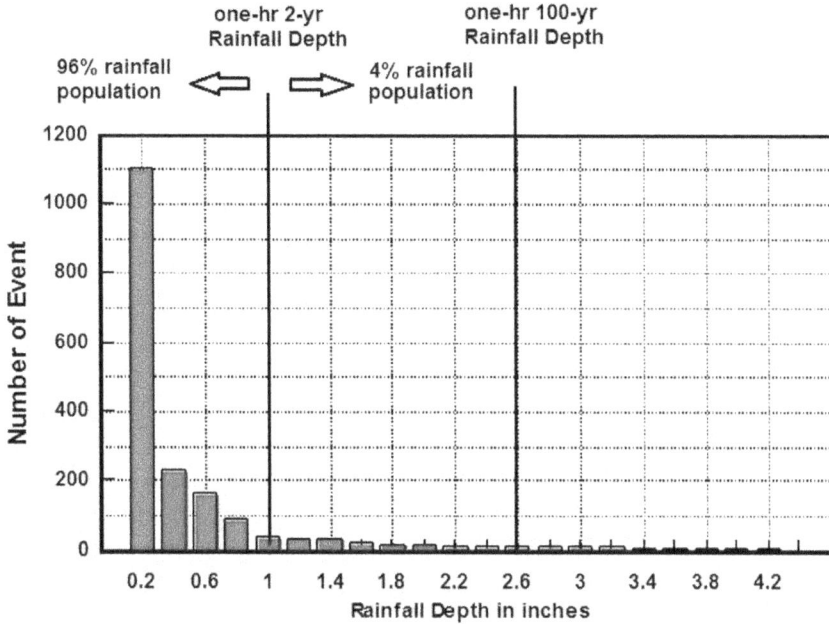

Figure 18.5 Rainfall depth distribution at Denver, Colorado

where $D(i)$ = i-th observed rainfall depth in [L] at the rain gauge, $d(i)$ = i-th on-ground rainfall depth in [L], and I_s = interception loss in [L], such as 0.05 to 0.1 inch (Guo and Urbonas 1996). Having the continuous rainfall record divided into individual storms, the statistics for event-depth, duration, and inter-event time can further be calculated as:

$$D_m = \frac{1}{N}\sum_{i=1}^{i=N}d(i) \tag{18.2}$$

$$S_D = \frac{1}{(N-1)}\left[\sum_{i=1}^{i=N}(d(i)-D_m)^2\right]^{\frac{1}{2}} \tag{18.3}$$

$$C_s = \frac{1}{S_D{}^3N(N-1)(N-2)}\left[\sum_{i=1}^{i=N}(d(i)-D_m)^3\right] \tag{18.4}$$

$$T_{im} = \frac{1}{N}\sum_{i=1}^{i=N}T_i(i) \tag{18.6}$$

$$T_{dm} = \frac{1}{N}\sum_{i=1}^{i=N}T_d(i) \tag{18.7}$$

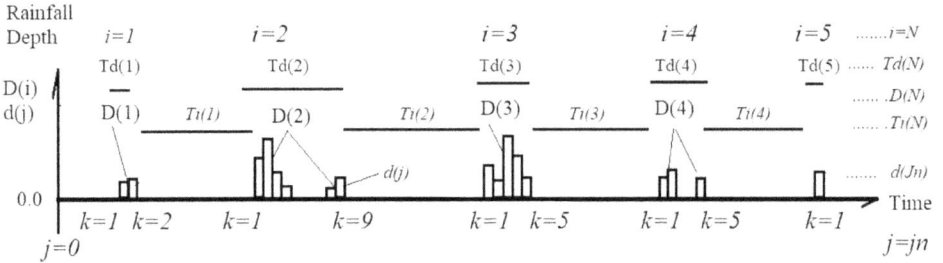

Figure 18.6 Rainfall event database generated from a continuous rainfall record

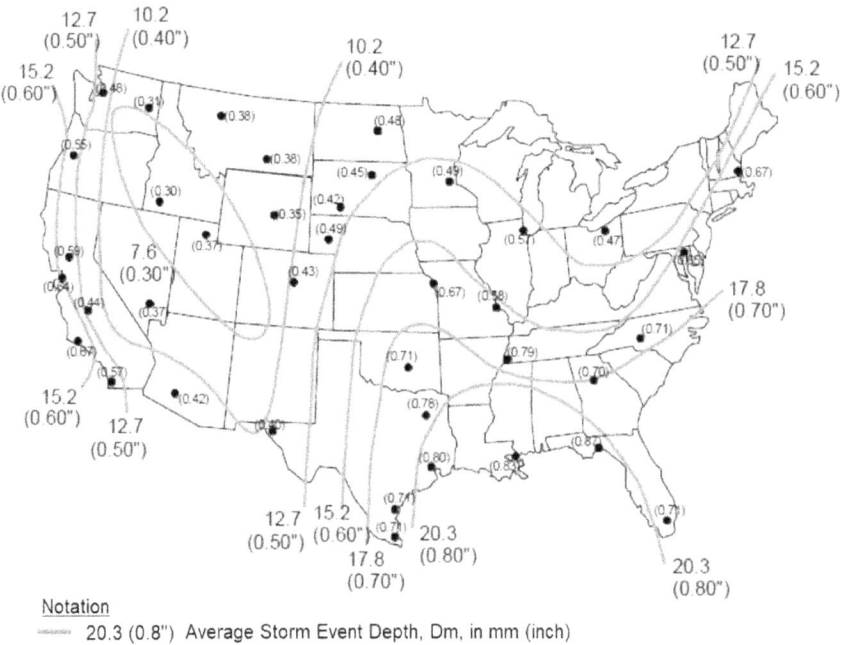

Notation

▬▬▬ 20.3 (0.8") Average Storm Event Depth, Dm, in mm (inch)

Figure 18.7 Average rainfall event depth for the US continent

where D_m = average event-depth in [L], N = total number of events in the continuous rainfall record, S_D = standard deviation in [L], C_s = skewness coefficient, T_{im} = average inter-event time in [T], $T_i(i)$ = i-th inter-event time in [T], $T_d(i)$ = i-th event duration in [T], T_{dm} = average event duration.

As demonstrated in Fig. 18.6, the continuous hourly rainfall depths, $d(j)$, from $j = 1$ to $j_n = 262800$ are divided into individual events expressed as d_i for the i-th rainfall event, and k = counter to determine the i-th event duration, T_{di}, and interevent time, T_i. Figs 18.7 and 18.8 show distributions of rainfall event-depth and duration. (Guo and Urbonas 2021).

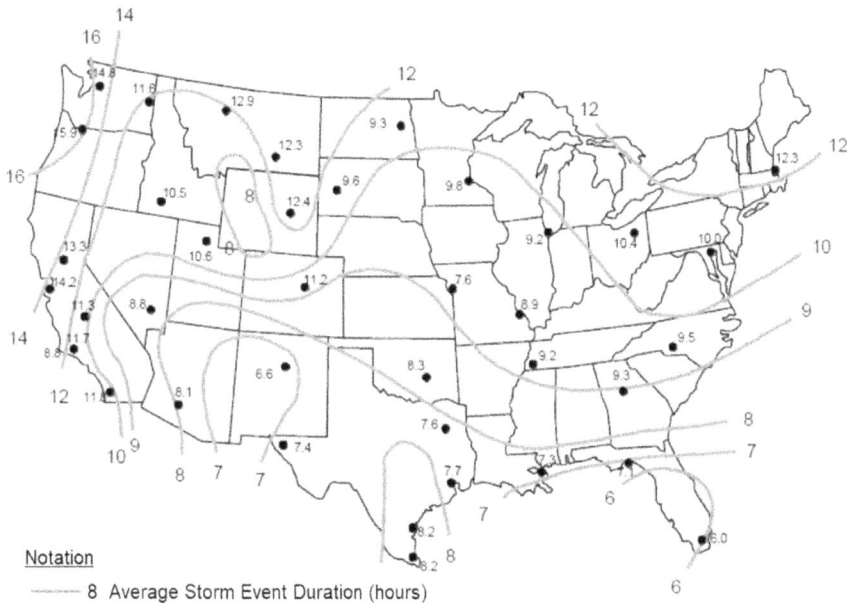

Figure 18.8 Average rainfall event duration for the US continent

18.4 CONCEPT OF RUNOFF CAPTURE

Referring to Fig. 18.5, a porous basin with a volume D would capture the rainfall events that produced an event depth less than or equal to D. As illustrated in Fig. 18.9, the runoff capture curve can be produced as an asymtotic curve from zero to unity. The area under the rainfall event-depth curve represents the runoff captured.

As illuminated in Fig. 18.9, the exponential model can be adopted to describe the distribution of the observed rainfall event-depths (Guo 2002). Taking into consideration the interception loss, the rainfall event-depth distribution is converted into a runoff capture curve as:

$$\text{Prob}\left[0 \leq \frac{D(i) - I_s}{D_m} \leq \frac{D}{D_m}\right] = \alpha = 1 - Ke^{\frac{-D}{D_m}} = 1 - \frac{m}{N} \tag{18.8}$$

$$K = e^{\frac{-I_s}{D_m}} \tag{18.9}$$

where $\text{Prob}\left[0 \leq \dfrac{D(i) - I_s}{D_m} \leq \dfrac{D}{D_m}\right]$ = probability distribution for rainfall event-depths, α = runoff capture rate, D = storage volume in [L per watershed] for the LID basin under design, m = the rank for D among the N observed events, I_s = hydrologic interception loss, such as 2 to 3 mm, K = constant related to the interception loss.

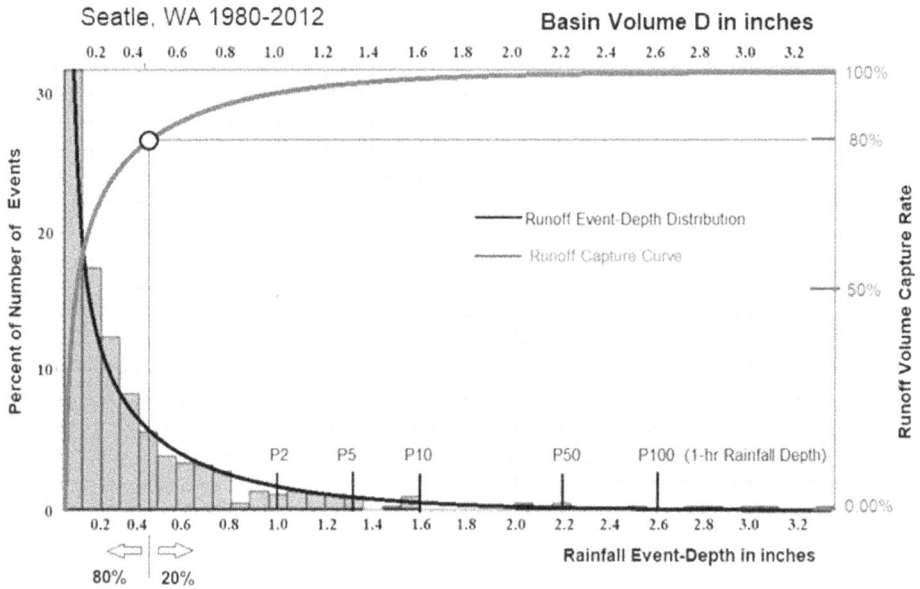

Figure 18.9 Runoff capture curve

Re-arranging Eq. 18.8 yields:

$$\frac{D}{D_m} = -\ln\left(\frac{1-\alpha}{K}\right) \tag{18.10}$$

where K = constant varying in a narrow range between 0.75 and 0.85 (Guo 2002). As shown in Fig. 18.10, Eq. 18.10 presents the non-exceedance probability curve for the rainfall event-depth distribution. On an impervious area, the runoff depth is equal to the rainfall depth. As a result, Eq. 18.10 may also represent the special case of a runoff capture curve for an impervious surface. Eq. 18.11 indicates that the higher the runoff capture rate, the bigger the storage volume. With such an increasing function, the optimal design volume can be identified using the diminishing return approach (Guo and Urbonas 1996). The optimal design volume is denoted as storm *water-quality capture volume* (WQCV), which has a runoff capture rate of approximately 80 to 85%. As a simplified recommendation, a runoff capture rate of $\alpha = 0.8$ should be accepted for WQCV (EPA Report 1986, Rossman 2010).

$$\frac{WQCV}{D_m} = -\ln\left(\frac{1-0.8}{0.8}\right) = 1.38 \quad \text{for } \alpha = 0.8 \text{ on an impervious surface} \tag{18.11}$$

Eq. 18.11 is a general expression for impervious surfaces. In fact, the basin's runoff coefficient, drain time, and rainfall duration are also crucial factors in the determination of the WQCV.

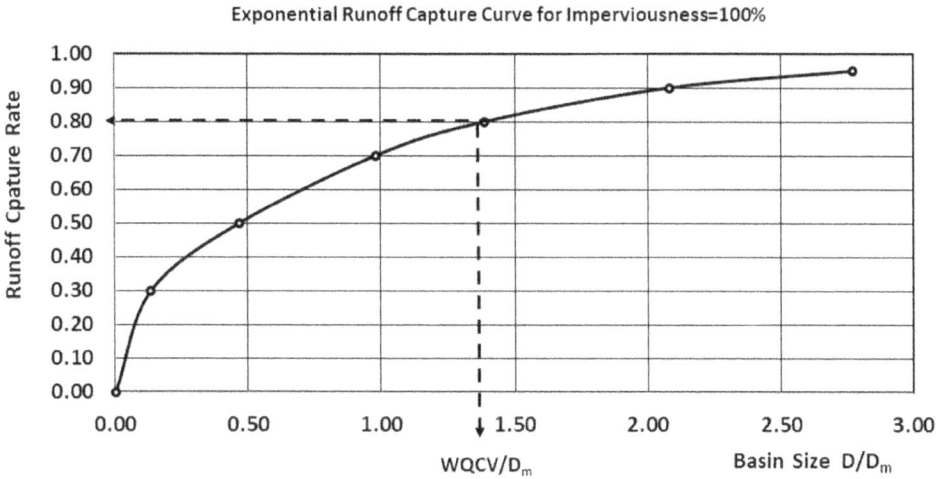

Figure 18.10 Exponential runoff capture curve for rainfall event depths

When the watershed imperviousness is 100%, the rainfall depths are equivalent to the runoff depths. Eq. 18.10 is a normalized one-parameter exponential runoff capture curve. Although Eq. 18.10 is simple and easy for engineering applications, it needs improvements to take into consideration the watershed imperviousness and basin's drain time.

18.5 RUNOFF CAPTURE ANALYSIS

18.5.1 Runoff volume capture analysis for single event

For convenience, all volumes used in the calculation of runoff capture are converted to the same units as rainfall depth, namely inches or millimeters per watershed. Applying the point rainfall-to-runoff volumetric approach, an on-ground event rainfall depth is converted to its runoff depth as:

$$P_R = Cd(i) \tag{18.12}$$

where P_R = event runoff volume in inch or mm per watershed and C = runoff coefficient. As illustrated in Fig. 18.11, during a storm event, the WQCB is loaded to its brimful and then overtopped. The maximum potential for runoff volume captured and treated by the WQCB is equal to the basin's storage volume, D_o, plus the runoff volume flowing through the basin during the storm duration, qT_d, as:

$$P_T = D_o + qT_d \tag{18.13}$$

where P_T = potential capacity in [L per watershed] that is considered captured and treated, D_o = basin's storage volume in [L per watershed], and T_d = rainfall event

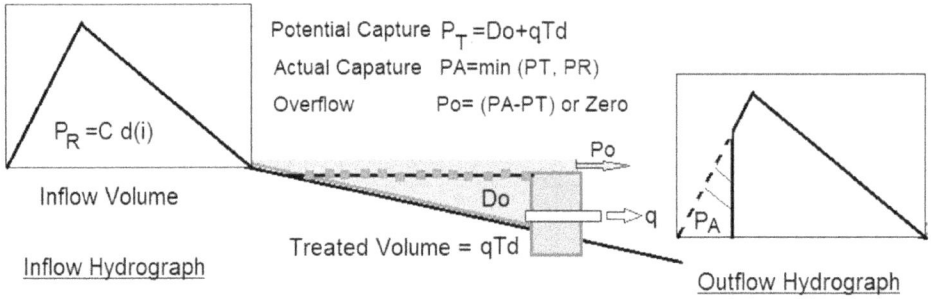

Potential Capture $P_T = Do + qTd$
Actual Capature $PA = min (PT, PR)$
Overflow $Po = (PA-PT)$ or Zero

Figure 18.11 Capture volume and outflow hydrograph

duration in hours. The product qT_d represents the runoff volume flowing through the WQCB during the event duration. The average flow release is determined as:

$$q = \frac{D_o}{T_e} \tag{18.14}$$

where q = *average flow release* from WQCB in [L/T] and T_e = basin's emptying (drain) time in hours. Since not every event can overtop the WQCB, there are two possibilities: (1) if the event runoff volume, P_R, is greater than the potential capacity, P_T, the excess runoff volume $(P_R - P_T)$ is the overflow without any treatment; and (2) if the runoff volume, $P_R \leq P_T$, the event is completely captured and treated. As result, the numerical procedure for determining the runoff capture volume, P_A, and overflow volume, P_o, is:

$$P_A = Min(P_R, P_T) \tag{18.15}$$

$$P_o = P_R - P_A \text{ if } P_o \geq 0 \text{ or } P_o = 0 \tag{18.16}$$

For this single event, the *runoff volume capture ratio* (RVCR), α, is defined as:

$$\alpha = \frac{P_A}{P_R} \tag{18.17}$$

Example 18.1: An urban catchment has a tributary area of 2.0 acres and a runoff coefficient of 0.85. Its WQCB in Fig. 18.12 is designed to have a storage volume of 0.07 acre-ft and a drain time of 6 hours. Determine the percentage of captured volume for the rainfall event of 1.0 inch over duration of 3.0 hours.

Solution: As illustrated in Fig. 18.12, a small, shallow porous basin only intercepts the early runoff on the rising inflow hydrograph. After the basin becomes full, the excess water overtops the basin outlet. The given event produces a runoff volume as:

$$P_R = 0.85(1.0 - 0.1) = 0.765 \text{ inch/watershed}$$

Figure 18.12 Illustration of runoff capture analysis

Let us convert the basin storage volume from acre-ft to inch/watershed as:

$$D_o = \frac{0.07 \times 12}{2.0} = 0.42 \, \text{inch/watershed}$$

The average release from this WQCB is:

$$q = \frac{0.42}{6.0} = 0.07 \, \text{inch/hr per watershed}$$

The basin's potential capacity in this case is:

$$P_T = 0.42 + 0.07 \times 3.0 = 0.63 \, \text{inch} < 0.765 \, \text{inch}$$

Because $P_R > P_T$, the basin was fully loaded and then spilled. The runoff volume capture ratio is calculated as:

P_A = min(0.765, 0.63) = 0.63 inch/watershed

P_o = 0.765 – 0.63 = 0.132 inch/watershed

α = 0.63/0.762 = 82%

The operation of a porous basin does not produce any significant attenuation effects on the peak flow.

Example 18.2: Continuing with Example 18.1, determine the percentage of captured volume for an event of 0.5 inch over a duration of 3.0 hours. Table 18.1 presents the details of the calculations.

Table 18.1 Solution for Example 18.2

Enter Watershed Information			
Watershed Area	A= 2.00	acres	
Runoff Coefficient	C= 0.85		
Incipient Runoff Depth	Is= 0.10	inch	
Enter Basin Information			
Storage Volume	Do= 0.42	inch	
Emptying Time	Te = 6.00	hours	
Average Release	q= 0.070	inch/hr	
Enter A Single Rainfall Event			
Enter Rainfall Depth	d(i) 0.50 inch $\implies$	3630.00	cubic feet
Enter Rainfall Duration	Td= 3.00	hours	
Runoff Volume Capture Analysis			
Basin's Max Potential Capacity	P_T= 0.630 inches	4573.80	cubic feet
Actual Rain Volume	P_R = 0.340 inch	2468.40	cubic feet
Runoff Capture Volume	P_A=MIN(P_T, P_R) 0.340 inch	2468.40	cubic feet
Spill-over Volume	Po = 0.000 inch	0.00	cubic feet
Runoff Capture Ratio	α=PA/PR 1.000		
Overflow Ratio	$1-\alpha$= 0.000		

For both cases, the average α = 90%. Examples 18.1 and 18.2 can be repeated for a long-term continuous rainfall record to determine the long-term runoff captured volume for a given WQCV.

18.5.2 Runoff volume capture analyses for long-term records

The long-term performance for a selected storage volume, D_o, and drain time, T_e, can be evaluated by the continuous rainfall record. Therefore, the total runoff volume generated from the watershed is summed up as:

$$P_{RT} = \sum_{i=1}^{i=n} P_{Ri} \tag{18.18}$$

where i = i-th event, n = number of events, and P_{RT} = total runoff volume in the record. The total runoff volume captured by the basin is:

$$P_{AT} = \sum_{i=1}^{i=n} P_{Ai} \tag{18.19}$$

where P_{AT} = total runoff volume captured through the period of the record. The overflow runoff volume, P_{OT}, is calculated as:

$$P_{OT} = \sum_{i=1}^{i=n} (P_{Ri} - P_{Ai}) \text{ if } P_{Ri} \geq P_{Ci}, \text{otherwise zero} \tag{18.20}$$

The over-all *runoff volume capture ratio* (RVCR) for the period of the record is defined as:

$$\alpha = \frac{P_{AT}}{P_{RT}}$$

(18.21)

where, α = RVCR ranging between zero and unity.

18.5.3 Runoff event capture ratio

Following the same approach as above, set the counter on the number of events that were completed intercepted without any overflow. The overall *runoff event capture ratio* (RECR), β, is defined as:

$$\beta = \frac{N_C}{N}$$

(18.22)

where, N_C = number of runoff events that were completely captured, and N = total number of events in the record. Both RVCR and RECR vary with watershed's runoff coefficient, basin's storage volume and drain time. An RVCR is sensitive to extreme events in the database, while a RECR is more sensitive to the hydrologic loss, I_s, which dictates the total number of events in the rainfall record. Although the difference between the RVCR and RECR is minor, they are not necessarily the same. There is no any rule of thumb to correlate one to the other.

18.6 DIMENSIONLESS RUNOFF CAPTURE FORMULA

As indicted in Examples 18.1 and 18.2, both the runoff volume and event capture ratios depend on the rainfall event duration, basin's drain time, and watershed imperviousness in terms of runoff coefficient. Based on the runoff capture analyses from some 40 major metro areas in the continental US, it was confirmed that the difference between α and β is negligible for engineering practices, and the empirical formula has been derived with a correlation coefficient $r^2 = 0.967$ as (Guo and Urbonas 2021):

$$\frac{D}{CD_m} = 0.466 \left(\frac{T_e}{T_d} \right)^{0.35} \left[-\ln\left(\frac{1-\alpha}{K} \right) \right] \text{ for } T_e = 12, 24, 48, \text{ to } 72 \text{ hr}$$

(18.23)

where D = design volume for porous basin in [L per watershed], C = watershed's runoff coefficient between zero to unity, T_e = basin's emptying time (drain time) in hours, T_d = local average event duration in hours, and α = selected runoff capture ratio between zero to unity. Using Eq. 18.23, the Exponential distribution, Eq. 18.10, is also converted from the rainfall event-depth distribution to the runoff depth distribution as:

$$\frac{D}{CD_m} = -\ln\left(\frac{1-\alpha}{K} \right) \text{ for } 0 \le \alpha \le 1.0 \text{ for } T_e = \text{extended to clogged}$$

(18.24)

As reported (Guo 2017), the runoff coefficients for frequent events can be related to the imperviousness area ratio as:

$$C = 0.858I_a^3 - 0.78I_a^2 + 0.774I_a + 0.04 \tag{18.25}$$

where I_a = watershed area imperviousness ratio between zero to unity. Eq. 18.23 is a regression equation for both runoff volume and event capture rates. The values of D_m and T_d can be read off from Figs 18.7 and 18.8. Eq. 18.23 applies to basin emptying times ranging from 12 to 72 hr. As shown in Fig. 18.13, when $T_e > 72$ hr, Eq. 18.23 approaches to Eq. 18.24 asymptotically. In practice, when the design information about the basin's operations or/and the local rainfall statistics are not readily available at the project site, the exponential runoff capture curve, Eq. 18.24, offers a conservative design.

18.7 EMPIRICAL FORMULA FOR WQCV

As per the US EPA's recommendations, the WQCV has an optimal runoff capture ratio of $\alpha = 0.8$. Applying the approach of diminishing returns to the runoff capture curves in Fig. 18.13 leads to the normalized WQCV as:

$$\frac{WQCV}{CD_m} = 0.88 + 0.01T_e (12 \leq T_e \leq 72) \tag{18.26}$$

As indicted in Eq. 18.26, the normalized WQCV ranges from 1.0 to 1.6.

Runoff Volume Capture Curve

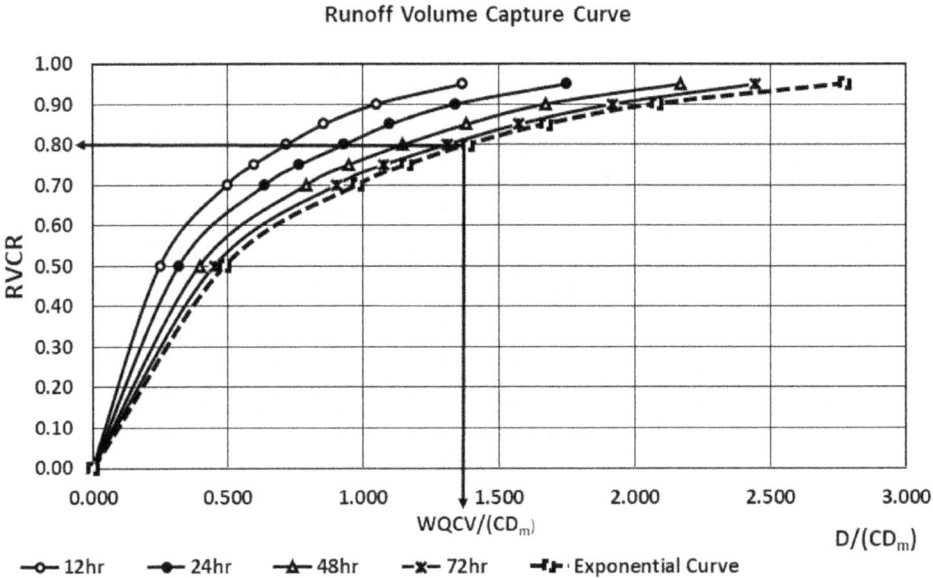

Figure 18.13 Normalized runoff volume capture curves for the US continent

Example 18.3: As shown in Fig. 18.14, the project site of 10 acres (4.0 hectares) is located in the City of Denver, Colorado. The site will be developed to have a watershed imperviousness ratio of 0.80. A porous basin is proposed to intercept 75% of the total area, or an area of 2.5 acres, directly drain into the street. Determine the size of the proposed LID basin with a drain time of 24 hr.

Solution: The runoff coefficient in this case can be derived from the proposed watershed imperviousness ratio of 0.8 as:

$$C = 0.858\,(0.8)^3 - 0.78(0.8)^2 + 0.774(0.8) + 0.04 = 0.6$$

With the drain time of 24 hours and $D_m = 0.41$ inch for the City of Denver (see Fig. 18.7), the WQCV in Eq. 18.26 is calculated as:

$$\frac{WQCV}{0.6 \times 0.41} = 0.88 + 0.01 \times 24 = 1.12. \text{ So, WOCV} = 0.275\,\text{inch}\,(7.0\,\text{mm}).$$

Under the specified area-interception ratio of 75%, the effective impervious area draining into the basin is 7.5 acres (10 acres × 0.75 = 7.5 acres). The required basin volume is determined as:

Basin volume = WQCV × area = 0.275/12 × 7.5 × 43560 = 7487 cubic ft.

Suggest a WQCB of 90 × 90 square ft with a water depth of 12 inches (1 ft or 30 cm).

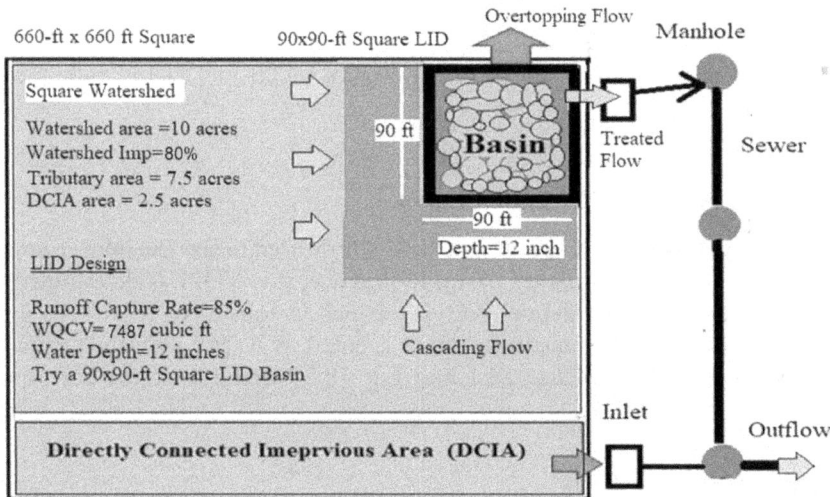

Figure 18.14 Porous basin designed for WQCV

It is important that the proposed porous basin is protected with grass buffers to filter out urban trash and debris. The basin must be equipped with a subdrain system to release seepage flow back to the sewer system and an overtopping weir to pass excess stormwater in the case of extreme events.

Example 18.4: A project site of 0.8 hectare (2.0 acres) is located in the City of Chicago, Illinois. The site will be developed to have a watershed imperviousness ratio of 0.8 or $C = 0.6$. Determine the size of the proposed LID basin with a drain time of 48 hr.

Solution: At the preliminary design stage, Eq. 18.24 is applicable to this case. With $\alpha = 0.8$, the basin's volume is estimated as:

$$\frac{D}{0.60 \times 13.2} = 1.174 \text{ for } \alpha = 0.8$$

So, $D = 9.3$ mm (0.366 inch) per watershed or a volume of 74.4 m³ (2657 ft³) is recommended for the basin design. With more design information, including the average rainfall event-depth, $D_m = 13.2$ mm (0.52 inch), from Fig. 18.7 and the event duration, $T_d = 9.3$ hr, from Fig. 18.8 for the Chicago area, the basin volume is further refined as:

$$\frac{D}{0.60 \times 13.2} = 0.446 \times \left(\frac{48}{9.3}\right)^{0.35} \left[-\ln\left(\frac{1-0.8}{0.8}\right)\right] = 1.147 \text{ for } \alpha = 0.8$$

For this case, $D = 9.1$ mm (0.358 inch) per watershed, or a volume of 72.8 m³ (2598.1 ft³). The WQCV at this project site is calculated as:

$$\frac{WQCV}{0.6 \times 13.2} = 0.88 + 0.01 \times 48 = 1.36.$$ So WQCV = 11.1 mm (0.437 inch) per watershed, or a volume of 88.8 m³ (3171 ft³).

18.8 WATER QUALITY PEAK FLOW

A water quality peak flow (WQPF) is the design flow used to size the inlet cross-section at the entrance of a water quality basin. In practice, the WQPF is determined by the water quality precipitation depth (WQPD), which is equivalent to a 4- to 6-month event. Since the storage volume in the basin is equal to WQCV, which is equal to the rainfall excess according to Eq. 18.1 and Eqs 18.23 and 18.24, the corresponding WQPD is determined as:

$$WQPD = WQCV / C + I_s \qquad (18.27)$$

The value of interception loss, I_s, is 0.05 to 0.1 inch. Usually, a LID basin is applied to a small urban catchment. As a result, the design rainfall intensity is determined as:

$$WQPI = \frac{C_1 \times WQPD}{(C_2 + T_c)^{C_3}}$$ (18.28)

where WQPI = water quality peak intensity in [mm/hr or inch/hr], T_c = time of concentration of the catchment in minutes, and C_1, C_2, and C_3 are constants used in the local rainfall intensity–duration–frequency (IDF) formula. The WQPF is further determined by the Rational method as:

$$WQPF = K \times C \times WQPI \times A$$ (18.29)

where WQPF = peak flow in cms or cfs, K = 1/360 for SI units or 1.0 for English units, A = catchment's tributary area in hectares or acres, and C = runoff coefficient.

Example 18.5: Continue with Example 18.3. Referring to Fig. 18.14, design the weir cross-section at the entrance of the proposed LID basin.

Solution: We know that C = 0.6 and WQCV = 0.275 inch from Example 18.3. From the catchment analysis, the time of concentration is T_c = 15 minutes. Using Eq. 18.27, the WQPD is calculated as:

$$WQPD = \frac{0.275}{0.6} + 0.1 = 0.558\,inch$$

It is noted that the 2-yr 1-hr precipitation depth at the project site is approximately 1.0 inch. The WQPD for this case is equivalent to a 4- to 6-month rainfall event. The rainfall IDF formula at the project site is given as:

$$WQPI = \frac{28.5 \times WQPD}{(10 + T_c)^{0.789}} = \frac{28.5 * 0.558}{(10 + 15)^{0.789}} = 1.25\,inch/hr$$

According to Eq. 18.29, the peak inflow to the basin is:

$$WQPF = 0.6 \times 1.25 \times 10 = 7.50\ cfs$$

Consider the weir coefficient of 3.0 for English units. As shown in Fig. 18.15, the basin is a 90-ft by 90-ft square in shape with a brimful depth of 1.0 ft. Set the water depth on top of the weir crest equal to 0.5 ft and the weir length to be 10 feet. The inflow capacity for this proposed weir is:

$$Q_p = C_w L_w H_w^{1.5} = 3.0 \times 10 \times 0.5^{1.5} = 10.6\ cfs > 7.5\ cfs\ (Ok!)$$

In operation, the proposed basin will be overtopped during an extreme event, and the basin will only intercept the early runoff volume up to WQCV.

Vertical View

water surface

WQPF   Hw=0.5 ft

⇨   1.0 ft
                                    Basin

Weir at Entrance

Top View

inlet channel   — 90 ft —

⇨   ⌐ 10 ft
                            Basin   90 ft

weir

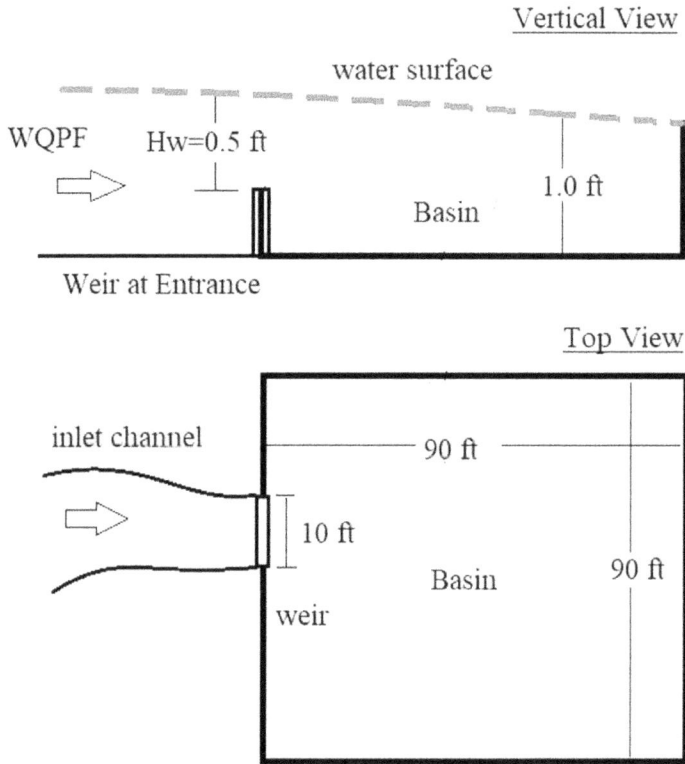

Figure 18.15 Vertical and top views of a proposed water quality basin

18.9 CONCLUSIONS

1. The simple exponential model in Eq. 18.24 is useful when detailed information is not readily available. Otherwise, Eq. 18.23 will be considered to include the on-site C, T_e, and T_d. Using Figs 18.7 and 18.8, a localized runoff capture curve can be constructed for LID basin designs. ,

2. The normalized governing equation, Eq. 18.23, is formulated with the key factors of D/CD_m, and T_e/T_d. For a specified capture rate, α, both RVCR and RECR demand a greater basin volume as the drain time increases. Having a drain time of longer than 72 hours, the synthetic RVCR and RECR curves asymptotically approach to the exponential runoff capture curve Eq. 18.24. In practice, the exponential runoff capture curve shows that the basin has a prolonged drain time due to clogging.

3. Fig. 18.13 presents different synthetic runoff capture curves derived for $T_e = 12$, 24, 48, and 72 hours. A RVCR is more sensitive than RECR to outliers in the rainfall data base. A RECR is a better indicator to quantify the urbanization impacts on a combined sewer overflow system. In general, the difference between the pair of (RVCR, RECR) is numerically negligible for the US continent.

4. A LID basin is designed with a pre-selected runoff capture rate. If the WQCV is chosen, the basin will have the optimal runoff capture rate of approximately 85%. The normalized, WQCV/CD_m, is varied with respect to basin's drain time, T_e, and can be derived by diminishing returns to optimize the capture rate (Guo and Urbonas 1996). Eq. 18.26 is the regression equation for calculating the ratio WQCV/CD_m.
5. The selection of the drain time depends on the construction of the basin's infiltrating media. In general, a drain time of 12 hours for a depth of 30 cm (12 inches) is recommended as the threshold condition for LID basin designs (Guo 2020). In other words, whenever the drain time of the basin becomes longer than 12 hours, it is time for basin repairs or/and sub-base replacement.

HOMEWORK

Q18-1 A site of 2 acres located in the City of Los Angles, California, is to be developed with I_a = 60%. At the preliminary stage, without considering T_d and T_e, estimate the storage volume with a runoff capture rate of 80% at this site and suggest a square basin with a depth of 30 cm. (Hint: Use Eq. 18.24.) (2) At the final stage, the project is updated with more detailed design information. Using Fig. 18.8, finalize the storage volume for T_e = 24 hours and α = 80%. (Hint: Use Eq. 18.23.) (3) Compare the storage volumes between (1) and (2) when T_e = 72 hours.

Solution:

Location of Basin Site		LA CA	
Avg Event Depth	Dm=	0.60 inches	15.24 mm
Avg Event Duration (or Te=Td)	Td=	12.00 hours	
Basin Empting (Drain) Time	Te=	24.00 hours	
Target Runoff Capture Rate	α =	0.80	
Runoff Incipient Depth	Is=	0.10 inches	2.54 mm
Watershed Area	A=	2.00 Acre	0.79 Hectares
Watershed Imp%	Imp%=	60.00 % Runoff C=	0.41
Value of K for Runoff Incipient Depth	K=	0.85	
Basin Volume			
D/CDm=	D/CDm	0.857 0.857	
D/Watershed	V=	0.21 inches	5.34 mm
Basin Storage Volume	V=	0.04 acre-ft	0.004 Hec-m
Basin Storage Volume	V= 1526.44 cufic ft		42.05 m³
Basin Geometry			
Enter Basin Brimfull depth	Y=	1.00 ft	0.30 m
Enter Basin Width	W=	39.00 ft	11.89 m
Basin Length	L=	39.14 ft	11.93 m

Q18-2 Continue with the WQCV = 0.21 inch in Q18-1. Knowing that C = 0.41, D_o = 0.21 inch, and T_e = 6 hours, your tasks are: (1) to estimate the average *runoff volume capture ratio* for the following cases given in pairs of (rainfall depth in inches, duration in hours): (1.5, 4.5), (1.0, 3), (0.75, 2), (0.15, 2), (1.0, 2). (2) to estimate the average *runoff event capture ratio*.

Q18-3   In China, the analyses of WQCV were performed for more than 30 metropolitan areas. Fig. Q18-3 presents the $WQCV_o$ in mm for C = 1.0. Determine the WQCV for a site of 3 hectares with C = 0.77 in the City of Beijing. Suggest a size of circular basin with a depth of 45 cm. (Hint: WQCV = C × $WQCV_o$)ure

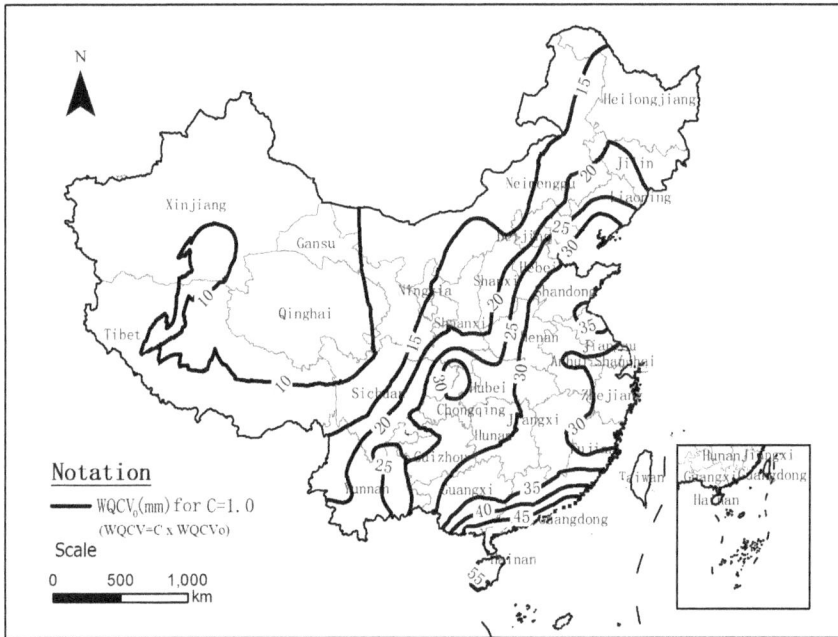

Figure Q18-3 WQCV$_o$ with *C* = 1 in China

REFERENCES

ASCE (1994). "Design and Construction of Urban Stormwater Management System", *American Society of Civil Engineers, Manuals and Reports of Engineering*, Practice No. 77, chapter 1.

ASCE WEF (1998). *Urban Runoff Quality Management*, WEF Manual of Practice no. 23, Alexandria, VA. and Reston, VA.

Booth, Derek B. and Jackson, Rhett C. (1997). "Urbanization of Aquatic Systems Degradation Thresholds, Stormwater Detention, and the Limits of Mitigation", *Journal of American Water Resources Association*, vol. 22, no. 5.

Davis, A.P. (2008). "Field Performance of Bioretention: Hydrology Impacts", *Journal of Hydrologic Engineering*, vol. 13, no. 2, pp. 90–95.

Driscoll, E.D., Palhegyi, G.E., Strecker, E.W., and Shelley, P.E. (1989). *Analysis of Storm Events Characteristics for Selected Rainfall Gauges throughout the United States*, US Environmental Protection Agency, Washington, DC.

EPA Report (1986). *Methodology for Analysis of Detention Basins for Control of Urban Runoff Quality*, US Environmental Protection Agency, EPA440/5-87-001, September.

Guo, J.C.Y. (1999). "Detention Basin Sizing for Small Urban Catchments", *ASCE Journal of Water Resources Planning and Management*, vol. 125, no. 6, November.

Guo, J.C.Y. (2002). "Overflow Risk of Storm Water BMP Basin Design", *ASCE Journal of Hydrologic Engineering*, vol. 7, no. 6, November.

Guo, J.C.Y. (2009). "Retrofitting Detention Basin for LID Design with a Water Quality Control Pool", *ASCE Journal of Irrigation and Drainage Engineering*, vol. 135, no. 6, October.

Guo, J.C.Y. (2013). "Green Concept in Stormwater Management", *Journal of Irrigation and Drainage Systems Engineering*, vol. 2, no. 3.

Guo, J.C.Y. (2017). *Urban Flood Mitigation and Stormwater Management*, CSC Publisher, New York.

Guo, J.C.Y. (2020). "Drain Time for Stormwater Porous Basin", *ASCE Journal of Hydrologic Engineering*, vol. 25, Issue 5, March.

Guo, J.C.Y. and Hughes, W. (2001). "Storage Volume and Overflow Risk for Infiltration Basin Design", *ASCE Journal of Irrigation and Drainage Engineering*, vol. 127, no. 3, May/June, pp. 170–176.

Guo, J.C.Y. and Urbonas, B. (1996). "Maximixed Detention Volume Determined by Runoff Capture Ratio", *Journal of Water Resources Planning and Management*, vol. 122, no. 1, pp. 33–39.

Guo, J.C.Y. and Urbonas, B. (2002). "Runoff Capture and Delivery Curves for Storm-water Quality Control Designs", *Journal of Water Resources Planning and Management*, vol. 128, no. 3, pp. 208–215.

Guo, J.C.Y. and Urbonas, B. (2021). "Stormwater Quality Capture Volume for Mitigating Urban Runoff Impacts", *ASCE Journal of Hydrologic Engineering*.

Guo, J.C.Y., Urbonas, B., and K. MacKenzie (2011). "The Case for a Water Quality Capture Volume for Stormwater BMP", *Journal of Stormwater*, October.

Guo, J.C.Y., Urbonas, B., and MacKenzie K. (2014). "Water Quality Capture Volume for LID and BMP Designs", *ASCE J of Hydrologic Engineering*, vol. 19, no. 4, April, pp. 682–686.

Guo, J.C.Y., Li, Q.L., Urbonas, B., and Wang, L.W. (2019). "Runoff Capture Methods Developed for Stormwater Low-Impact Development Designs", *ASCE J of Hydrologic Engineering*, vol. 24, no. 4, April.

Guo, J.C.Y., Wang, W.L., and Li, J.Q. (2020). "Cascading Flow System for Urban Drainage Design", *ASCE Journal of Hydrologic Engineering*, vol. 25, no. 7, July.

Montgomery, D.C. and Runger, G.C. (2007). *Applied Statistics and Probability for engineers*, John Wiley & Sons, Hoboken, NJ.

NCDC (2013). *Climate Data Online*, National Climatic Data Center, www.ncdc.noaa.gov/cdo-web/datasets

Rossman, L.A. (2010). *Storm Water Management Model*, Ohio, EPA/600/R-05/040.

Ryan, M.A. (2011). *The Clean Water Act Handbook*, 3rd Edition, ABA Publishing, Chicago, IL.

UDFCD (2010). *Urban Storm Water Drainage Criteria Manual*, Vol. 3, Urban Drainage and Flood Control District, Denver, CO.

WQ-COSM (2020). "Water Quality Capture Optimization and Statistics Model", www.urbanwatersheds.org

Chapter 19

Porous basin

Urbanization results in more impervious areas. The reduction of soil infiltration leads to increases in runoff volume and peak flow. Overland flows on urban pavements are loaded with solids and pollutants. Pollutant sources include debris, dirt, chemicals, and contaminants from streets, open areas, business districts, residential divisions, and industrial parks. The increase in storm runoff is closely related to the area ratio of imperviousness (EPA 1983). Using the Colorado urban hydrograph procedure (CUHP), a sensitivity study of runoff volume and flow to the area imperviousness was conducted for a 100-yr storm event. Considering that the baseline case has an area imperviousness of 5%, as shown in Fig. 19.1, the peak discharge increases 3.84 times and the runoff volume increases 1.61 times when the watershed imperviousness increases to from 5 to 90%.

Increases of flood flows during extreme events (Q-problem) trigger concerns about traffic safety on the street, while increases in frequent runoff volumes (V-problem) undermine the wellbeing of the water environment. Stormwater management is aimed at reducing the negative impacts of urbanization on the receiving water bodies. Stormwater best management practices (BMPs) suggest: (1) *upstream source control systems* using low-impact-development (LID) devices, (2) *mid-stream collection and delivery systems* using sewers, streets and channels, and (3) *downstream detention and treatment systems* using detention, retention, and wetland facilities.

The concept of *low-impact-development* (LID) was evolved from the concept of preserving the pre-development watershed regime. The effort is two-fold developed to alleviate both *Q* and *V* problems respectively: (1) *to reduce post-development peak flows* in order to sustain the level of protection from flood potentials, (2) *to reduce the increased runoff volume* to protect the water environment. Flood mitigation and stormwater management are a regional effort starting from upstream watershed improvement to downstream extended detention strategy. With the latest innovative concepts, various types of porous basins and permeable pavers are invented to take advantage of soil infiltration and vegetal filtering processes to settle solids and also to dispose of runoff volume. In this chapter, design criteria and procedures for various LID devices are introduced. In general, a LID device should be placed upstream as a pollution source control and sized for the stormwater runoff capture volume (WQCV) for on-site runoff volume disposal.

DOI: 10.1201/9781003284239-19

Urbanization Impact

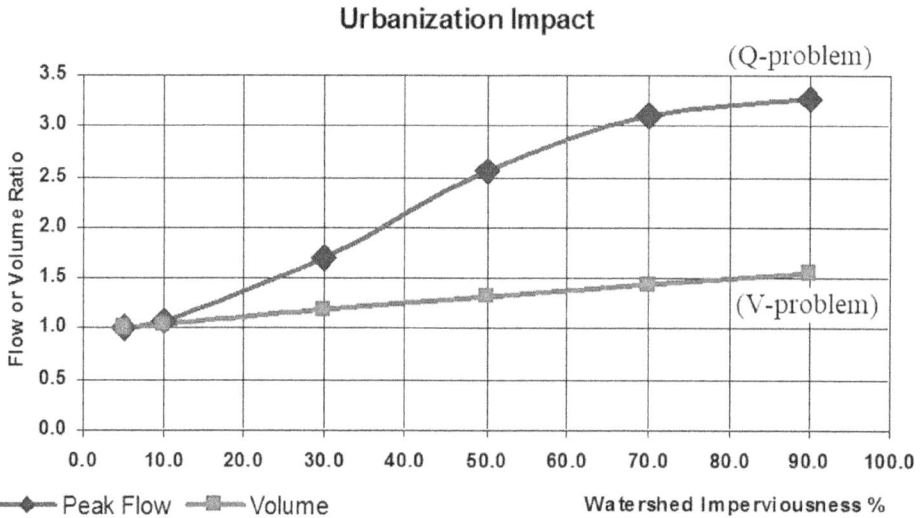

Figure 19.1 Urban impact on runoff flows and volumes

(a) Two Separate Flow System

(b) Cascading Flow System

Figure 19.2 Cascading flow system

19.1 LID SITE PLAN

A LID layout is designed to spread runoff flows generated from the upper impervious surface onto the lower pervious area, such as vegetation beds and porous landscaping areas, for additional infiltration benefits and water-quality enhancement. Fig. 19.2 presents a comparison between the conventional *separate flow system* and the innovative *cascading flow system* under the concept of LID. A separate flow system consists of two flow paths (1) a storm drain for the impervious area, and (2) a grass swale for the pervious areas. In contrast, a cascading flow system has only one flow path, which collects the flows from the impervious areas and then runs through the pervious area before reaching the street gutter.

(a) Rain Garden   (b) Infiltration Swale

Figure 19.3 Examples of LID designs

It is essential that a LID unit is placed at the source of the runoff, such as roof downspouts, outfall points of parking lots, outlets of sport fields, and exits of residential areas. As recommended, a LID site should be laid out with the following steps (USWDCM 2010):

1. to decrease the impervious areas at the project site,
2. to minimize the directly connected impervious area (MDCIA),
3. to decentralize runoff flows and volumes, and
4. to dispose runoff into a hydrologically functional landscape such as rain gardens, bio-detention systems, filter/buffer strips, grassed swales, and infiltration trenches in Fig. 19.3 (Guo, Urbonas, and MacKenzie in 2014).

In an urban setting, the Q-problem of stormwater is directly related to extreme events for increased peak flow, high flow velocity, and extended inundation. Q-problems are considered to be a public safety issue that can be alleviated by flood mitigation methods. The *V-problems* are more of a public health issue, directly related to the increased runoff volumes from frequent events. V-problems can be improved by stormwater on-site quality control through infiltration and filtering processes. As illustrated in Fig. 19.4, a conventional drainage system consists of: (1) *storm sewers* to collect flow from *minor events* (2 to 5-yr events) and (2) *street gutters* to pass the flow from *major events* (10- to 100-yr events). In the 1980s, stormwater detention was implemented to reduce the outflow from 100-year events. After the year 2000, urban renewal projects began to improve drainage systems with: (1) a regional extended detention basin (EDB) placed at the outfall to provide a full-spectrum outflow release control (from *micro, minor* and *major events*), and (2) LID devices sized up to *micro events* (3 to 6-month events) and placed immediately upstream of street inlets for on-site runoff volume disposal, and water-quality controls.

As shown in Fig. 19.5, the micro–minor–major (3M) cascading flow system provides a full-spectrum runoff release control on both runoff flows (Q-problem) and volumes (V-problem) (Guo 2013).

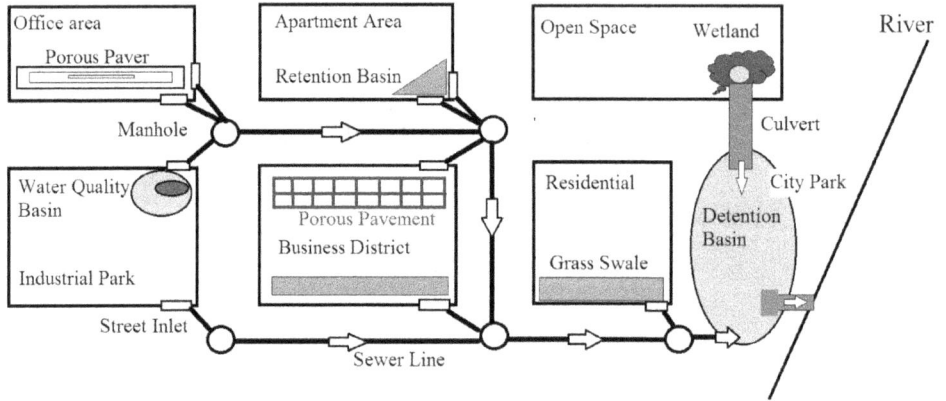

Figure 19.4 Micro–minor–major (3M) cascading flow systems

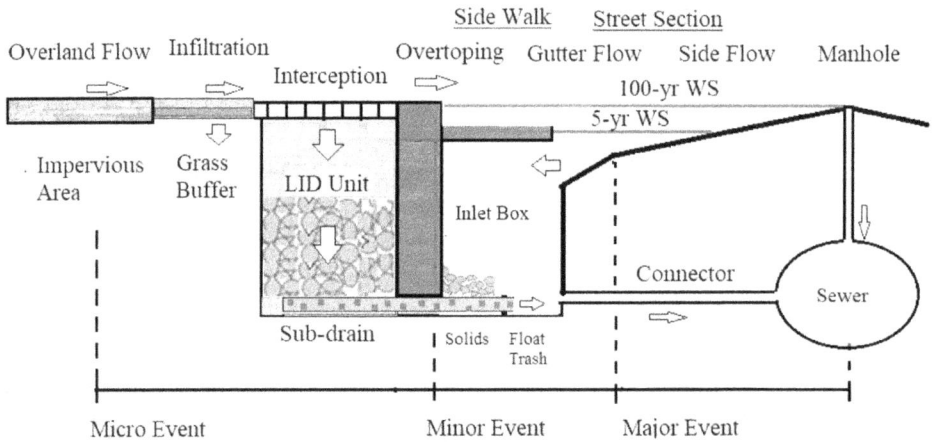

Figure 19.5 Examples of LID sites

19.2 INCENTIVE INDEX FOR LID SITE

A LID site is a flow-path dependent layout. As shown in Fig. 19.6, the receiving pervious area in a cascading flow system is covered with native soils, grass, or/and structured porous pavers. For example, the use of modular-block porous pavement or reinforced turf in low-traffic zones, such as parking areas and infrequently used service drives such as fire lanes, alleys, or sidewalks, can significantly reduce the site imperviousness. Before sizing a WQCB, it is important that the site is laid with cascading flows to reduce the runoff volume. This practice can significantly reduce the size of the downstream storm sewers and detention basins.

19.2.1 Area-weighted imperviousness

An urban catchment is composed of impervious and pervious areas. For modeling convenience, an irregular catchment shall be converted into its equivalent rectangular

(a) Gravel Infiltration Basin   (b) Sandy Infiltration Bed

Figure 19.6 Receiving porous areas

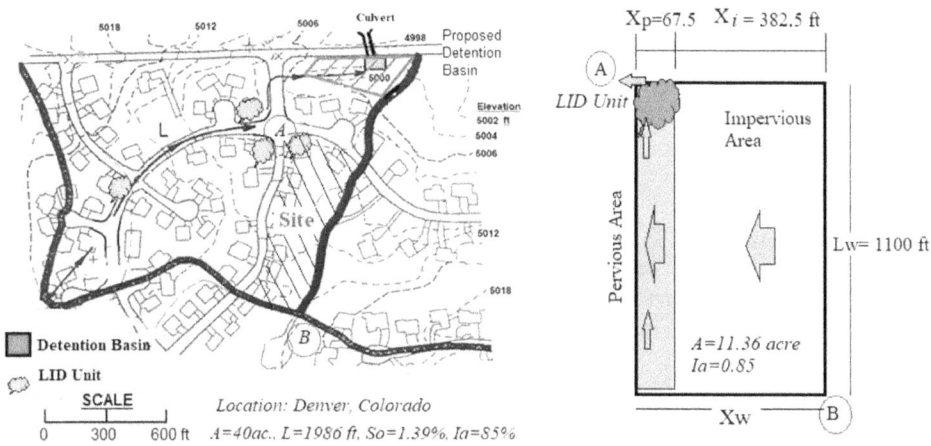

Figure 19.7 Flow systems at the project site

sloping plane. The watershed in Example 6.1 is divided into four zones, of approximately 10 acres per zone. The shaded area in Fig. 19.7 is converted into a cascading flow system in which the upper impervious area is drained onto the lower pervious area (SWMM5, Rossman 2009).

At the outfall point, the resultant hydrograph is the sum of these two overland flows. The relationship between the three flow lengths is:

$$X_w = X_I + X_P \tag{19.1}$$

where X_w = total flow length in [L], X_I = upper flow length on the impervious plane in [L], and X_P = lower flow length on the pervious plane in [L]. Normalizing Eq. 19.1 with X_w, we have

$$I_a = \frac{X_1}{X_w} = \frac{A_I}{A_T} \tag{19.2}$$

$$\frac{X_P}{X_w} = \frac{A_P}{A_T} = 1 - I_a \tag{19.3}$$

$$A_T = A_I + A_P \tag{19.4}$$

where I_a = area imperviousness percentage, A_I = impervious area in [L²]. A_p = pervious area in [L²], and A_T = total area in [L²].

19.2.2 Volume-weighted imperviousness

Under a cascading flow system, the additional infiltration benefit can be added to the flow path. Using Eqs 19.2 and 19.4, the relationship between the receiving pervious area and the source impervious area is derived as:

$$I_a = \frac{A_I}{A_I + A_P} = \frac{A_r}{A_r + 1} \tag{19.5}$$

$$A_r = \frac{A_I}{A_P} \tag{19.6}$$

where A_r = area ratio between the upstream impervious area and downstream pervious area. At a LID site, the *area imperviousness* is replaced with the *effective imperviousness*, I_E, which is defined by the runoff volume-weighted ratio as (Guo 2008):

$$V_T = I_E V_I + (1 - I_E) V_P \tag{19.7}$$

$$V_I = P A_T \tag{19.8}$$

$$V_P = (P - F) A_T \tag{19.9}$$

where V_T = runoff volume in [L³] from the cascading plane, V_I = runoff volume in [L³] as if the entire plane were impervious, V_P = runoff volume in [L³] as if the entire plane were pervious, I = rainfall depth in [L], and F = soil infiltration amount in [L]. Rearranging Eq. 19.7 yields:

$$I_E = \frac{V_T - V_P}{V_I - V_P} \tag{19.10}$$

With an additional infiltration loss on the receiving pervious area, the effective imperviousness, I_E, must be numerically less than the area imperviousness I_a. Let the pavement-area-reduction factor (PARF), K, be defined as (Blackler and Guo in 2012 and 2013):

$$I_E = K I_a \tag{19.11}$$

In engineering practice, the PARF serves as an indicator on the effectiveness of a cascading flow system. The PARF can be used to evaluate various alternatives of LID designs. The PARF is proportional to the ratio of soil infiltration rate to rainfall intensity, and the ratio of impervious to pervious areas (Guo 2008).

$$K = fct\left(\frac{F}{P}, A_r\right) = fct\left(\frac{F}{P}, I_a\right) \tag{19.12}$$

Under the assumption that the pervious area would have a 100% interception of the runoff flow generated from the upstream impervious area, Fig. 19.8 was produced based on numerical simulations using the EPA SWMM computer model (Guo 2008). As expected, the higher F/P ratio, the lower the PARF, and the higher the A_r ratio, the higher the PARF.

As the ratio of f/I increases, the runoff flow generated from the pervious area is diminished. As the ratio of I_a increases, the impervious area dominates the generation of runoff flows. For instance, under $f/I = 2.0$, the high infiltration rate consumes the surface runoff from the pervious area; as a result, the runoff flow is mainly produced from the impervious area at the site.

Example 19.1: The catchment in Fig. 19.9 has a drainage area of 10 acres (4.0 hectares, or an equivalent square of 200 m by 200 m). The area imperviousness percentage at the site is 85%. Considering a rainfall duration of one hour and an infiltration depth of $F = 1.0$ inch, estimate the effective imperviousness percentage for the cascading flow

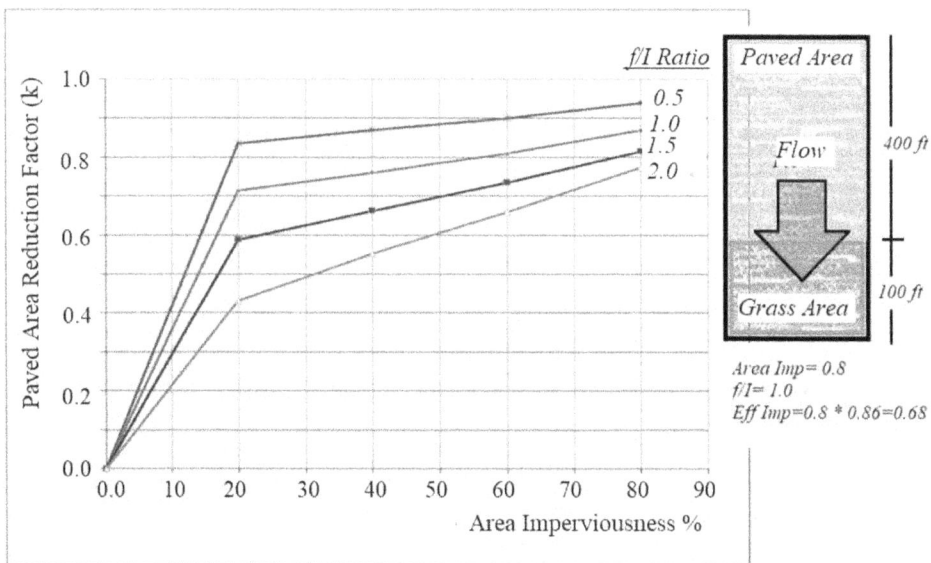

Figure 19.8 Paved area reduction factor

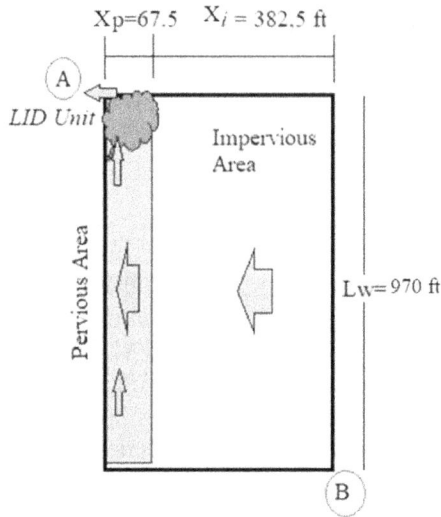

Figure 19.9 Example LID layout

system under a 1-hr, 2-yr event that has a precipitation depth of 1.0 inch. Repeat the above for a 1-hr, 100-yr storm event with a precipitation depth of 2.0 inches.

Solution: For this case, $X_I = 382.5$ ft and $X_P = 67.5$ ft are used for numerical simulations.

$$A_r = \frac{A_I}{A_P} = \frac{382.5 \times 970}{67.50 \times 970} = 5.67$$

$$I_a = \frac{A_r}{A_r + 1} = \frac{5.67}{5.67 + 1} = 0.85$$

At the project site, the 1-hr, 2-yr event has a ratio of $F/P = f/I = 1.0$. From Fig. 19.8, $K = 0.88$. As a result, this project site has an effective imperviousness of $I_E = 0.88 \times 0.85 = 0.75$. Under the 1-hr, 100-yr event with $F/P = f/I = 0.5$, the corresponding $K = 0.99$, i.e. it has no impact on a 100-yr event. This is a typical characteristic of the LID effect. The effect of LID cascading flow is significant for frequent events, but it diminishes for extreme events.

19.3 CASCADING DRAINAGE PATTERN

As shown in Fig. 19.10, a LID unit is sized to intercept up to the water-quality capture volume (WQCV) generated from the tributary area. The size of the LID device depends on its tributary area and impervious percentage. As a rule of thumb, the area ratio between the tributary area and the LID unit is 5 to 10, depending on the cascading layout. In practice, a LID device may collect 70 to 100% of the runoff flows generated from the impervious area. A small portion of the impervious area may bypass the LID device to become a directly connected impervious area (DCIA). Any and all

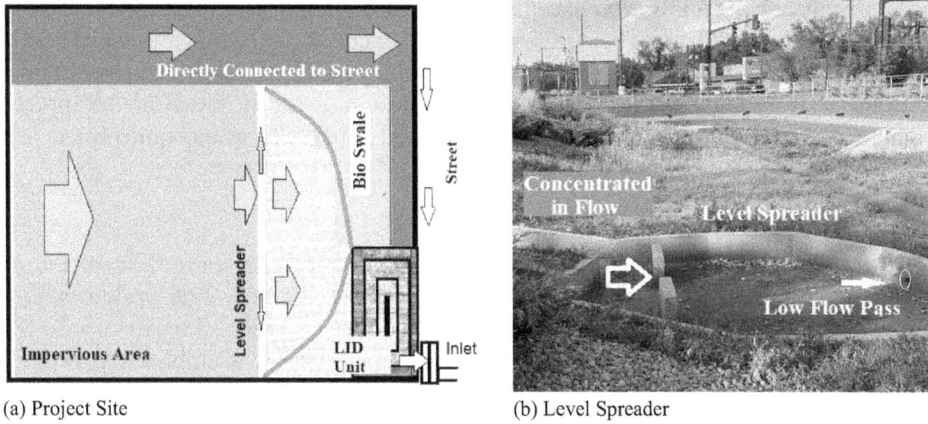

(a) Project Site   (b) Level Spreader

Figure 19.10 Site planning using a cascading drainage system

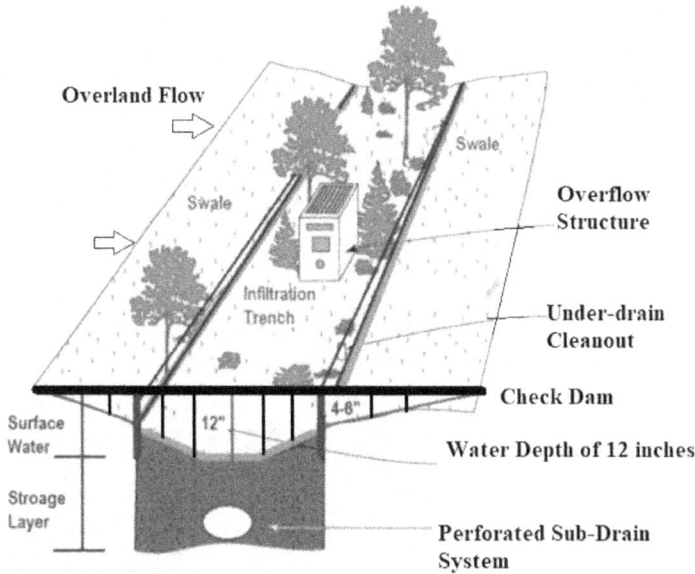

Figure 19.11 Bio-swale to collect overland flows

concentrated flows from the impervious area should drain into a *level spreader*, which is shallow pool with 4 to 6-inch berms. Flows overtopping the berms create unit-width overland flows evenly spread onto the infiltration bed. Overland flows on a pervious area may become erodible as the flow depth increases. A bio-swale, shown in Fig. 19.11, is recommended to be the collector channel, which guides the overland flows to move towards the downstream LID unit for further filtering and infiltration processes.

A bio-swale is built similar to an earth trench with a shallow depth (24–36 inches) and a wide bottom (10–20 ft). The native soils may be replaced with engineered filtering media to enhance the infiltration benefits. To slow down the flow movement, a bio-swale is built on a grade of 0.5 to 1%. On steep ground, a check dam, shown in Fig. 19.11, is built across the bio-swale to create a backwater pool, which serves as a porous basin to infiltrate the runoff volume. It is crucial that the drain time for the bio-swale be approximately 6 to 12 hours to avoid public environmental issues such as mosquito beds.

19.4 TYPES OF LID DEVICE

There are two major types of LID devices: (1) *porous pavement* and (2) *porous basin*. A porous pavement is built on a flat surface to enhance the flow interception, while a porous basin should be constructed with a pool 12 to 18 inches (30 to 40 cm) deep to store the WQCV. Underneath the porous bottom, the sub-base is assembled with multiple layers of filtering media to filter the water flow for better quality. Ideally, a LID unit will have a drain time of 12 to 24 hours, depending on the local regulations of water rights.

19.4.1 Porous basins

Applications of porous basins include a *bioretention basin* (BRB), *rain garden* (RG), and *porous landscaping detention basin* (PLDB). Although the general structure among these porous basins is similar, there are some minor differences. The PLDB in Fig. 19.12 is composed of a porous basin with a water depth up to 12 to 15 inches (30 to 38 cm) and multiple sub-layers of filtering media. Beneath the porous basin bottom, the filtering system consists of multiple layers, including: an upper layer of 15 to 18-inch (38 to 45 cm) sand-mix for filtering process, and a lower layer of 8 to 12-inch (20 to 30 cm) gravel serving as a subsurface reservoir (Hsieh and Davis 2005). Because the

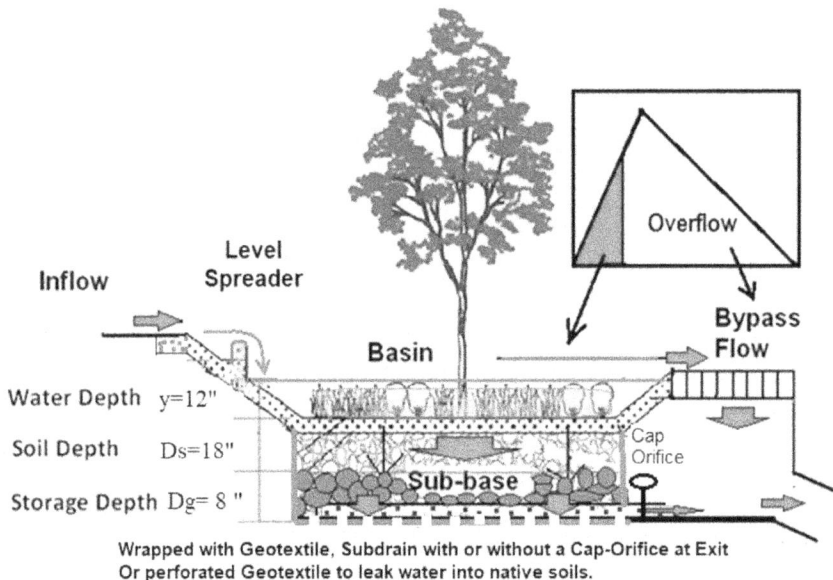

Figure 19.12 Layout of a landscaping detention basin (Rain Garden)

gravel layer drains faster than the sand layer, a suction pressure is induced to accelerate the seepage flow through the sand layer. An RG, shown in Fig. 19.13, does not have the gravel layer. A geotextile fabric is used to wrap the RG unit if no leak into the native soils is allowed; otherwise, a perforated fabric is preferred. A PLDB should have thicker layers of sand-mix and gravel to accommodate the growth of vegetation roots. Care must be taken when choosing bushes and plants for a PLDB.

As illustrated in Fig. 19.14, if the local native soils do not drain well, it is necessary to replace them with sandy and loamy media. A layer of geotextile fabric is selected

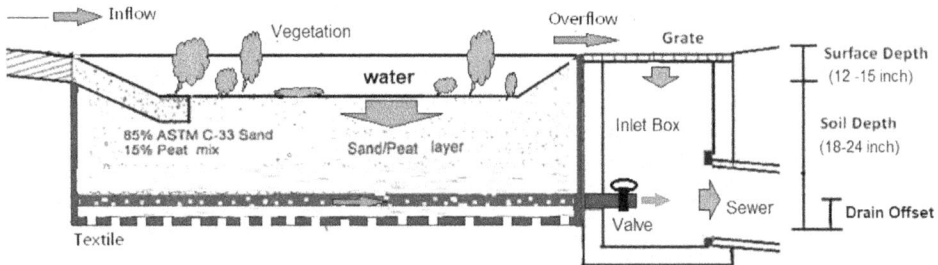

Wrapped with Geotextile, Subdrain with or without a Cap-Orifice at Exit
Or perforated Geotextile to leak water into native soils.

Figure 19.13 Basic layout of a rain garden

Figure 19.14 Construction stages for a porous basin

to control the contact between stormwater and groundwater. If the stormwater flows may carry heavy metals, it is necessary to maintain no contact with the surrounding soils, and a perforated subdrain system must be installed and tied to the downstream inlet box. The excavated area should be backfilled with topsoil for vegetation and planting in the basin.

A LID site is sensitive to the flow path. Care must be taken when setting up the elevations along the flow path. As shown in Fig. 19.15, the overland flows are guided to reach the porous basin. Stormwater flows around the WQCB and enters the basin at the low point. A WQCB has a capacity up to a depth of 12–15 inches (30–45 cm). After the basin is full, any and all excess stormwater will overtop the berm and drain into the inlet box.

The WQCV in Fig. 19.16 serves a tributary area that is more than 20 times the size of the porous basin. The stormwater stored in the basin will be infiltrated into the subsurface reservoir, where the seepage flow is filtered, stored, and gradually released through the perforated subdrain pipe into the inlet box. The infiltrating rate on the porous bottom represents the inflow to enter the subsurface filtering layers, while the seepage flow into the subdrain pipe represents the outflow. The operation of a porous basin is controlled by either the infiltrating rate or the seepage rate, whichever is smaller (Guo 2012). During an event, all the aggregate voids must be filled up with infiltrating water before the seepage flow can be fully developed through the saturated medium. If the subsurface seepage flow cannot sustain the infiltrating flow, the water

Figure 19.15 Examples of porous pavements and basins

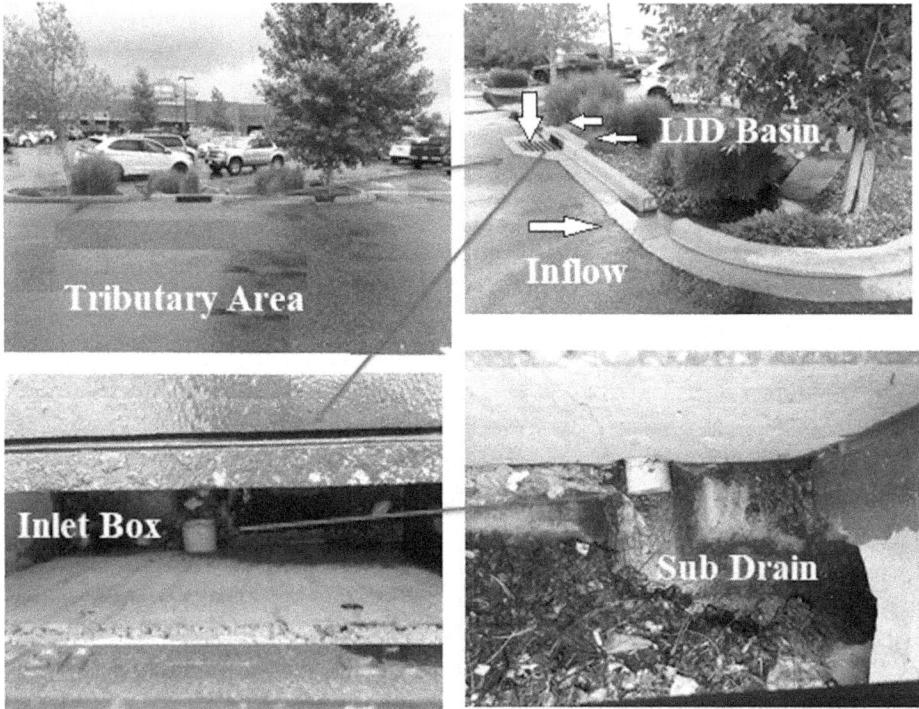

Figure 19.16 Example of a porous basin with a subdrain system

mounding will be built up to balance the inflow and outflow rates. This phenomenon is manifested by standing water remaining in the porous basin. Therefore, it is advisable that the subsurface geometry beneath the basin be designed to provide an adequate hydraulic gradient in order to sustain the continuity of flow.

19.4.2 Porous pavement

Porous pavements, shown in Fig. 19.17, are suitable for low-traffic areas, including patios, walkways, driveways, fire lanes, and parking spaces. As illustrated in Fig. 19.17, a porous pavement is formed with 4- to 6-inch permeable asphalt, concrete, tiles or pavers on a gently sloped ground (<1%). Underneath the porous pavement, it is preferred to have two layers of filtering medium. The upper layer is built with a 4 to 8-inch rock and gravel layer, while the lower layer is filled with a sand-mix medium. A porous pavement unit may be wrapped with a geotextile fabric if no water leak is preferred. A subdrain system using flexible pipes 2- to 4-inch in diameter shall be installed to collect and deliver the clean water to the downstream manhole. Otherwise, a perforated geotextile fabric is used to allow local groundwater recharge at a rate of 0.5 to 1.0 inch/hr.

The effectiveness of a porous pavement system highly depends on the absorption of surface runoff. A level spreader using a 1 to 2-inch berm should be installed at the lower end of the pavement surface. Excess stormwater overtops the berm. Underneath

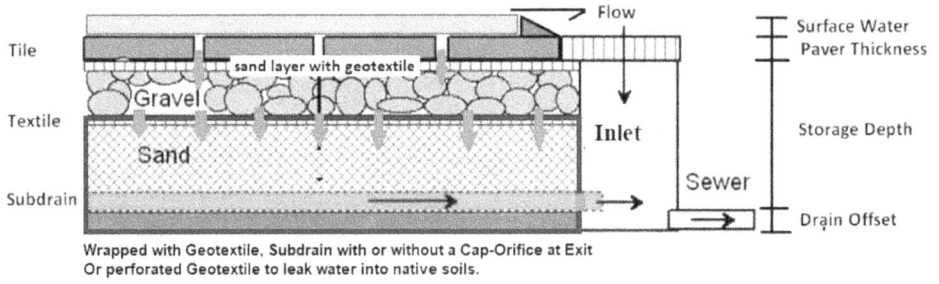

Figure 19.17 Basic structure of a porous pavement

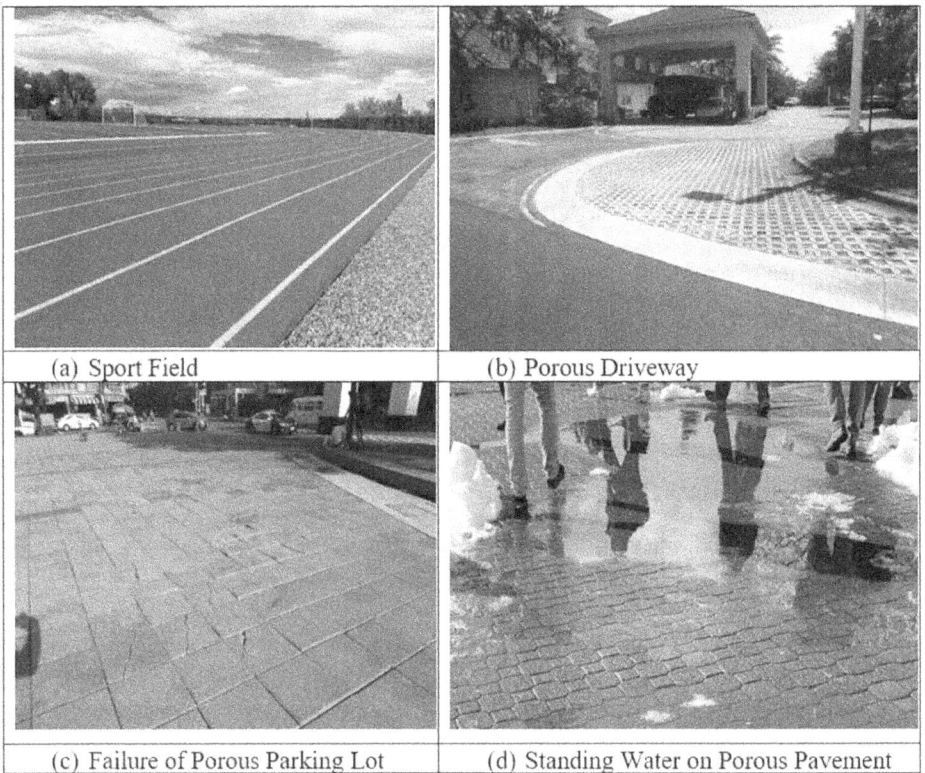

Figure 19.18 Examples of porous pavements

the permeable pavers exists a thick layer of rock and gravel serving as a subsurface reservoir to absorb the surface water. The stored water volume infiltrates into the sand layer for filtering process. After the sand layer becomes saturated, the seepage flow is then collected into the subdrain pipe or the native soils for groundwater recharge.

Fig. 19.18 presents examples of porous pavements. It is important to place the pavers on the flow path and at the low point. Porous pavements are not suitable for heavy traffic loading areas. Regular maintenance is required to avoid clogging problems.

19.5 DESIGN VOLUME FOR A POROUS BASIN

19.5.1 WQCV for basin sizing

An RG is designed as an on-site stormwater disposal facility. The storage volume for an RG is often sized for the water quality capture volume (WQCV) or the equivalent to the natural depression storage volume ranging from 0.25 to 1.0 inch. As mentioned in Chapter 16, WQCV is directly related to the local *average event rainfall depth*, D_m, and the runoff capture rate. The *runoff volume capture ratio* (RVCR) is preferred if a continuous probability model is used to portray the rainfall distribution, or the runoff event capture ratio (RECR) should be employed if the runoff discrete database is used. Applying the Exponential distribution to the rainfall database, the RVCR is derived as (Guo and Urbonas 2002):

$$C_v = 1 - \alpha\, e^{\frac{-D_o}{CD_m}} \tag{19.13}$$

$$\alpha = e^{\frac{-I_s}{D_m}} \tag{19.14}$$

where C_v = RVCR based on the Exponential distribution, numerically $0 \le C_v \le 1.0$, D_m = local average rainfall event-depth in [L], D_o = WQCV in [L]/watershed determined by the Exponential distribution, C = runoff coefficient of the tributary catchment to the RG, I_s = interception loss in [L], such as 0.05 to 0.1 inch, and α = constant related to the rainfall interception loss (Guo and Urbonas in 2002). For a pre-selected C_v, the value of D_o can be determined using Eq. 19.13. Next, the storage volume for the RG's surface storage basin is:

$$V_o = D_o A \tag{19.15}$$

where V_o = WQCV in [L³] and A = catchment area in [L²] of the tributary to the RG basin. Safety is always a concern when designing an RG. Often, the water depth in an RG is set to be 12 to 15 inches (30 to 38 cm). With a pre-selected basin depth, the basin's cross-sectional area is determined as:

$$A_B = \frac{V_o}{Y_o} \tag{19.16}$$

where A_B = average cross-section area in [L²], and Y_o = basin depth in [L] of 12 to 15 inches. To enhance the infiltrating process, the basin bottom should be on a flat to mild slope ($\le 1.0\%$).

Example 19.2: Continued from Example 19.1. The catchment in Fig. 19.9 has a drainage area of 10 acres (4.0 hectares). The catchment is located at the project site, where $D_m = 0.41$ inch. The effective imperviousness for this catchment is determined to be $I_e = 75\%$. Design the surface basin for this RG unit.

Solution: Consider $I_s = 0.1$ inch and $D_m = 0.41$ inch. The value of α is calculated as:

$$\alpha = e^{\frac{-I_s}{D_m}} = e^{\frac{-0.1}{0.41}} = 0.784$$

As illustrated in Example 19.1, the area imperviousness for this case is 0.85. Applying the PARF to the area imperviousness, the effective imperviousness is determined to be 0.75. The corresponding runoff coefficient is:

$$C = 0.86I_a^3 - 0.78I_a^2 + 0.77I_a + 0.04 = 0.86 \times 0.75^3 - 0.78 \times 0.75$$
$$+ 0.77 \times 0.75 + 0.04 = 0.54$$

Setting $C_v = 0.8$, the value for D_o is calculated as:

$$C_v = 1 - \alpha\, e^{\frac{-D_o}{CD_m}} = 1 - 0.784 e^{\frac{-D_o}{0.54 \times 0.41}} = 0.8 \text{ So}, D_o = 0.30 \text{ inch}$$

$$V_o = D_o A_B = \left(\frac{0.30}{12}\right) \text{ft} \times 10 \text{ acre} = 0.25 \text{ acre-ft}$$

Let $Y_o = 12$ inches. The basin's cross-sectional area is 0.25 acre or 40 by 270 sq ft.

19.5.2 Pore storage capacity in filtering layers

Most soil properties are related to the *soil moisture content* that represents the percentage of the pore volume in the sand layer that has been filled with water. The layer of sand becomes saturated when the moisture content is equal to its porosity. During a storm event, the available pore volume in a sand layer is defined as:

$$D_1 = H_1(\theta_1 - \theta_o) \tag{19.17}$$

where D_1 = equivalent water depth in the sand layer in [L], H_1 = thickness in [L] of the upper sand layer, θ_1 = saturated moisture content in the sand layer such as 0.30 to 0.35 for sand, and θ_o = initial moisture content. It is noted that the saturated moisture content cannot exceed the porosity of the sand layer, and the initial moisture content cannot be below the wilting point, as shown in Fig. 19.19.

Similarly, the equivalent water depth in the gravel layer is calculated as:

$$D_2 = H_2(\theta_2 - \theta_o) \tag{19.18}$$

where D_2 = equivalent water depth in [L] in the gravel layer, H_2 = thickness in [L] of lower gravel layer, θ_2 = saturated moisture content in the gravel layer, such as 0.40 to 0.45 for gravel, and θ_o = initial moisture content, such as 0.05 to 0.1. The total excavated depth for the LID unit is the sum of:

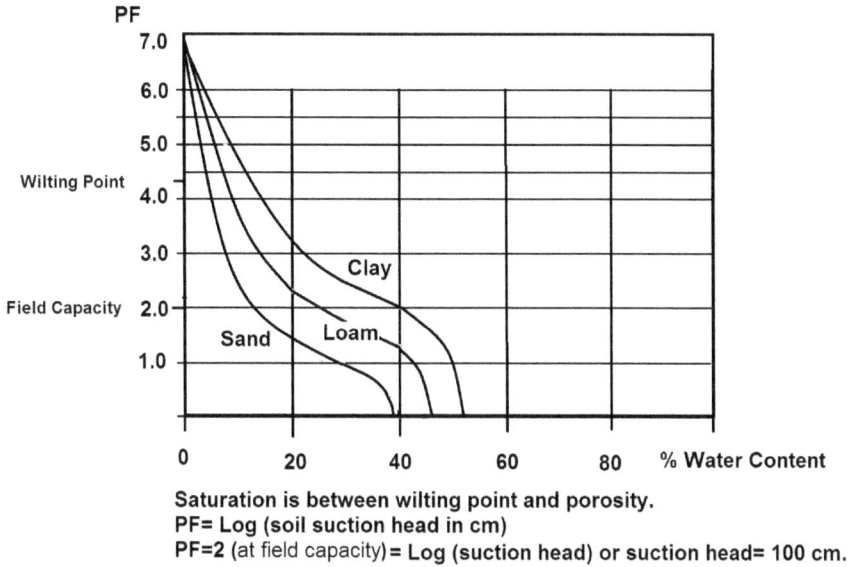

Saturation is between wilting point and porosity.
PF= Log (soil suction head in cm)
PF=2 (at field capacity)= Log (suction head) or suction head= 100 cm.

Figure 19.19 Soil properties and water contents

$$D_T = Y_o + H_1 + H_2 \tag{19.19}$$

where D_T = excavated depth in [L]. If the excavated depth will reach or close to the local groundwater table, we have to change the design to reduce the storage volume. For instance, two smaller LID units could be used for a shallower excavated depth.

Example 19.3: A tree box in Fig. 19.20 is designed to capture stormwater from an impervious area of 0.25 acres. Knowing that WQCV = 0.4 inch, determine the dimensions of the tree box.

Solution: Referring to Fig. 19.20, the WQCV from a 0.25-acre area is:

$$V = \frac{0.4}{12} \times 0.25 \times 43560 = 363 \text{ ft}^3$$

The tree box is a three-layer system that has: H_w = 12 inches, H_s = 18 inches, and H_g = 8 inches. The subsurface storage volume per unit area is summed as:

$$H_1 (\theta_1 - \theta_o) + H_2 (\theta_2 - \theta_o) = 18 \times (0.35 - 0.1) + 8.0 \times (0.45 - 0.10) = 7.3 \text{ inches}$$

$$D_T = Y_o + H_1 + H_2 = 12 + 18 + 8 = 38 \text{ inches}$$

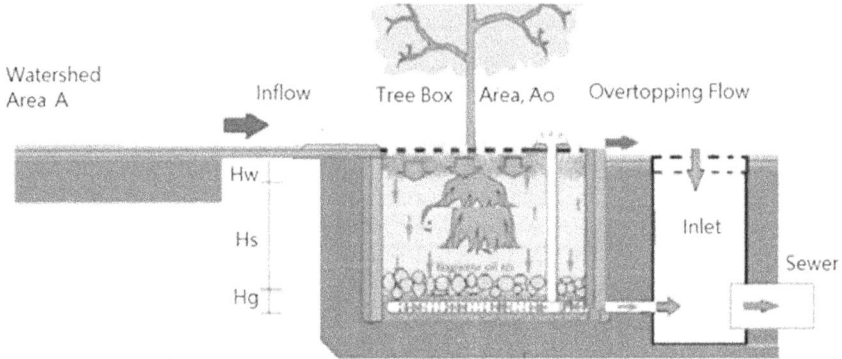

Figure 19.20 Tree box

The cross-sectional area of the tree box is:

$$\text{Area} = \frac{363}{(12+7.3)/12} = 227\,\text{sq ft}$$

During the water loading process, the sub-base layers act as a reservoir, which has a storage capacity of 7.3 inches over a vertical distance of 26 inches below the porous bed. As soon as the layer becomes saturated, the porous bed is hydraulically connected to the subdrain pipe through the saturated sub-base layers. As a result, the sub-base layers become a vertical tube to transmit water from the surface basin into the subdrain. Such a seepage flow is dominated by either the seepage process through the filtering layers or the orifice flow at the subdrain exit, whichever is smaller.

19.6 SEEPAGE FLOW AND DRAIN TIME

Drain time is critically important to the operation of a LID unit because it controls the enhancement of water quality. Based on the characteristics of urban pollutants, the drain time for a porous basin is usually set to be 12 to 24 hours (USDCM 2010). The average infiltration rate is:

$$f = \frac{Y_o}{T_o} \tag{19.20}$$

Where f = water infiltrating rate in [L/T], such as mm/hr or inch/hr, and T_o = design drain time in [T], such as hours. As a dual flow system above and below the porous bed, the flow movement through a porous bed can be analyzed using the principle of continuity between the infiltrating water flow on the porous bed and the seepage flow through the sub-base media (Guo, Kocman, and Ramaswami 2009). As illustrated in Fig. 19.21, under the assumption of steady state, the infiltrating flow rate described by Horton's formula must be equal to the seepage flows determined by Darcy's law as:

$$Q = fA_B = K_s I_s A_B = K_g I_g A_B \tag{19.21}$$

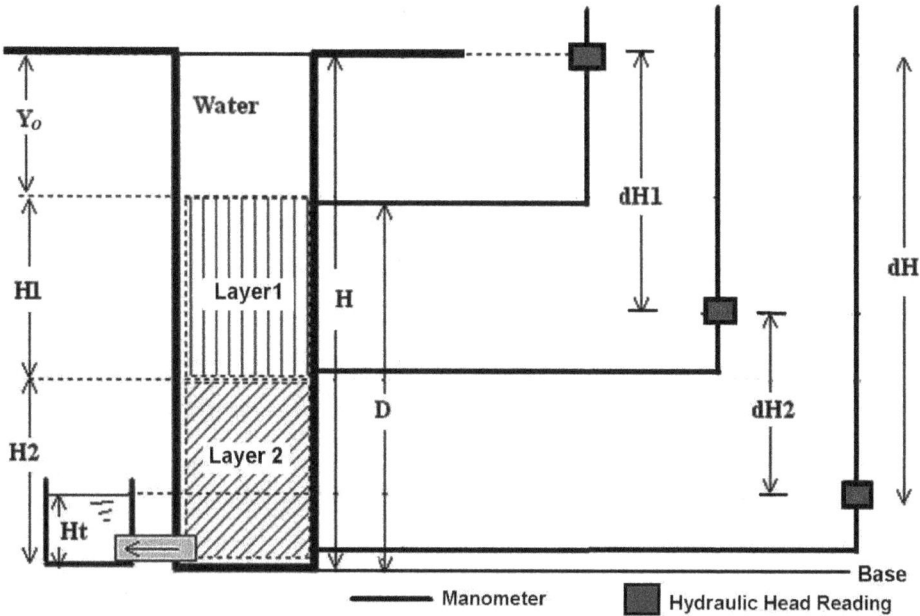

Figure 19.21 Seepage flow through filtering media

Table 19.1 Hydraulic conductivity coefficients for sub-base filtering media

Sub-base Filtering Material	Hydraulic Conductivity Under Fresh Conditions inch/hr	Hydraulic Conductivity Under Clogged Conditions inch/hr
Sand-mix	2.50	1.0
Gravel	25.0	1.0

Note: 1 inch = 25.4 mm.

where Q = flow released from LID unit [L^3/T], A_B = porous bed's bottom area [L^2], K_s = hydraulic conductivity coefficient in [L/T] for a sand-mix layer such as 2.5 inch/hr, I_s = hydraulic gradient through the sand layer, K_g = hydraulic conductivity coefficient in [L/T] for a gravel layer, such as 25 inch/hr, and I_g = hydraulic gradient through a gravel layer.

The hydraulic conductivity coefficients for sand-mix and gravel layers from laboratory tests (Kocman, Guo, and Ramaswami in 2012) are summarized in Table 19.1.

Referring to Fig. 19.12, the flow is driven by the available head in the system as:

$$H = Y_o + H_1 + H_2 \qquad (19.22)$$

where H = total hydraulic head in [L], H_1 = thickness of the upper layer in [L], and H_2 = thickness of the lower layer in [L]. It is noted that the upper layer for a porous basin is filled with sand-mix, while the porous pavement requires the upper layer to be gravel. The energy losses for the seepage flow through the sand and gravel layers are calculated as:

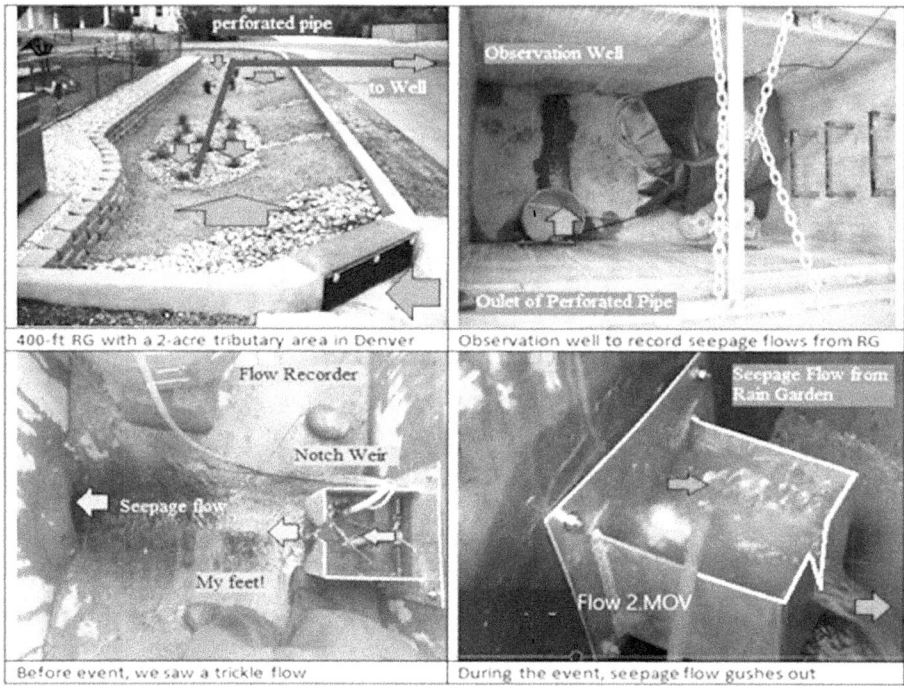

Figure 19.22 Outlet of a perforated pipe with and without a cap orifice

$$\Delta H_1 = I_1 H_1 \tag{19.23}$$

$$\Delta H_2 = I_2 H_2 \tag{19.24}$$

As a result, the residual head applied to the underdrain pipe is:

$$\Delta H = H - \Delta H_1 - \Delta H_2 \tag{19.25}$$

where ΔH_1 = energy loss in the upper layer, ΔH_2 = energy loss in the lower layer, and ΔH = total loss through both filtering layers. The design of the porous basin or pavement involves an uncertainty in how to select the design parameters. For instance, the infiltration rate through the sand-mix layer varies from 10 to 15 inch/hr for a newly constructed basin, 3 to 5 inch/hr for a matured porous bed, and 1.0 inch/hr or less for a clogged basin (Kocman, Guo, and Ramaswami in 2012). Infiltration of less than 1.0 inch/hr will result in such a prolonged inundation that the infiltrating bed needs to be replaced. For design, a moderate infiltration rate of 3.0 to 5.0 inch/hr is selected to meet the design criteria for both water quality and quantity control. In operation, it the challenge is how to mimic pre-development flow release when the infiltration rate is decays over time. In practice, a cap orifice, shown in Fig. 19.22, is employed to regulate the flow release at the exit of the underdrain pipe.

19.6.1 Case 1: Without a cap orifice

Without a cap orifice, the perforated underdrain pipe is directly connected to a down-stream manhole. At the underdrain exit, the flow pressure drops to atmospheric pressure. As a result, the infiltration rate in Eq. 19.25 must satisfy the balance of energy as:

$$\Delta H = H - \Delta H_1 - \Delta H_2 = 0 \text{ without a cap orifice} \tag{19.26}$$

It implies that without a cap orifice, the available headwater in the system dictates the flow capacity. As a result, the rain garden may be drained at a release rate higher than pre-development conditions. Its operation may have a drain time shorter than the required residence time for stormwater filtering and solid settlement.

19.6.2 Case 2: With a cap orifice

To regulate the flow release, a cap orifice can be installed at the exit of the perforated underdrain pipe. A cap orifice backs up the flow system to cause saturation in the lower layer. In doing so, the flow release is regulated with the cap orifice. To satisfy the principle of conservation of energy, the friction loss through the underdrain pipe is computed as:

$$\Delta H_N = kL\frac{N^2 Q^2}{D^{(16/3)}} \tag{19.27}$$

where ΔH_N = friction loss in [L] through the circular underdrain pipe, L = pipe length in [L], D = diameter in [L] of the underdrain pipe, N = Manning's roughness coefficient, and k = 4.65 for units of feet-second or 10.28 for units of meter-second. The cross-section area for the required cap orifice is calculated as:

$$A_o = \frac{Q}{C_d\sqrt{2g(H - \Delta H_1 - \Delta H_2 - \Delta H_N)}} \text{ with a cap} \tag{19.28}$$

where A_o = opening area of the cap orifice in [L²], C_d = discharge coefficient, and g = gravitational acceleration in [L/T²]. In practice, the cap orifice must have a diameter smaller than the underdrain pipe.

Example 19.4: A rain garden is designed to have an infiltration basin and two-layered filtering system. The infiltration bed for a rain garden has a flat area as A_B = 500 ft². Referring to Fig. 19.12, the dimensions of filtering system are: Y = 12 inches, H_1 = 18 inches, H_2 = 8 inches. The infiltration rate for the filtering media is estimated to decay from 10.0 to 1.0 inch/hr. The hydraulic conductivity is 2.5 inch/hr for the upper sand layer and 25.0 inch/hr for the lower gravel layer. Without a cap orifice, the flow rate released from this rain garden is determined using a trial-and-error procedure. Let us start with a guessed infiltrating rate of 5.0 inch/hr.

$$Q = fA_B = \frac{5.0}{12 \times 3600} \times 500 = 0.058 \text{ cfs}$$

The energy gradients through the two filtering layers are computed as:

$$I_1 = \frac{f}{K_1} = \frac{5.0}{2.5} = 2.0 \quad \text{for the sand layer}$$

$$I_2 = \frac{f}{K_2} = \frac{5.0}{25.0} = 0.2 \quad \text{for the gravel layer}$$

The energy losses through the sand and gravel layers are calculated as:

$$\Delta H_1 = I_1 H_1 = 2.0 \times 18 = 36.0 \text{ inches}$$

$$\Delta H_2 = I_2 H_2 = 0.2 \times 8 = 1.6 \text{ inch}$$

$$H = Y_o + H_1 + H_2 = 12.0 + 18.0 + 8.0 = 38.0 \text{ inches}$$

$$\Delta H = H - \Delta H_1 - \Delta H_2 = 38.0 - 36.0 - 1.6 = 0.4 \text{ inches (close to zero)}$$

With $f = 5.0$ inch/hr, the total energy loss is 37.6 inches, compared to the total available energy of 38 inches in the system. Therefore, the unregulated flow rate through the rain garden is determined to be 5.1 inch/hr after the second iteration.

Based on the pre-development conditions at the project site, the flow release from this rain garden is not allowed to exceed 3.0 inch/hr. The task is to design a cap orifice that will reduce the flow release from 5.1 to 3.0 inch/hr. Repeating the above procedure with $f = 3.0$ inch/hr, the cap orifice is determined as:

$$Q = fA_B = \frac{3.0}{12 \times 3600} \times 500 = 0.035 \ cfs$$

The energy gradients through the two filtering layers are computed as:

$$I_1 = \frac{f}{K_1} = \frac{3.0}{2.5} = 1.2 \text{ for the sand layer}$$

$$I_2 = \frac{f}{K_2} = \frac{3.0}{25.0} = 0.12 \text{ for the gravel layer}$$

The energy losses through the sand and gravel layers are calculated as:

$$\Delta H_1 = I_1 H_1 = 1.2 \times 18 = 21.6 \text{ inches}$$
$$\Delta H_2 = I_2 H_2 = 0.12 \times 8 = 0.96 \text{ inches}$$

Considering the underdrain pipe is described as: $D = 4$ inches, $L = 25$ feet, and $N = 0.012$, the friction loss through the underdrain pipe is:

$$\Delta H_N = 4.62 L \frac{N^2 Q^2}{D^{(16/3)}} = 4.62 \times 25 \times \frac{0.012^2 \times 0.035^2}{(4/12)^{(16/3)}} = 0.007 \text{ft} = 0.084 \text{ inch}$$

With $C_d = 0.70$, the cross-sectional area for the cap orifice is calculated as:

$$A_o = \frac{0.035}{0.70\sqrt{2 \times 32.2(38 - 21.6 - 0.96 - 0.084)/12}} = 0.0055\,\text{sq ft, or one inch in-diameter.}$$

The design example reveals that ΔH_2 for the lower gravel layer and ΔH_N through the short subdrain pipe are numerically negligible in comparison with the ΔH_1 that the sand-mix layer dictates because the conductivity of the sand-mix is much smaller than that of gravel. As a result, Eq. 19.28 is reduced to and normalized as:

$$\frac{A_o}{A_B} = \frac{F_f}{C_d\sqrt{2\left(1 - \dfrac{\Delta H_1}{H}\right)}} \tag{19.29}$$

$$F_f = \frac{f}{\sqrt{gH}} \tag{19.30}$$

$$\frac{\Delta H_1}{H} = \frac{f}{K_1}\frac{H_1}{H} \tag{19.31}$$

where F_f = infiltration Froude number, and the subscript of 1 represents the variables for the sand layer. Eq. 19.29 indicates that this system is characterized by the infiltration flow Froude number. Considering that $C_d = 0.7$, Eq. 19.29 is converted into the design chart in Fig. 19.23.

For instance, the design example has an infiltration Froude number of:

$$F_f = \frac{f}{\sqrt{gH}} = \frac{\dfrac{3.0}{12 \times 3600}}{\sqrt{32.2 \times \dfrac{38}{12}}} = 6.88E - 06$$

$$\frac{\Delta H_1}{H_t} = \frac{f}{K_1}\frac{H_1}{H} = \frac{3.0}{2.5}\frac{18}{38} = 0.57$$

From Fig. 19.23, the area ratio is found to be 10.5×10^{-6} or $A_o = 0.0055$ sq ft for the cap orifice.

19.7 DRY TIME OF SUB-BASE

As discussed above, the *drain time* is defined as the time taken to deplete the water depth in the storage basin. The *dry time* is the time taken to drip water out of the sand and gravel layers by gravity. In comparison, the dry time is dominated by the sand layer. The depletion process in a basin is an unsteady flow that is subject to varying

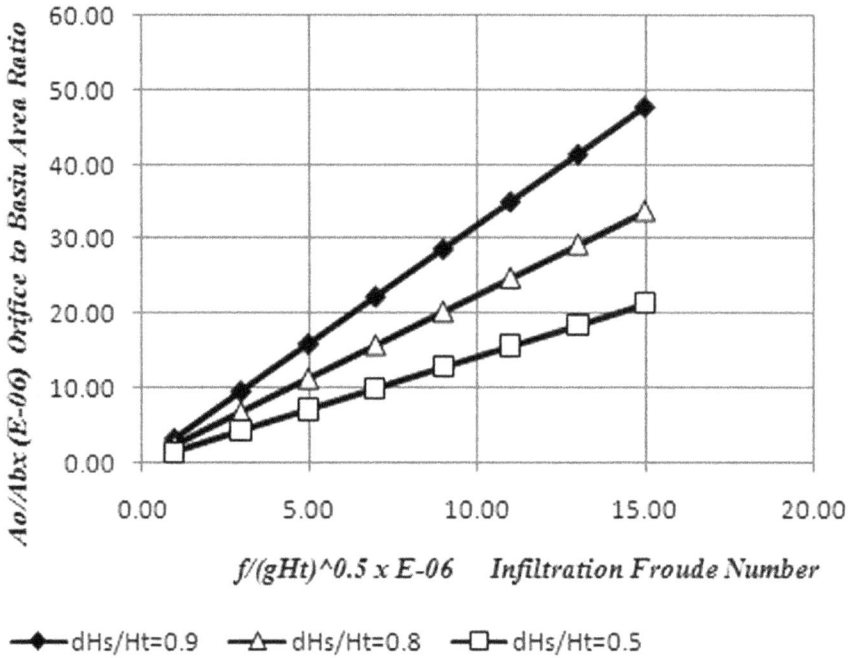

Figure 19.23 Cap orifice for a rain garden

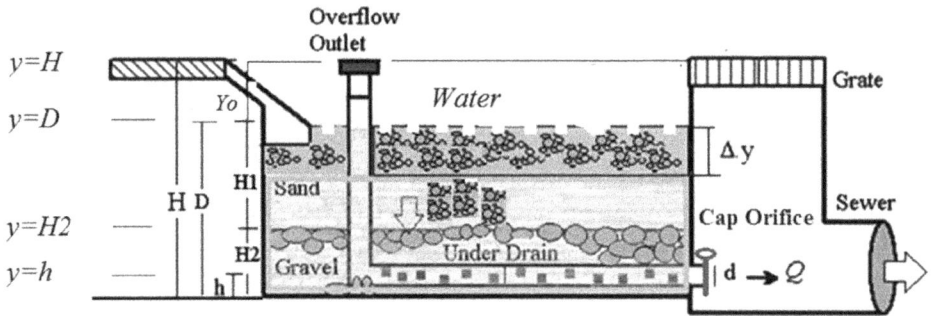

Figure 19.24 Depletion of water volume through a rain garden

headwater. Under the assumption that the seepage flow is faster than the orifice flow, the depletion process is then dominated by orifice hydraulics. Referring to Fig. 19.24, the continuity between the volume in the storage basin and the flow released through the orifice is:

$$A_B \Delta D = Q \Delta t = C_d A_o \sqrt{2g(D-h)} \Delta t \qquad (19.32)$$

where A_B = porous bed area in [L^2], ΔD = recession depth in [L], Q = flow release in [L^3/T] through the subdrain, Δt = time step, C_d = orifice coefficient for the subdrain outlet, A_o = opening area in [L^2] as the cross-sectional area of subdrain, g = gravitational acceleration in [L/T^2], H = total headwater in [L], Y_o = water depth in the storage basin in [L], D = total saturated depth in [L] in the filtering layers, including H_1 in [L] for the sand-mix layer and H_2 in [L] for the gravel layer, and h = height in [L] at the center line of the subdrain.

Integrating Eq. 19.32 from H to D yields the drain time, T_{DW}, for the storage basin as:

$$T_{DW} = \frac{2A_B[(H-h)^{1/2} - (D-h)^{1/2}]}{C_d A_o \sqrt{2g}} \big/ 3600 \text{ (hours)} \tag{19.33}$$

After emptying the storage basin, the LID unit continues depleting water content from the saturated filtering layer. The water volume in the sand-mix layer depends on the porosity. As a result, Eq. 19.32 is modified to:

$$n_s A_B \Delta D = Q\Delta t = C_d A_o \sqrt{2g(D-h)}\Delta t \tag{19.34}$$

where n_s = porosity of the sand-mix. Integrating Eq. 19.34 from D to H_2 yields the dry time, T_{DS}, for the sand layer as:

$$T_{DS} = \frac{2n_s A_B[(D-h)^{1/2} - (H_2-h)^{1/2}]}{C_d A_o \sqrt{2g}} \big/ 3600 \text{ (hours) for the sand layer} \tag{19.35}$$

After the sand layer becomes dry, the dry time for the gravel layer, T_{DG}, is calculated from H_2 to h using the porosity of the gravel, n_G, as:

$$T_{DG} = \frac{2n_G A_B(H_2-h)^{1/2}}{C_d A_o \sqrt{2g}} \big/ 3600 \text{ (hours) for the gravel layer} \tag{19.36}$$

For a simple case, the filtering system has only a layer of sand-mix. Grouping all system parameters together, Eq. 19.35 is reduced to:

$$C_f = \frac{C_d A_o \sqrt{2g}}{n_s A_B} = \frac{2(D-h)^{1/2}}{T_{DS}} \text{ (inch}^{0.5} \text{ / hr)} \tag{19.37}$$

where C_f = flow coefficient, which is a required input parameter when using the EPA SWMM-LID computer model. Eq. 19.37 converts the dry time into a depth-based ratio.

It is important to understand that Eq. 19.20 is applicable to conditions under a constant headwater or during the peak period of a runoff event that keeps the storage basin being filled. Therefore, both the seepage flow and the orifice release can be reasonably estimated under a constant headwater. Such a constant infiltration rate as in Eq. 19.20 is a simplified approach for design. In contrast, Eq. 19.33 is only applicable during the depletion process. The drain time is estimated under a varied headwater after rain has ceased. Eqs 19.33 to 19.37 tend to underestimate the drain and dry times because friction losses through the sand and gravel layers are ignored. Of course, a calibrated orifice coefficient can be introduced to compensate for friction losses if sufficient data is available. As a rule of thumb, the orifice coefficient for Eqs 19.33 to 19.37 is reduced to the range of 0.3 to 0.4 without a cap orifice. When the rain garden is operated with a cap orifice, the entire sand and gravel layers remain so saturated that friction losses can be ignored. Such a depletion process is similar to a water tank. As a result, with a cap orifice, no reduction to the orifice coefficient is recommended.

Example 19.5: Referring to Fig. 19.12, the WQCB has a water depth of 12 inches in the storage basin and a subsurface filtering sand layer of 26 inches. The subdrain does not have a cap orifice. The design parameters are: $A_B = 1500$ sq feet, $n_s = 0.35$, $H = 38$ inches, $D = 26$ inches, $d = 2.0$ inches, $h = 1.0$ inch, and $C_d = 0.35$ (reduced to compensate for friction losses). Determine the drain time for 12 inches of water and the dry time for the 26-inch sand layer.

Solution: The drain time is determined for water to recede from $H = 38$ inches to $D = 26$ inches. Applying Eq. 19.33 to the storage basin yields:

$$T_{DW} = \frac{2 \times 1500 \times \left[\left[(\frac{38}{12} - \frac{1}{12})^{\frac{1}{2}} - (\frac{26}{12} - \frac{1}{12})^{\frac{1}{2}} \right] \right]}{0.35 \times \left[3.14 \times (\frac{2.0}{12})^2 \right] / 4 \times \sqrt{2 \times 32.2}} / 3600 = 4.26 \text{ hr}$$

After the storage basin is emptied, the saturated filtering depth is 26 inches. The dry time is determined to dry the sand layer from $D = 26$ inches to $h = 1.0$ inches as:

$$T_{DS} = \frac{2 \times 0.35 \times 1500 \times \left[\left(\frac{26}{12} - \frac{1}{12} \right) \right]^{\frac{1}{2}}}{0.35 \times [3.14 \times (2.0/12)^2 / 4] \times \sqrt{2 \times 32.2}} / 3600 = 6.88 \text{ hr}$$

The corresponding flow coefficient for the sand layer is:

$$C_f = \frac{2(D - h)^{1/2}}{T_{DS}} = \frac{2 \times (26 - 1)^{1/2}}{3.44} = 2.91 \text{ inch}^{0.5} / \text{hr}$$

Example 19.6: For the purposes of rain harvest, the 48-inch circular tank shown in Fig. 19.25 is used to store storm runoff to a depth of 60 inches. The underdrain on this tank is controlled by an orifice with $d = 1.0$ inches and $h = 0.5$ inch. Determine the drain time.

(a) Industrial Rain Tank (b) Household Rain Tank

Figure 19.25 Rain tank

Solution: Emptying a rain tank does not involve any filtering medium. Therefore, the assumption of no friction is acceptable. Thus, Eq. 19.33 shall be used with $A_B = 12.6$ sq ft, $C_d = 0.70$, and $d = 1.0$ inch to calculate the drain time from $H = 60$ inch to $h = 0.5$ inch as:

$$T_{DW} = \frac{2 \times 1.0 \times 12.6 \times \{[(60-0.5)/12]^{\frac{1}{2}} - (0.5-0.5)/12]^{\frac{1}{2}}\}}{0.7 \times [3.14 \times (1/12)^2 / 4] \times \sqrt{2 \times 32.2}} / 3600 = 0.51 \text{ hour}$$

The operation of a rain tank can be synchronized with local weather forecasting. The tank will carry a slow flow release during dry days and then be reloaded during wet days. A storage tank can be emptied much faster than a rain garden. The time of delay in the operation of rain harvest is an important factor that should observe local water rights regulations.

19.8 CLOGGING EFFECT AND LIFECYLE

Over years, a clogging effect will develop in an infiltration-based LID unit. Sediment deposits in a LID unit accumulate on top of the porous bed and gradually form a layer of hardened cake that clogs the pores on the porous bed. The clogging effect may also migrate into the upper filtering layer to cause a reduction in infiltration rate. With a reduced infiltration rate, the drain time becomes prolonged. As recommended, the rain garden needs to be repaired or replaced after the clogged infiltration rate becomes less than 1.0 inch/hr (UDFCD 2010). Such a low infiltration rate results in inundation in the infiltration bed of longer than 12 hours. Prolonged standing water is considered to be a hazard to the public. To be conservative, the design infiltration rate for LID units is set to a minimum infiltration rate of 1.0 inch/hr. In operation, the LID unit needs to be replaced when the infiltration rate decays to such a minimal rate.

In the laboratory, the clogging effect was studied by continuously adding sediment-laden stormwater onto a porous bed. After many cycles of wet-dry applications, the decay of infiltration rate versus the sediment load was investigated and plotted as shown Fig. 19.26. An empirical equation was derived from the data regression analysis as (Kocman, Guo, Ramaswami in 2012):

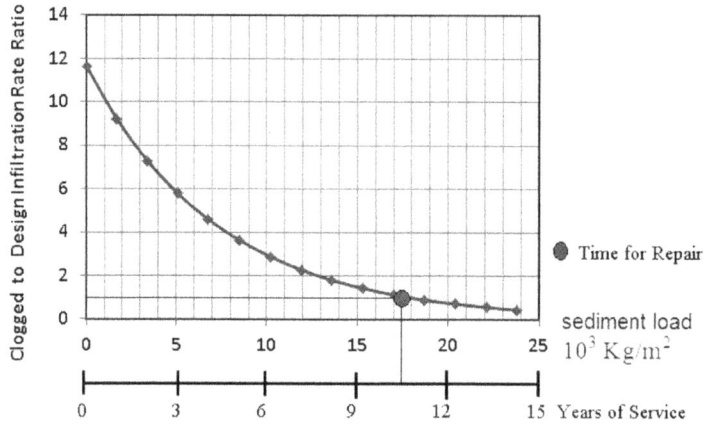

Figure 19.26 Clogging effect through lifecycle of a LID unit

$$\frac{f_T}{f} = f_1 e^{-0.1369 L_s} \tag{19.38}$$

where f_T = clogged infiltration rate [L/T] measured after the sediment load, L_s, in [10^3 kg/m^2] applied to a unit area of porous bed, f_1 = first year infiltration rate in [L/T], such as 10 to 12 inch/hr, and f = design infiltration rate in [L/T], which is the minimal rate acceptable for operation, such as 1.0 inch/hr. The first-year infiltration rate can be greater than 12 inch/hr and then gradually decays to the design infiltration rate of 1.0 inch/hr over years of service.

In practice, it is more meaningful if the reduction in infiltration rate can be converted into the years of service for the LID unit under design. The annual sediment yield generated from the tributary area to the LID unit is estimated by the annual event mean concentration of sediment and the annual runoff volume as:

$$V_o = CP_o A \tag{19.39}$$

$$L_o = C_S V_o \tag{19.40}$$

where V_o = annual runoff volume in [L^3/yr], C = runoff coefficient representing the land use in the tributary area, P_o = annual rainfall depth in [L], A = tributary area in [L^2], C_S = mean event sediment concentration [M/L^3], and L_o = annual sediment load generated from the tributary area [M/yr]. It is important that the annual snowfall depth in the high country is excluded from the annual rainfall depth, because the sediment yield is better related to rainfall-runoff in the thunderstorm season than snowmelt runoff in the early spring. In practice, the RG is designed to intercept the stormwater generated from the tributary area. Aided by Eqs 19.39 and 19.40, the annual sediment load added to the RG is estimated as:

$$L_B = \frac{L_o}{A_B} = C_S CP_o \frac{A}{A_B} \tag{19.41}$$

where A_B = porous bed area or basin bottom area in $[L^2]$ and L_B = annual unit-area sediment load in RG [M/L²/year]. Using Fig. 19.26, a cumulative sediment load, L_s, can be converted into the RG's years of service as:

$$N = \frac{L_S}{L_B} \qquad (19.42)$$

where N = RG's years of service and L_S = cumulative sediment load into the RG $[M/L^2]$.

Example 19.7: A PLDB is built at the outfall point of a parking lot. The area ratio of the parking lot to the PLDB is 20 to 1. The event mean sediment concentration in the urban runoff is approximately 240 mg/L. The annual precipitation at the project site is 400 mm (15.7 inches). Aided by Eq. 19.41 with C_S = 240 mg/L (equivalent to mg/kg), A/A_B = 20, C = 0.9 for a parking lot, and P_o = 0.4 m, the annual unit-area sediment load, L_B, is calculated as:

$$L_B = (240 \text{ mg/kg}) \times (1000 \text{ kg/m}^3) \times (0.9 \times 20 \times 0.4 \text{ m}) = 1.728 \times 10^3 \text{ kg/m}^2$$

According to Fig. 19.26, a PLDB needs a replacement when the clogged infiltration rate decays to 1.0 inch/hr or the cumulative unit-area sediment load (L_S) is close to 19.5 (10^3 kg/m²). For this case, this PLDB is expected to have a service of approximate 10 years before an over-all repair or replacement. Fig. 19.27 presents a plugged porous bed due to the tremendous sediment load generated from an adjacent construction site. The cake layer on the bed was removed with a turbo-power vacuum cleaner.

(a) Plugged Porous Bed (*f<1 inch/hr*)   (b) Vacuum Clean to Remove Cake Layer

Figure 19.27 Repair of a clogged porous bed

19.9 EVALUATION OF LID PERFORMANCE

The ultimate goal of a LID is to preserve the watershed hydrologic regime. From the hydrologic point of view, it is an effort to mimic the pre-development watershed hydrologic conditions. Using the runoff flow-frequency curve as a basis, a 100-yr single-layer detention basin can reduce the post-development 100-yr peak flow to the pre-development flow as shown in Curve 2 in Fig. 19.28, but it fails to control frequent events. The innovative green concept in stormwater management is to apply various mitigation measures convert Curve 2 to Curve 4. For instance, replacing a one-layer 100-yr detention basin with a 3M extended detention basin to control 2-, 10-, and 100-yr events, Curve 1 is improved to perform as Curve 3. As reported (Guo and Urbanos 1996), extreme events only make up the top 4–6% of the runoff flow population. It means that another 94–96% of small runoff flows would trickle through the basin without adequate detention processes. Under the concept of the LID, on-site infiltration and filtering devices are recommended to intercept up to 80% of runoff events. With a LID's storage capacity of WQCV, Curve G can be achieved. In comparison, Curve 3 presents a flood mitigation plan to solve the Q-problem, while Curve G provides a full-spectrum control to solve the additional V-problem.

The Q-problem is represented by the changes in the *flow-frequency curve* that depicts the relationship between the flow and its probability of recurrence. Preservation of a flow-frequency curve implies that the development does not change the expected flood damage. In practice, a well-designed stormwater detention basin may reduce the peak flow as illustrated in Fig. 19.29, but the stored water volume will have to be released at a higher rate for a long time. Prolonged high flows do induce erosion to the stream bed and scour to the channel banks. These are typical changes in stream morphology. It is important to recognize the fact that a detention basin provides a solution to Q-problems by reducing the peak flow, but it does not reduce the runoff volume at all. Therefore, the V-problems can only be

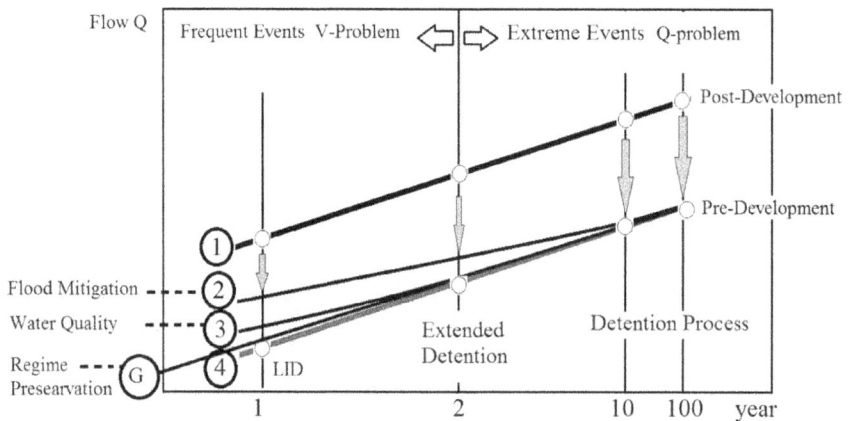

Figure 19.28 Preservation of the flow–frequency relationship

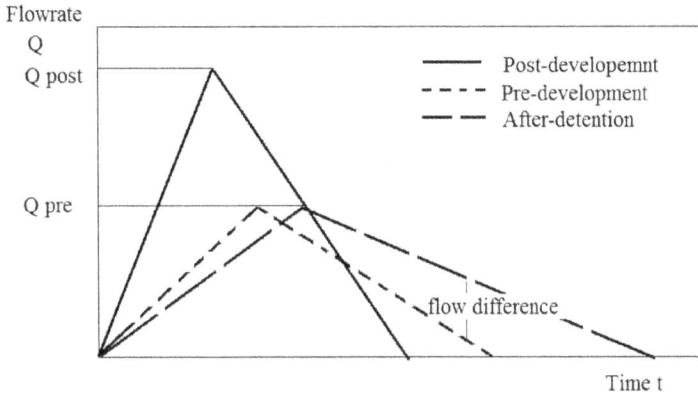

Figure 19.29 Prolonged high flows after detention

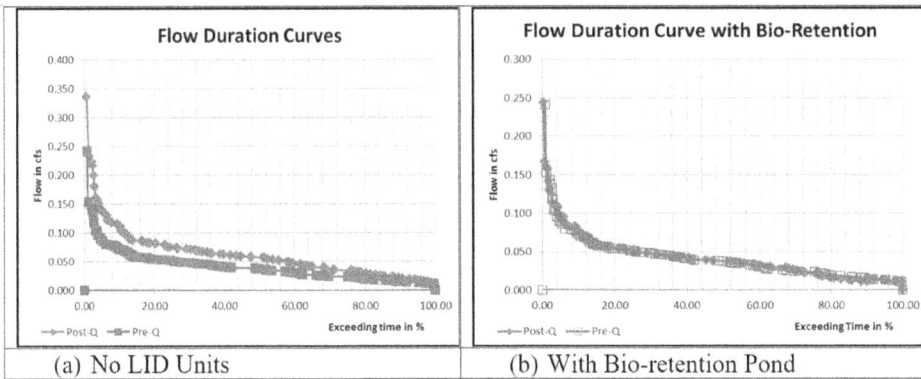

Figure 19.30 Preservation of a flow–duration curve

alleviated with infiltration basins that are designed to reduce runoff volumes (EPA 1983, EPA 2006).

The V-problem can be quantified by the *flow-duration curve*, which describes the percentage of time that flows exceed a given magnitude. The area under a duration curve represents the expected runoff volume. Therefore, on top of the *flow-frequency curve*, it is suggested that the flow-duration curve be preserved as well. Fig. 19.30 presents a case study in which a one-acre lot was developed for residential use with an impervious percentage of 75%. The changes in the flow-duration relationship are quantified by the runoff simulation using the 1-hr, 30-yr continuous precipitation record. The gap between the pre- and post- flow-duration curves represents the increased runoff volume. Adding a 300-ft^2 bioretention pond to the site, the computer simulation indicates that the infiltration capacity of the bioretention pond is sufficient to preserve the flow-duration curve.

An urban renewal project often involves adding LID facilities as a *micro drainage system* that intercepts 3- to 6-month events. With the infiltration capability in a

LID facility, the amount of WQCV is transferred into the soil media that can be either for the purpose of groundwater recharge or a delayed release into the storm sewer system as a base flow. Storm drains and sewers are designed to serve as the *minor drainage system* that conveys 2- to 5-yr peak runoff flows. During 10- to 100-yr events, the excess stormwater will overflow into street gutters as the *major drainage system*. The 3M (*micro, minor,* and *major*) *drainage system* should be laid out as a cascading flow system that begins with LID devices for upstream source control and ends with drainage facilities for downstream flood control (Guo 2013).

19.10 CONCLUSIONS

The effectiveness of a stormwater LID site is flow-path dependent. It is necessary to develop cascading flows to drain storm runoff from the upstream impervious surface onto the downstream pervious areas for more infiltration. Based on the impervious-to-pervious-area ratio, the effective imperviousness is determined and then used to estimate the required WQCV for the project site.

A porous basin such as a BRB, RG, or PLDB should be designed using the concept of a hydrologic system to take both the on-surface and sub-surface flows into consideration. The filtering layers beneath the porous bed should be structured to completely consume the hydraulic head available in the system. In practice, it is important that the infiltration rate on the porous bed and the seepage rate through the sand layer are properly evaluated for the design conditions to avoid undesirable prolonged standing water in the basin. The drain time for the LID unit is defined as how fast the water depth in the basin can be emptied, while the dry time is calculated as to how fast the sub-base becomes unsaturated again. The drain time is approximated by the infiltration rate through the porous bed and the dry time is calculated based on the seep flow rate through the subdrain pipes. Use the local annual runoff amount and event mean concentration to predict the decay of the infiltration rate when the LID unit is becoming clogged. When the infiltration rate is reduced to the design rate, such as 1 inch/hr, the LID unit needs to be replaced.

As mentioned above, LID units are designed to reduce runoff volumes from frequent events. As a result, any large events will overtop the LID unit. As shown in Fig. 19.31 the reduction percentage on peak flows depends on the area ratio of the LID unit to its tributary catchment, and its effectiveness is diminished from small to extreme events. The more LID units, the less the storm drains. In design, it is a challenge to determine the tradeoff between the LID unit and the storm sewer. Nevertheless, a LID unit will reduce the size of the storm drain.

On top of flow reduction, a LID unit captures many small events and reduces infiltrating flows into combined sewer systems. This effect is a significant improvement to the combined sewer overflow (CSO) system when facing how to reduce the number of overflows from a combined sewer into the downstream water body. Numerically, the reduction on the number of overflows from a CSO system can be well modeled with a long-term rainfall-runoff simulation using the EPA SWMM5.

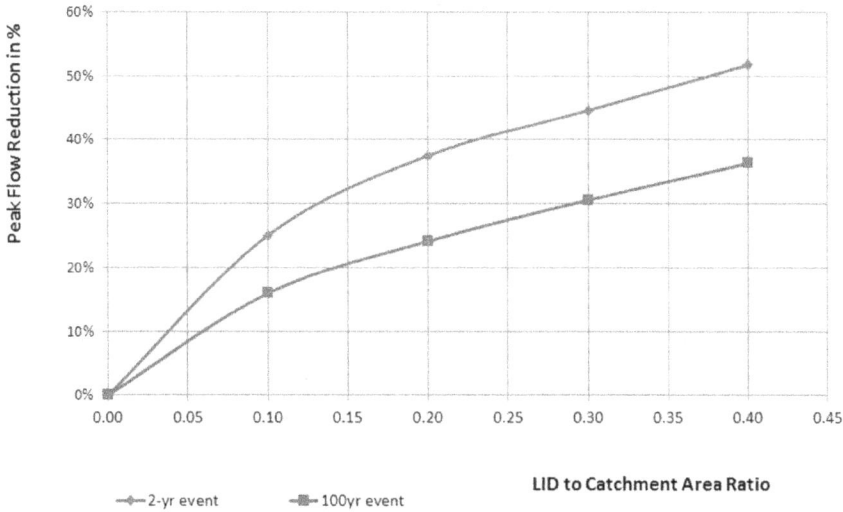

Figure 19.31 Reduction of peak flow using a rain garden as a LID unit

HOMEWORK

Q19-1 Fig. Q19-1 presents two drainage systems: a separate flow system and a cascading flow system. (1) Determine the area-imperviousness, I_a, for the separate flow system. (2) Under $f/I = 1.0$, estimate the effective imperiousness, I_E, for the cascading flow system. (3) With a runoff capture rate of $\alpha = 0.8$, determine the two WQCVs respectively for I_a and I_E as the watershed's impervious ratio. (4) Explain why there is a difference between the two WQCVs.

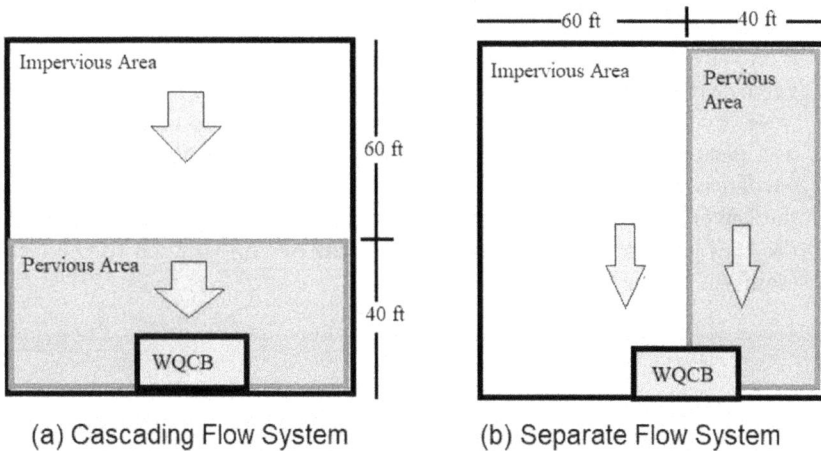

(a) Cascading Flow System   (b) Separate Flow System

Figure Q19-1 Cascading flow and separate flow systems

Q19-2 The WQCB is built for the cascading flow system in Q19-1(a). The storage depth in the basin is 12 inches. Determine the surface area for the proposed WQCB and estimate the peak-flow reduction percentages for 2- and 100-yr events respectively. (Hint: Fig. 19.31.)

Q19-3 As illustrated in Fig. Q19-3, the WQCB is designed to capture a volume of 45 mm from a site of 40 m by 40 m. The WQCB has a depth of 0.225 m on an area of 10 m by 40 m. Verify whether the proposed system in Fig. Q19-3 is adequate for both the storage capacity in the on-surface basin and the pore storage volume in the sub-surface sand and gravel layers.

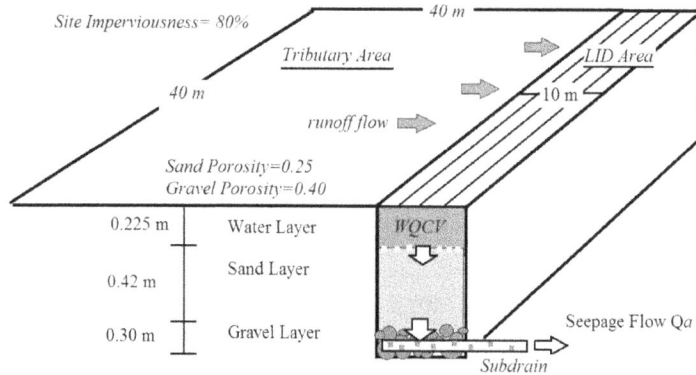

Figure Q19-3 Layout of a LID infiltration basin

Q19-4 The WQCB in Q19-3 is built at the outfall of a parking lot. The area ratio of the parking lot to the rain garden is 4 to 1. The event mean sediment concentration in the local runoff flows is approximately 200 mg/L. The annual precipitation at the project site is 600 mm (23.6 inches). (A) Determine the annual sediment amount loaded into the rain garden. (B) Estimate the lifetime of service for this rain garden.

Q19-5 A catchment is drained with a combined overflow sewer system into the river. There are two measures to reduce overflow events into the river: a rain garden at the entrance of the sewer system and a flow diversion near the sewer exit. Fig. Q19.5 presents the distribution of 144 runoff flows observed in a period of 3 years. (1) Count the number of overflow events if the rain garden is installed to intercept runoff flows up to 3 cfs at the entrance of the sewer system. (2) Count the number of the overflow events with both the rain garden and flow diversion to transfer up to 2 cfs to the treatment facility.

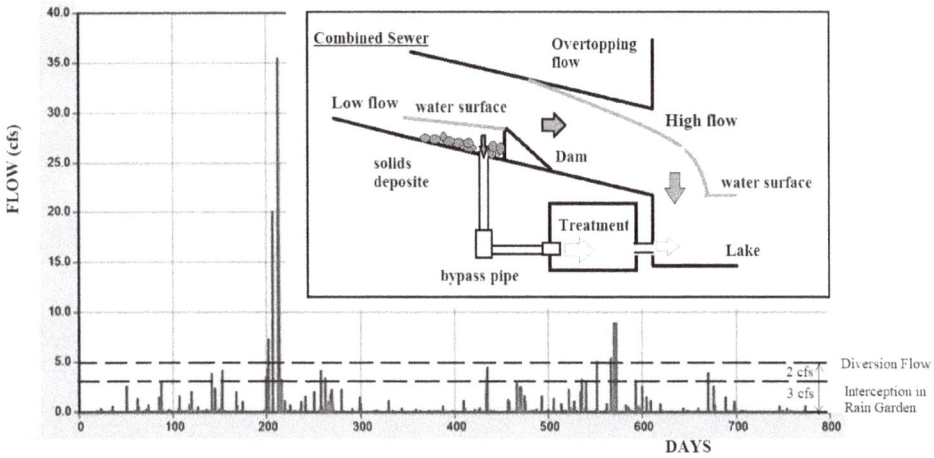

Figure Q19-5 Distribution of runoff flows for a CSO system

Q19-6 Referring to Fig. Q19-6, the rain garden is built with a basin that has a storage depth of 12-inch on a porous area of 1500 sq feet (Y_o = 12 inches), a sand layer (H_s = 18 inches), a gravel layer (H_g = 8 inches), and a subdrain pipe (4-inch in diameter for a length of 300-ft). Assume the conductivity coefficients are 0.25 inch/hr for the sand layer and 25.0 inch/hr for the gravel layer. Determine the infiltration rate when the basin is loaded with a water depth of 12 inches.

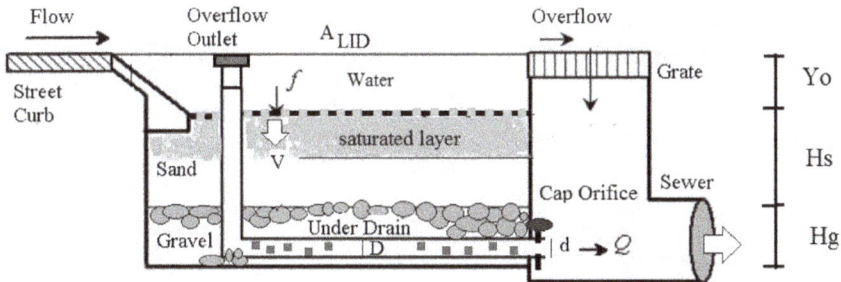

Figure Q19-6 Illustration of a rain garden for seepage flow analysis

Solution:

A) Rain Garden Flow System

Surface Area for LID Unit	A-LID 1,500.0	sq ft
Water depth in the surface basin	Y_o = 12.00	inches

B) Sub-Base Geometry for Two-Layered LID Basin

Enter Initial Soil Moisture Content	θ_o = 10.00	percent
Enter Thickness of Upper Sand Layer	H_s = 18.00	inches
Enter Hydraulic Conductivity of Sand Layer	K_s = 2.50	inch/hr
Enter Porosity for Upper Sand Layer	θ_s = 35.00	percent
Enter Thickness of Lower Gravel Layer	H_g = 8.00	inches
Enter Conductivity of Lower Gravel Layer	K_g = 25.00	inch/hr
Enter Porosity for Lower Gravel Layer	θ_g = 40.00	percent
Enter Subdrain Pipe Diameter	D = 4.00	inches
Enter Subdrain Manning's Roughness	N = 0.025	
Enter the Length of Subdrain Pipe	L = 300.0	feet
Enter Diameter for Cap Orifice	D_o = 4.00	inches
Enter Discharge Coef for Cap Orifice	C_d = 0.60	
Total storage depth = $Y_o + H_s(\theta_s - \theta_o) + H_g(\theta_g - \theta_o)$	*H*-storage = 18.90	inches

C) Enter the Design Infiltration Rate

Equivalent Water Depth = $Y_o + H_g \times \theta_g + H_s \times \theta_s$	*D*-water = 21.47	inches
	Guess f = 4.07	inch/hr
Seepage Flow through Surface Area = $f \times$ A-LID	*Q*-in = 0.1413	cfs
Total Headwater Depth available = $Y_o + H_g + H_s$	H = 38.00	inches
Energy Loss through Upper Layer = Is $\times$ Hs = f/Ks $\times$ Hs	ΔH_s = 29.30	nches >0
Energy Loss through Lower Layer = Ig $\times$ Hg = f/ Kg $\times$ Hg	ΔH_g = 1.30	inches >0
Subdrain Pipe Flowing Full Velocity = Q/A	V = 1.619	fps
Energy Slope for Flowing Full = $(NV^{)2}/ (2.22R^{4/3})$	S_e = 0.0201	ft/ft
Energy headwater for orifice = H-ΔHg-ΔHs-LS$_e$	Δho = 1.36	inches >0
Cap Orifice Flow = $C_d \pi D_o^2/4 \times (2g \, \Delta ho)^{0.5}$	*Q*-out = 0.1414	cfs
Check if Δq = Q$_{in}$ – Q$_{out}$	Δq = 0.000	Close to zero
Drain Time for the Basin on Surface = Y_o/f	*T*-drain = 2.95	hr
Dry Time for Soil Layers = $(\theta_s \times H_s + \theta_g \times H_g)/f$	*T*-dry = 2.33	'hr
Sum of Drain and Dry Times =	*T*-total = 5.28	'hr

For this case, the system drains very fast. It takes a drain time of 2.95 hours to deplete the water depth of 12 inches.

Q19-7 Referring to Fig. Q19-6, the rain garden was designed to have a drain time of 12 hours. A cap orifice with a diameter of 0.92 inches ($d = 0.92$ inches) is installed at the exit of the subdrain pipe. Prove that the drain time is close to 12 hours.

Solution:

A) Rain Garden Flow System

Surface Area for LID Unit	A-LID 1,500.0 Square feet
Water depth in the surface basin	$Y_o = 12.00$ 'inches

B) Sub-Base Geometry for Two-Layered LID Basin

Enter Initial Soil Moisture Content	$\theta_o = 0.10$
Enter Thickness of Upper Sand Layer	$H_s = 18.00$ inches
Enter Hydraulic Conductivity of Sand Layer	$K_s = 2.50$ inch/hr
Enter Porosity for Upper Sand Layer	$\theta_s = 35.00$ percent
Enter Thickness of Lower Gravel Layer	$H_g = 8.00$ inches
Enter Conductivity of Lower Gravel Layer	$K_g = 25.00$ inch/hr
Enter Porosity for Lower Gravel Layer	$\theta_g = 40.00$ percent
Enter Subdrain Pipe Diameter	$D = 4.00$ inches
Enter Subdrain Manning's Roughness	$N = 0.025$
Enter the Length of Subdrain Pipe	$L = 300.0$ feet
Enter Diameter for Cap Orifice	$D_o = 0.92$ inches
Enter Discharge Coef for Cap Orifice	$C_d = 0.60$
Total storage depth $= Y_o + H_s(\theta_s - \theta_o) + H_g(\theta_g - \theta_o)$	H-storage $= 18.90$ inches

C) Enter the Design Infiltration Rate

Equivalent Water Depth $= Y_o + H_g \times \theta_g + H_s \times \theta_s$	D-water $= 21.47$ inches
	Guess $f = 1.01$ inch/hr
Seepage Flow through Surface Area $= f \times$ A-LID	Q-in $= 0.0351$ cfs
Total Headwater Depth available $= Y_o + H_g + H_s$	$H = 38.00$ inches
Energy Loss through Upper Layer $= I_sH_s = f/K_s \times$ Hs	$\Delta H_s = 7.27$ Inches>0
Energy Loss through Lower Layer $= I_g \times H_g = f/K_g \times$ Hg	$\Delta H_g = 0.32$ inches >0
Subdrain Pipe Flowing Full Velocity $= Q/A$	$V = 0.402$ fps
Energy Slope for Flowing Full $= (NV^2/(2.22R^{4/3})$	$S_e = 0.0012$ ft/ft
Energy headwater for orifice $= H - \Delta H_g - \Delta H_s - LS_e$	$\Delta H_o = 30.03$ inches >0
Cap Orifice Flow $= C_d\, P_i\, D_o^2/4 \times (2g\, \Delta H_o)^{0.5}$	Q-out $= 0.0352$ cfs
Check if $\Delta q = Q_{in} - Q_{out}$	$\Delta q = 0.000$ Close to zero
Drain Time for the Basin on Surface $= Y_o/f$	T-drain $= 11.88$ hr
Dry Time for Soil Layers $= (\theta_s \times H_s + \theta_g \times H_g)/f$	T-dry $= 9.41$ 'hr
Sum of Drain and Dry Times $=$	T-total $= 21.29$ 'hr

For this case, the proposed cap orifice of 0.92 inch in diameter will reduce the flow release down to $f = 1.01$ inch/hr and extend the drain time to 11.88 hours. Throughout the lifetime of service, the infiltration rate continues decaying from its highest rate down to the clogged rate. It is necessary to operate a cap orifice to maintain such a low release rate that water quality enhancement is warranted.

REFERENCES

Blackler, G. and Guo, J.C.Y. (2012). "Field Test of Paved Area Reduction Factors using a Storm Water Management Model and Water Quality Test Site", *ASCE Journal of Irrigation and Drainage Engineering*, vol. 17, no. 8, August.

Blackler, G. and Guo, J.C.Y. (2013). "Paved Area Reduction Factors under Temporally Varied Rainfall and Infiltration", *ASCE Journal of Irrigation and Drainage Engineering*, vol. 139, no. 2, February 1.

Booth, D.B. and Jackson, R.C. (1997). "Urbanization of Aquatic Systems Degradation Thresholds, Stormwater Detention, and the Limits of Mitigation", *Journal of American Water Resources Association*, vol. 22, no. 5.

Davis, A.P. (2007). "Field Performance of Bioretention: Water Quality", *Environmental Engineering Science*, vol. 24, no. 8, 1048.

EPA (1983). *Results of the Nationwide Urban Runoff Program*, Final Report, US Environmental Protection Agency, NTIS no. PB84-185545, Washington, DC.

EPA (2006). "*National Recommended Water Quality Criteria*", Office of Water, Washington, DC.

Earles, T., Guo, J., MacKenzie, K., Clary, J., and Tillack, S. (2010). "A Non-Dimensional Modeling Approach for Evaluation of Low Impact Development from Water Quality to Flood Control", *Low Impact Development*, pp. 362–371.

Guo, J.C.Y. (2007). "Stormwater Detention and Retention LID Systems", *Journal of Urban Water Management*, July.

Guo, J.C.Y. (2008). "Runoff Volume-Based Imperviousness Developed for Storm Water BMP and LID Designs", *ASCE Journal of Irrigation and Drainage Engineering*, vol. 134, no. 2, April.

Guo, J.C.Y. (2010). "Preservation of Watershed Regime for Low Impact Development using (LID) Detention", *ASCE Journal of Engineering Hydrology*, vol. 15, no. 1, January.

Guo, J.C.Y. (2012). "Cap-orifice as a Flow Regulator for Rain Garden Design", *ASCE J of Irrigation and Drainage Engineering*, vol. 138, no. 2, February.

Guo, J.C.Y. (2013). "Green Concept in Stormwater Management", *Journal of Irrigation and Drainage Systems Engineering*, vol. 2, no. 3.

Guo, J.C.Y. and Urbonas, B. (1996). "Maximized Detention Volume Determined by Runoff Capture Rate", *ASCE Journal of Water Resources Planning and Management*, vol. 122, no. 1, January.

Guo, J.C.Y. and Urbonas, B. (2002). "Runoff Capture and Delivery Curves for Storm Water Quality Control Designs", *ASCE Journal of Water Resources Planning and Management*, vol. 128, no. 3, May/June.

Guo, J.C.Y., Kocman, S., and Ramaswami, A. (2009). "Design of Two-Layered Porous Landscaping LID Basin", *ASCE Journal of Environ Engineering*, vol. 145, no. 12, December.

Guo, J.C.Y., Blackler, E.G., Earles, A., and MacKenzie, K. (2010). "Effective Imperviousness as Incentive Index for Stormwater LID Designs" *ASCE Journal of Irrigation and Drainage Engineering*, vol. 136, no. 12, December.

Guo, J.C.Y., Urbonas, B., and MacKenzie, K. (2011). "The Case for a Water Quality Capture Volume for Stormwater BMP", *Journal of Stormwater*, October.

Guo, J.C.Y. Urbonas, B. and MacKenzie K. (2014). "Water Quality Capture Volume for LID and BMP Designs", *ASCE J of Hydrologic Engineering*, vol. 19, no. 4, April, pp. 682–686.

Hsieh, C.H. and Davis, A.P. (2005). "Evaluation and optimization of bioretention media for treatment of urban storm water runoff", *Journal of Environmental Engineering*, vol. 131, no. 11, pp. 1521–1531.

Hunt, W.F., Jarrett, A.R., Smith, J.T., and Sharkey, L.J. (2006). "Evaluating bioretention hydrology and nutrient removal at three field sites in North Carolina", *Journal of Irrigation and Drainage Engineering*, vol. 132, no. 6, pp. 600–608.

Kim, H., Seagren, E.A., and Davis, A.P. (2003). "Engineered Bioretention for Removal of Nitrate from Stormwater Runoff", *Water Environment Research*, vol. 75, no. 4, pp. 335–367.

Kocman, S.M., Guo, J.C.Y., and Ramaswami, A. (2012). "Waste-Incorporated Subbase for Porous Landscape Detention Basin", *ASCE Journal of Environmental Engineering*, vol. 137, no. 10, October.

Lee, J.G. and Heaney, J.P. (2003). "Estimation of Urban Imperviousness and its Impacts on Storm Water Systems", *Journal of Water Resources Planning and Management*, vol. 129, no. 5, pp. 419–426.

Rossman L.A. (2009). "Storm Water Management Model User's Manual Version 5.0", Water Supply and Water Resources Division National Risk Management Research Laboratory, Cincinnati, OH, p. 34.

Thompson, A.M., Paul, A.C., and Balster, N.J. (2008). "Physical and hydraulic properties engineered soil media for bioretention basins", *ASCE*, vol. 51, no. 2, pp. 499–514.

UDFCD (2010). *Urban Storm Drainage Criteria Manual, Volume 3: Best Management Practices*, Urban Drainage and Flood Control District, Denver, CO.

Chapter 20

Stormwater regional planning

20.1 DESIGN FLOW AS LOADING TO SYSTEM

Tributary catchments along a regional waterway, such as sewer trunk lines and channels, take several stages to be developed. Usually, the land parcels in an urban catchment are randomly developed over a period of time. This implies that urban runoff flows continuously vary with respect to time until the entire watershed has been fully developed. As a rule of thumb, downstream properties must provide adequate easement to pass incoming flood flows from upstream areas, while no flood damage is allowed to be transferred to downstream properties in case of overdevelopment in the upstream areas. Before a regional development is initiated, an RDMP report should be announced to define the locations and dimensions of the regional drainage systems and facilities. After a number of years, such a report needs updating to reflect the ongoing urbanization process and changes made to the original RDMP. Therefore, the RDMP is not only applicable to a regional development but also to urban renewal projects. A renewal project tends to have more challenges in coping with downstream and upstream existing drainage systems within limited land resources. Often an interim drainage facility is needed to accommodate the situation when an upsized sewer drains into the existing, undersized sewer.

The hydrologic analyses in a RDMP present design flows at design points (nodes) along the drainage network (links). The design point in Fig. 20.1 is assigned to the location where the design flow must be known for sizing the future hydraulic structure. As expected, there are inflows (links) merged at the design point (node). In case that the direct release will overload the downstream trunk line, then the design point should be converted into a storage facility to reduce outflows. Therefore, the continuity of flows at a design point is written as:

$$Q_o \geq \sum_{k=1}^{k=n} Q_k \tag{20.1}$$

where Q_o = allowable outflow in $[L^3/T]$ to the downstream waterway, Q_k = inflow to the design point from the k-th branch in $[L^3/T]$, n = number of incoming branches. In case that Eq. 20.1 is true, the system is a direct release; otherwise the design point is a strategic location for stormwater detention.

DOI: 10.1201/9781003284239-20

$$Q_o >= (Q_1+Q_2); \; n=2 \qquad\qquad Q_o < (Q_1 +Q_2+Q_3); \; n=3$$

Waterway

Waterway

Allowable ⇧ Q_o
Inflow

⇧ Q_o

V Detention
Basin

⇧ Outflow

○ Node

⇧   ⇧ Inflow

Q_1 Q_2

○

⇧   ⇧   ⇧

Q_1 Q_2 Q_3

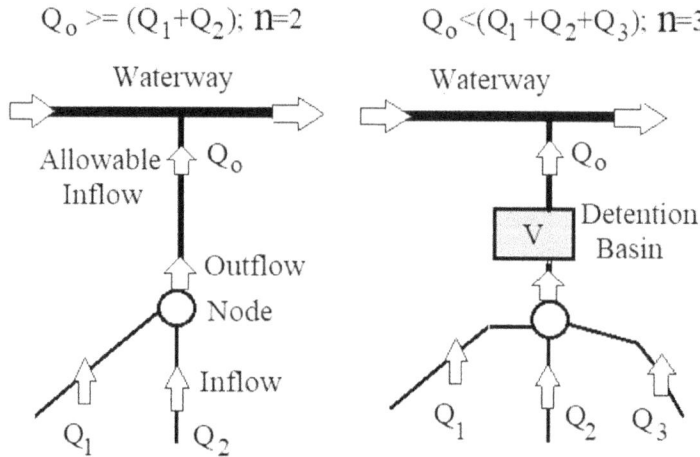

Figure 20.1 Design point with and without a detention basin

The outflow released from a node is a flow loading into the downstream waterway. Based on the downstream critical flow capacity and tailwater condition, an RDMP presents a legitimate spatial and temporal distribution of runoff flows and volumes allocated along the waterway. To mitigate adverse impacts due to continuous urban developments over a period of time, it is essential to develop a stable spatial distribution for incoming runoff flows from tributary catchments. In other words, the RDMP presents the best strategy using a combination of conveyance and storage facilities to maintain the stable runoff flow loadings to the waterway.

20.2 RUNOFF FLOW LOADING APPROACH

The runoff flow loading approach (ROFLA) is an algorithm used to determine the spatial and temporal distributions along a waterway. As illustrated in Fig. 20.2, along a waterway, the runoff flows and volumes increase in the downstream direction as the tributary areas increase. This fact implies that there exists a spatial flow loading distribution along the waterway or the changes of flow from the station, x, to the downstream station at $x+dx$, where dx is the increased distance in the direction of the flow. Similarly, we also experienced that the runoff flows and volumes are increased at a design point as its tributary area becomes more developed over a period of time. This fact implies that there exists a temporal flow loading distribution at any point along the waterway. In general, the Rational method is suitable for a small urban renewal project that covers a tributary area less than 100 hectares. Use the Rational method to illustrate the concept of ROFLA as:

$$Q = CIA \qquad\qquad (20.2)$$

where Q = flow rate in [L^3/T] generated from the tributary area into the waterway, C = runoff coefficient representing the urbanization conditions, I = rainfall intensity in

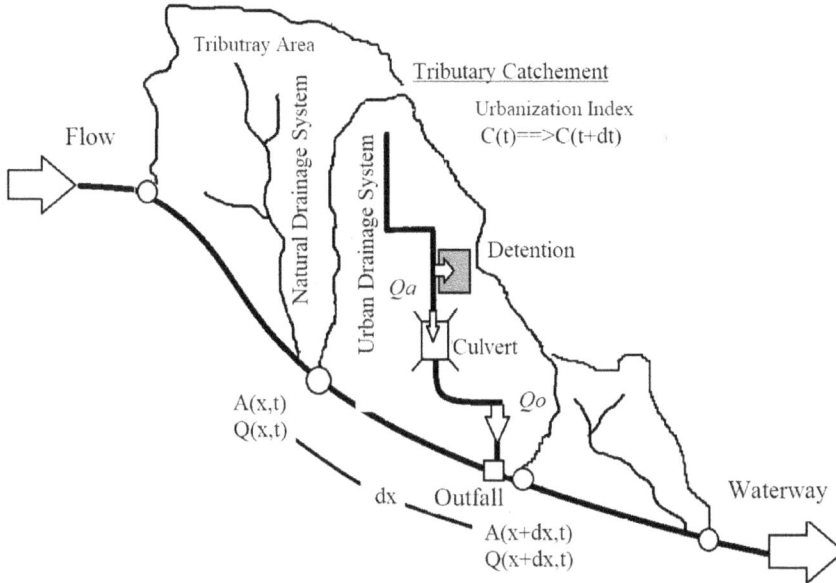

Figure 20.2 Illustration of runoff flow loading along a waterway

[L/T], and A = tributary area in [L²]. Along a waterway, the flow changes with respect to time and location. As a result, the variation of flow, $Q(t,x)$, is written as:

$$Q(t + dt, x + dx) = Q(t,x) + dQ(t,x)$$ (20.3)

$$dQ(t,x) = dQ_t + dQ_x = \frac{\partial Q}{\partial t} dt + \frac{\partial Q}{\partial x} dx$$ (20.4)

where t = time in [T], x = distance or station in [L], dQ_t = flow increase in [L³/T] over dt in [T], dQ_x = flow variation in [L³/T] from dx in [L], and dQ = total variation in flow in [L³/T]. Substituting Eq. 20.2 into Eq. 20.4 yields:

$$dQ_t = IA\frac{\partial C}{\partial t} dt = IA[C(t+dt) - C(t)] = IA(C_a - C_o) \quad \text{at location } x$$ (20.5)

$$dQ_x = IC\frac{\partial A}{\partial x} dx = IC_o[A(x+dx) - A(x)] \quad \text{at time } t$$ (20.6)

where C_o = pre-development runoff coefficient at time t, and C_a = post-development runoff coefficient at time $t+dt$. Eq. 20.5 is derived based on the assumption that the rainfall statistics and tributary area do not vary with respect to time. For instance, during the period of urbanization, the 100-yr precipitation depth remains unchanged and the tributary catchment's boundaries remain the same. Similarly, the assumption for Eq. 20.6 is that the rainfall statistics and runoff coefficient are not changed in space.

For instance, the point rainfall depth at the center of the catchment remains unchanged and the land use over the catchment is homogeneous. The numerical integration of Eq. 20.6 is the *flow loading spatial distribution* of $Q(t, x)$ along the waterway from x to $x+dx$. It reflects the fact that the greater the tributary area, the higher the runoff flow. The integration of Eq. 20.5 is the *flow loading temporal distribution* at the catchment's outlet, representing the increase in runoff flows from the catchment due to progressive urbanization over a period of time.

The hydrologic analyses performed in the RDMP include both pre- and post-development conditions. The negative impact of urbanization is quantified by the increased runoff flow. Such an increased runoff flow should be mitigated with various stormwater facilities, including low-impact-development (LID) devices placed at runoff source points or/and a detention basin installed at the outfall point (Guo 2017). The ultimate goal of an RDMP is to preserve the flow loading along the waterway, or the flow increase in Eq. 20.5 should be nullified through the stormwater storage facilities.

20.3 CONVEYANCE AND STORAGE FACILITIES

Often the pre-development conditions serve as the basis to define the allowable flow release from a tributary catchment. Under the proposed conditions, the peak flow generated from the tributary catchment is increased according to the proposed land use. To be safe, the stormwater conveyance facilities through the catchment should be sized for the post-development peak flow as:

$$Q_a = C_a I A \quad \text{for conveyance facility design} \tag{20.7}$$

where Q_a = post-development peak flow in [L³/T]. The design rainfall intensity is selected based on a critical rainfall duration such as the time of concentration of the catchment. As a common practice, the rainfall intensity is determined with the local rainfall intensity–duration–frequency (IIDF) formula. The total rainfall depth is calculated as:

$$P = I \, T \tag{20.8}$$

where P = design rainfall depth in [L] and T = rainfall duration in [T]. The corresponding rainfall volume for Eq. 20.7 is calculated as:

$$V_a = C_a A P \tag{20.9}$$

where V_a = post-development runoff volume in [L³]. Although the drainage system should be sized for post-development conditions, the allowable release at the outfall should be defined by pre-development conditions as:

$$Q_o = C_o I A \tag{20.10}$$

$$V_o = C_o A P \tag{20.11}$$

where Q_o = pre-development peak flow in [L³/T] and V_o = pre-development runoff volume in [L³]. Using Eqs 20.8, 20.9, and 20.11, the required detention volume to reduce the post- to pre-development peak flow is:

$$dV = V_a - V_o = IA(C_a - C_o)T = PA(C_a - C_o)$$
(20.12)

where dV = required detention volume in [L³]. Eq. 20.12 quantifies the required detention volume to nullify $dQ_t = 0$ in Eq. 20.5. Eq. 20.12 exactly agrees with the ultimate goal set forth in the concept of LID (Guo 2010). The runoff coefficients in Eq. 20.12 are directly related to the precipitation depth and imperviousness in the catchment as (Guo and Mackenzie 2014):

$$C_a = \left(1 - \frac{D_{vi}}{P}\right)I_a + \left(1 - \frac{D_{vp}}{P} - \frac{F}{P}\right)(1 - I_a)$$
(20.13)

$$C_o = \left(1 - \frac{D_{vi}}{P}\right)I_o + \left(1 - \frac{D_{vp}}{P} - \frac{F}{P}\right)(1 - I_o)$$
(20.14)

where D_{vi} = depression loss in [L] on the impervious area, such as 0.1 inch (2.5 mm), D_{vp} = depression loss in [L] on the pervious area, such as 0.4 inch (10 mm), I_a = post-development imperviousness ratio, and I_o = pre-development imperviousness ratio, F = infiltration loss in [L], such as 1.8 inches (45.7 mm) for Soil A, 1.0 inch (25.4 mm) for Soil B, and 0.88 inch (22.4 mm) for Soils C/D. Using Eqs 20.13 and 20.14), the difference between the pre- and post-runoff coefficients is:

$$\Delta C = \left(\frac{D_{vp} - D_{vi}}{P} + \frac{F}{P}\right)(I_a - I_o)$$
(20.15)

Substituting Eq. 20.15 into Eq. 20.12 yields:

$$dV = (D_{vp} - D_{vi} + F)(I_a - I_o)A \quad \text{for the detention basin design}$$
(20.16)

Eq. 20.16 suggests that the detention volume is equal to the hydrologic losses weighted with the difference between the pre- and post-impervious ratios applied to the entire catchment area. Eqs 20.7 and 20.16 are utilized to design the permanent stormwater conveyance and detention systems.

20.4 INTERIM FACILITIES FOR URBAN RENEWAL

As the population and economics grow in an urban area, the demands on land resources may trigger urbanization processes at a faster pace than the city has planned for. As a result, many urban areas have been developed ahead of their RMDP. Without an RMDP, land use and traffic routes are not well coordinated but are randomly developed in time and in space. Flood problems are induced and accumulated through under-sized drainage systems. As public safety becomes a serious concern, an urban renewal project is initiated. The major tasks in the RMDP for an urban renewal project are to identify existing flood problems, to upgrade drainage systems, to provide adequate easement, to pass flood flows, and to allocate land resources for stormwater storage facilities. As always, the challenges include how to choose various alternatives to establish positive hydraulic and energy grade lines along the sewer-line corridors, how to allocate adequate open space for stormwater detention, and how to enhance the use of open space for multiple purposes. During the construction phase, traffic detour and stormwater interim facilities are always needed to accommodate temporary services.

20.4.1 Temporary channel and off-stream storage systems

As illustrated in Fig. 20.3, the newly upsized channel collects the post-development flow generated from the urban renewal project site. This channel drains into the downstream existing waterway. At the confluence, the flow released from the upsized new channel needs to be regulated with a culvert. The culvert's capacity is subject to the critical capacity in the waterway. To be conservative, the most restricted tailwater depth at the confluence point is subject to the required freeboard height, such as 1.0 ft (30 cm) for a 10-year event, and 0.5 ft (15 cm) for a 100-yr event (MacKenzie 2010). Therefore, the maximal flow release through the culvert into the downstream waterway is estimated as:

$$Q_m = A_m \sqrt{\frac{1}{1 + K_e + K_n}} \sqrt{2g(H_n - H_m)} \tag{20.17}$$

$$K_n = \left[\frac{2g}{k^2} \frac{N^2}{R^{\frac{4}{3}}} L\right] \tag{20.18}$$

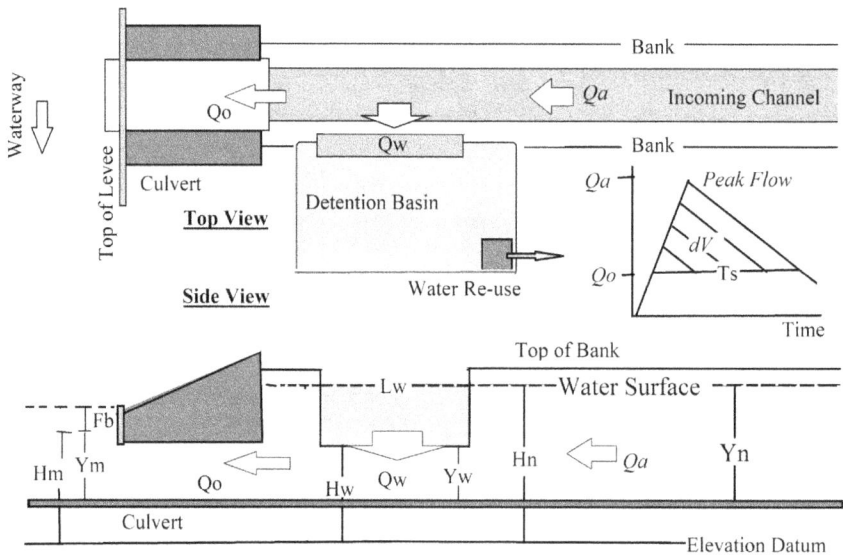

Figure 20.3 Temporary channel and off-stream storage system

where H_m = tailwater elevation in waterway in [L], F_b = freeboard height in [L], H_n = elevation of the energy grade line in [L] at conduit entrance, Q_m = maximal flow release in [L³/T], A_m = conduit cross-sectional area in [L²], K_n = friction loss coefficient, K_e = sum of minor loss coefficients, including exit, entrance, and bend losses, k = 1 for SI units or 1.486 for English units, L = length of conduit in [L], N = Manning's roughness coefficient, R = full-flow hydraulic radius of conduit in [L], and g = gravitational acceleration in [L/T²]. The goal of the ROFLA is to make sure that the incoming flow from the tributary catchment will not overload the receiving waterway. Therefore, Eq. 20.17 must further be compared with the critical capacities defined by other parameters associated with the downstream waterway, including the backwater effect, tidal effect, restrictions due to highway bridges, sediment movement, and changes in flow regime. With an adequate evaluation of the downstream waterway's capacity, the allowable release or equivalent pre-development flow is finalized as:

$$Q_o = \min(Q_m, Q_c) \qquad (20.19)$$

If $Q_o > Q_a$, a direct release, otherwise a detention release $\qquad (20.20)$

where Q_o = allowable flow release in [L³/T] equivalent to pre-development flow rate, Q_m = maximal capacity in [L³/T] defined by levees and freeboard along the downstream waterway, and Q_c = critical flow capacity in [L³/T] determined by constraints associated with the downstream waterway's conveyance capacity. Next, if the allowable flow release, Q_o, is greater than the post-development peak flow, Q_a, in Eq. 20.20, it is a case of direct release; otherwise, a temporarily storge facility is needed to reduce the peak flow. In a compact urban area, land is limited and expensive. As a result, the off-stream detention shown in Fig. 20.3 is the best choice because it only intercepts and stores the peak flow volume (Guo 2012). Observing the local regulations on water rights, the stored water may be released for water reuse within the allowable time frame.

As illustrated in Fig. 20.3, the incoming channel acts as a pool to split the incoming flow into straight-through flow into the downstream culvert and diverted flow into the off-stream basin. As a temporary facility, a side weir installed on top of channel's bank is the most economic device to transfer excess stormwater from the channel into the storage basin. The amount of flow volume diverted into the basin depends on the reliable headwater depth on top of the weir crest. As a result, the downstream conduit is sized to have its headwater depth slightly higher than the normal depth in the channel. The backwater profile of the M-1 curve is created to provide a long and stable headwater, starting from the culvert entrance to back up through the length of the side weir. The post-development peak flow in the channel, as calculated in Eq. 20.7, is divided into the straight-thru flow equal to the allowable, Q_o, as determined in Eq. 20.19, and the diverted flow, Q_w. The continuity equation states as:

$$Q_a = Q_w + Q_o \qquad (20.21)$$

The diverted flow through the side weir flow is calculated as:

$$Q_w = \frac{2}{3}C_o\sqrt{2g}L_w(Y_n - Y_w)^{1.5} \quad \text{for a bank top side weir design} \qquad (20.22)$$

where Q_w = diverted flow in $[L^3/T]$, C_o = orifice coefficient such as 0.6 to 0.7, L_w = length of the side weir in $[L]$, Y_n = normal depth in $[L]$ in the channel determined for the post-development flow, Q_a, and Y_w = height of the weir crest in $[L]$ above the channel floor. In practice, the length of the side weir is solved for the pre-selected height of the weir crest. Referring to Fig. 20.3, the storage volume is approximated by the triangular volume during the peak time as:

$$dV = 0.5(Q_a - Q_o)T_s \quad \text{for the storage volume design} \tag{20.23}$$

where dV = storage volume in $[L^3]$ and T_s = peak time for flow diversion in $[T]$. In practice, it is necessary to further confirm using the hydrograph routing method to refine Eq. 20.23 (Guo 2004).

Fig. 20.4 presents an example located near South Logan Street and Highway 285, Denver, Colorado. The Harvard Gulch is overloaded with runoff flows from an over-developed area. The existing culvert under South Logan Street is undersized. The excess stormwater is diverted into adjacent open space for temporary storage. Downstream is a vacant land parcel under development. Before the permanent drainage system is built, the culvert is tied into an overflow structure to spread flow into the vacant open space as an interim drainage system.

An off-stream storage basin can also be utilized to ease the inconsistency between the upstream and downstream underground sewer sizes. As illustrated in Fig. 20.5,

Figure 20.4 Detention-culvert-overflow system

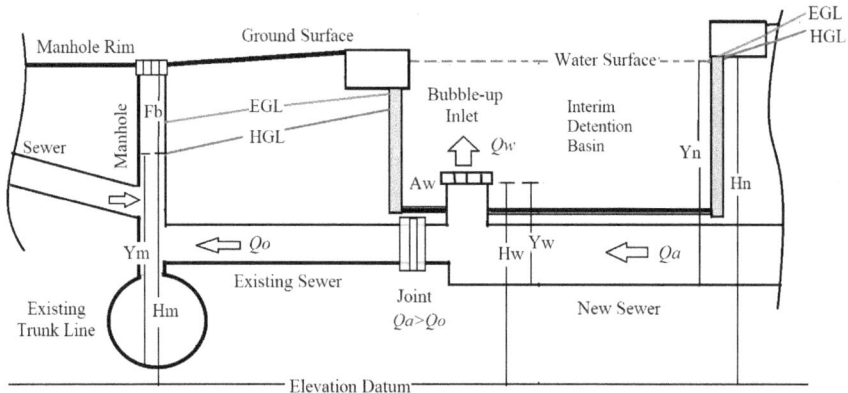

Figure 20.5 Temporary sewer and bubble-up storage basin system

the incoming sewer line is upsized to carry the post-development peak flow. However, the existing downstream trunk line was built to pass the pre-development peak flow. Upstream of the connection of these two sewer conduits, a bubble-up inlet can be installed to divert the excess stormwater by the hydraulic grade line (HGL) into open space on the ground. The energy grade line (EGL) between the temporary bubble-up storage basin and the outfall manhole is designed to satisfy Eq. 20.17. Furthermore, Eq. 20.19 is applied to finalize the allowable flow released into the downstream trunk line. The diverted flow through the bubble-up inlet is calculated as:

$$Q_w = C_o A_w \sqrt{2g(H_n - H_w)} \quad \text{for the bubble-up flow diversion design} \qquad (20.24)$$

where Q_w = bubble-up flow in $[L^3/T]$, A_w = bubble-up flow area in $[L^2]$, H_n = HGL elevation in $[L]$ of the incoming sewer flow, and H_w = elevation in $[L]$ of the horizontal orifice area. With a pre-selected elevation, H_o, the bubble-up flow area is determined by Eq. 20.24. Repeat Eq. 20.23 to size the storage volume.

Since the HGL and EGL along a sewer line are not stable variables for designs, it is necessary to verify the system's performance using numerical simulations. Computer models such as EPA SWMM (2005) provide a dynamic wave approach to apply both the continuity and energy principles to confirm the performance of the sewer-storage system under design.

Fig. 20.6 is an example of bubble-up inlets used during the interim period at the project site near South Jones Blvd and Interstate Highway I-225, Las Vegas, Nevada. During an extreme event, the incoming sewer lines are fully pressurized because the existing trunkline downstream is undersized. The excess stormwater is then spilled into adjacent vacant lands for temporary detention. The downstream trunk line will be upsized upon the completion of upstream developments.

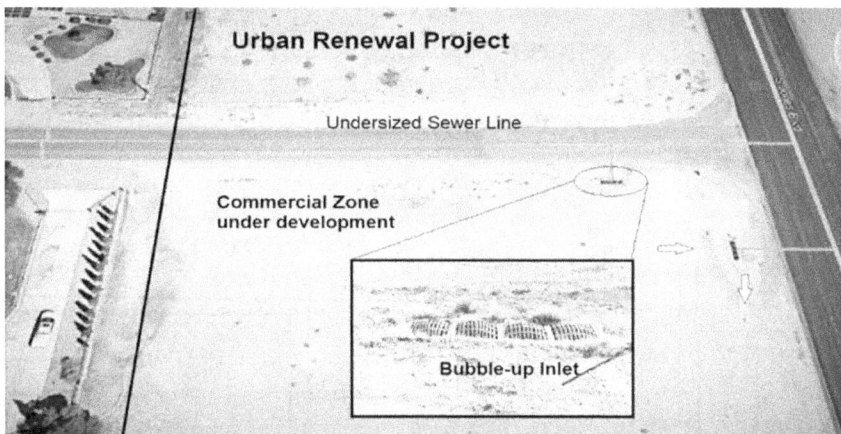

Figure 20.6 Bubble-up inlets built along Jones Blvd., Las Vegas, Nevada

20.4.3 Temporary flood gate and pump systems

Under the conditions of a high tide in the downstream waterway, the incoming channel in Fig. 20.7 needs a flood gate on the culvert exit to prevent reverse flows from the downstream waterway. As soon as the flood gate is shut down by the high tide, the channel flow will be diverted into the sump tank through the flat equalizers installed on the channel bottom. With the pre-selected height, H_w in Fig. 20.7, the diameter of these circular equalizers can be determined with Eq. 20.24. The pump system is operated with start-up and shut-off depths. The pump's characteristic curve dictates the flow-depth relationship in the sump tank. During the recession of flood flows, the flood gate is opened again and the stored water in the sump tank flows back into the channel. The operation of a flood gate depends on the weight and geometry of the gate, location of the hinge, and differential between the water surfaces as:

$$\Delta Y = (Y_n - Y_x) \geq \frac{W_s \sin \theta (S_s - 1)}{\gamma_w} \frac{}{S_s} \quad \text{for flood gate design} \qquad (20.25)$$

and ΔY = water depth difference in [L], Y_n = headwater depth upstream of the culvert, Y_x = recession water depth in channel in [L], W_s = steel gate's unit weight per area in [pounds/ft² or Newton/m²], γ_w = specific weight of water in [pounds/ft³ or Newton/m³], S_s = specific gravity of a steel flood gate, such as 7.5, and θ = minimal inclined angle to open the flood gate, such as 45 degrees. Eq. 20.25 provides guidance on how to select and operate the flood gate in Fig. 20.8. Details of the flood gate mechanism can be found elsewhere (Guo 2012). The filling–depletion loading cycle in the sump tank is described with the stage-outflow curve representing the combination of pump and gravity flows. Regardless of any specifics, the same principle is observed in which the pump flow should not exceed the allowable.

Through continuous efforts in urban renewal, a functional drainage network can be implemented with the proposed local and regional drainage facilities. Although

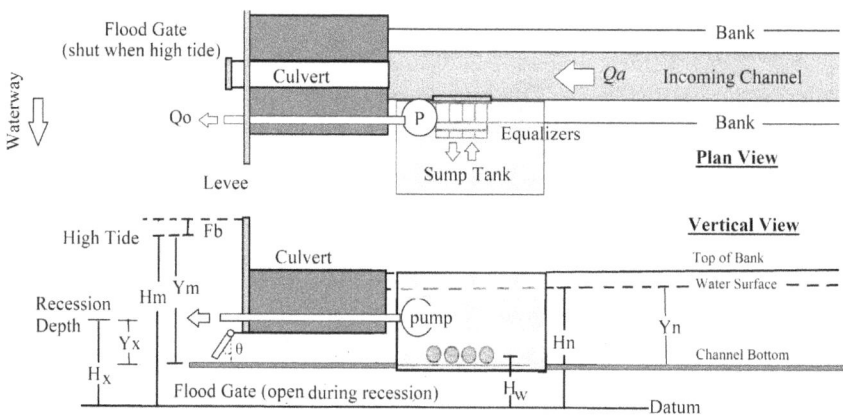

Figure 20.7 Temporary flood gate and pump station

(a) Pipe Outlet with Flood Gates

(b) Pipes for off-stream detention

Figure 20.8 Horizontal pipes serving as an equalizer to balance water surfaces

temporary storage facilities are aimed at a short length of service, adequate and frequent maintenance is still required for reliable performance.

20.5 CONCLUSIONS

1. Based on the concept of the *runoff flow loading approach* (ROFLA), both runoff flow spatial and temporal distributions are derived for a waterway under planning and design. The spatial distribution represents the increases in runoff flows as the tributary area to the waterway increases downstream, while the temporal distribution represents the changes in runoff flows over a period of time during the progressive urbanization process in the tributary catchments. Rather than being intuitive, the ROFLA derived in this study justifies the basic legal requests on the regional runoff management.

2. When interim drainage facilities are used to ease the inconsistencies between urban renewal projects, it is critically important to understand that the design event and construction material for a temporary drainage structure depend on the period of service. To be economic, re-used pipes and pumps are a rational choice for a short detour service (Guo 1998).

3. All flooding problems are case dependent, the general approach discussed provides a basis for deriving specific solutions, according to the given constraints and conditions. For instance, an off-stream detention basin can be replaced with on-stream detention if the floodplain can be widened or the waterway can be re-routed into an adjacent open space (Guo 2004). If the incoming sewer can be drained into a basin such as a parking lot, then the flow system can be driven by the gravity rather than the flow's EGL and/or HGL. Mostly, an on-stream basin will intercept the initial runoff volume for water-quality treatment, whereas the off-stream basin will be the most economic because it only intercepts the peak flow volume.

4. With the latest developments in low-impact-development (LID), all infiltration-based drainage devices can be placed upstream of street inlets for pollutant source control. All LID facilities should be sized to intercept the early runoff volume up to the runoff capture volume. The detention basin placed at the

system outfall should be designed to cope with the peak runoff volume for peak-flow reduction.

5. The kinematic wave (KW) approach is preferred for modeling stormwater movements in the planning stage because it is a conservative approach based on the continuity principle using gravity as the driving force. At the final stage, it is necessary to verify the system's performance using the dynamic wave (DW) approach, which simulates the flows with both continuity and energy principles for the given boundary and tailwater conditions.

REFERENCES

EPA SWMM (2005). *Storm Water Management Model Version 5*, Water Supply and Water Resources Division, National Risk Management Research Laboratory, Cincinnati, OH.

Guo, J.C.Y. (2004). "Hydrology-Based Approach to Storm Water Detention Design Using New Routing Schemes", *ASCE Journal of Hydrologic Engineering*, vol. 9, no. 4, July/August.

Guo, J.C.Y. (2010). "Preservation of Watershed Regime for Low Impact Development using (LID) Detention", *ASCE Journal of Engineering Hydrology*, vol. 15, no. 1, January.

Guo, J.C.Y. (2012). "Off-stream Detention Design for Stormwater Management", *ASCE Journal of Irrigation and Drainage Engineering*, vol. 138, no. 4, April 1.

Guo, J.C.Y. (2017). *Urban Flood Mitigation and Stormwater Management*, CSC Publisher, New York.

Guo, J.C.Y. and Mackenzie, K. (2014). "Modeling Consistency for small to large watershed studies", *ASCE Journal of Hydrologic Engineering*, vol. 19, no. 8, August, 04014009-1–7.

MacKenzie, K. (2010). *Urban Stormwater Drainage Criteria Manual*, Urban Drainage and Flood Control District / Mile High Flood District, Denver, CO.

Index

For Product Safety Concerns and Information please contact our EU
representative GPSR@taylorandfrancis.com
Taylor & Francis Verlag GmbH, Kaufingerstraße 24, 80331 München, Germany